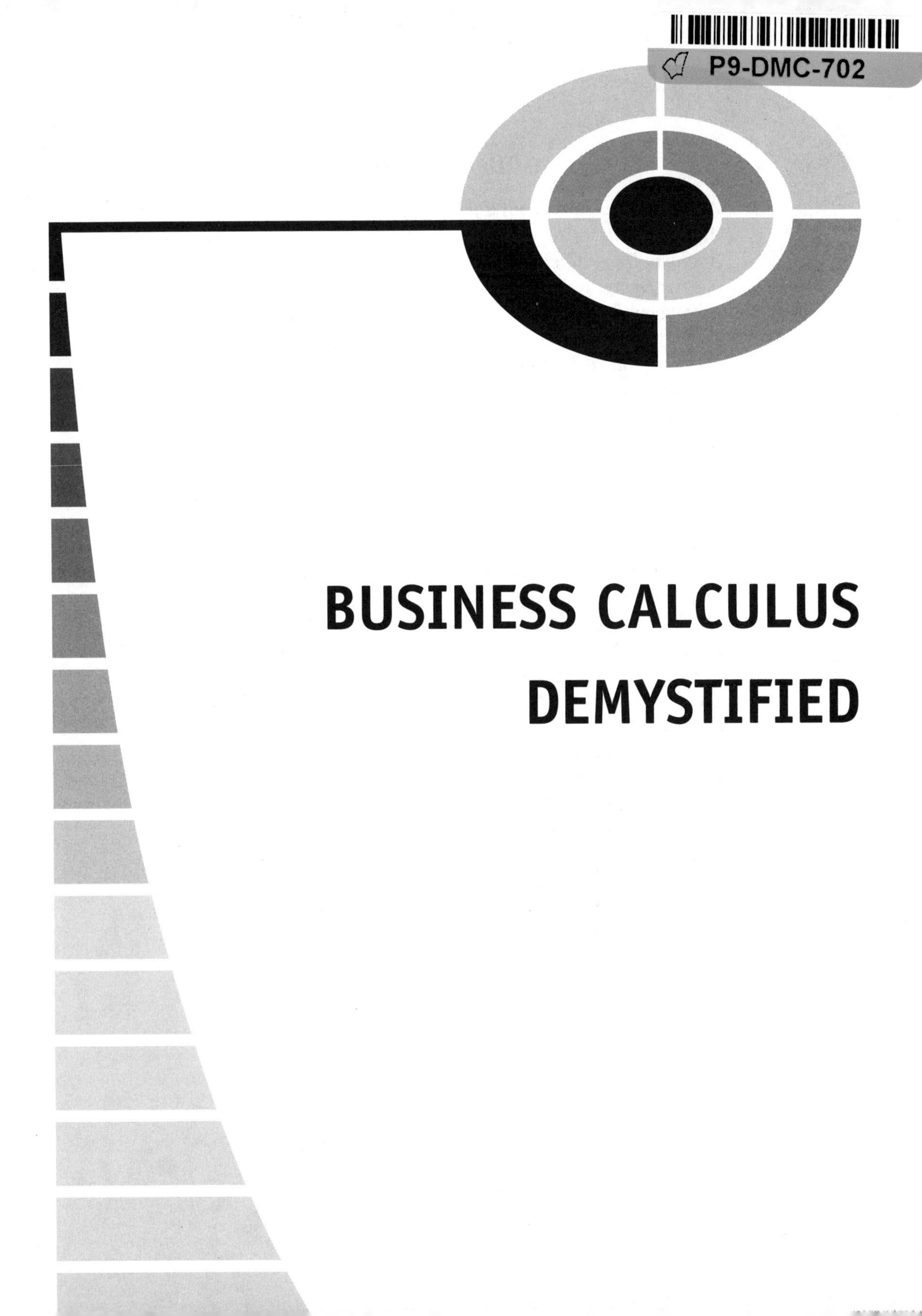

BUSINESS CALCULUS DEMYSTIFIED

Demystified Series

Advanced Statistics Demystified
Algebra Demystified
Anatomy Demystified
asp.net Demystified
Astronomy Demystified
Biology Demystified
Business Calculus Demystified
Business Statistics Demystified
C++ Demystified
Calculus Demystified
Chemistry Demystified
College Algebra Demystified
Data Structures Demystified
Databases Demystified
Differential Equations Demystified
Digital Electronics Demystified
Earth Science Demystified
Electricity Demystified
Electronics Demystified
Environmental Science Demystified
Everyday Math Demystified
Genetics Demystified
Geometry Demystified
Home Networking Demystified
Investing Demystified
Java Demystified
JavaScript Demystified
Linear Algebra Demystified
Macroeconomics Demystified
Math Proofs Demystified
Math Word Problems Demystified
Medical Terminology Demystified
Meteorology Demystified
Microbiology Demystified
OOP Demystified
Options Demystified
Organic Chemistry Demystified
Personal Computing Demystified
Pharmacology Demystified
Physics Demystified
Physiology Demystified
Pre-Algebra Demystified
Precalculus Demystified
Probability Demystified
Project Management Demystified
Quality Management Demystified
Quantum Mechanics Demystified
Relativity Demystified
Robotics Demystified
Six Sigma Demystified
sql Demystified
Statistics Demystified
Trigonometry Demystified
uml Demystified
Visual Basic 2005 Demystified
Visual C# 2005 Demystified
xml Demystified

BUSINESS CALCULUS DEMYSTIFIED

RHONDA HUETTENMUELLER

New York Chicago San Francisco Lisbon London
Madrid Mexico City Milan New Delhi San Juan
Seoul Singapore Sydney Toronto

The McGraw-Hill Companies

2 3 4 5 6 7 8 9 0 DOC/DOC 0 1 0 9 8 7

ISBN 0-07-145157-9

The sponsoring editor for this book was Judy Bass and the production supervisor was Richard C. Ruzycka. It was set in Times Roman by Keyword Publishing Services Ltd. The art director for the cover was Margaret Webster-Shapiro; the cover designer was Handel Low.

Printed and bound by RR Donnelley.

McGraw-Hill books are available at special quantity discounts to use as premiums and sales promotions, or for use in corporate training programs. For more information, please write to the Director of Special Sales, McGraw-Hill Professional, Two Penn Plaza, New York, NY 10121-2298. Or contact your local bookstore.

This book is printed on recycled, acid-free paper containing a minimum of 50% recycled, de-inked fiber.

CONTENTS

PREFACE

This book was written to help you solve problems and understand concepts covered in a business calculus course. To make the material easy to absorb, only one idea is covered in each section. Examples and solutions are given in detail so that you will not be distracted by missing algebra and/or calculus steps. Topics that students find difficult are written with extra care.

Each section contains an explanation of a concept along with worked out examples. At the end of each section is a set of practice problems to help you master the computations, and solutions are given in detail. Each chapter ends with a chapter test so that you can see how well you have learned the material, and there is a final exam at the end of the book.

If you have recently taken an algebra course, you can probably skip the algebra review at the beginning of the book. The material in Chapters 1 and 2 lay the foundation for the concept of the derivative, which is introduced in Chapter 3. The formulas in Chapter 4 are used throughout the book and should be memorized. Calculus techniques and other formulas are covered in Chapters 6, 7, 8, 9, and 11. Calculus can solve many business problems, such as finding the price (or quantity) that maximizes revenue, finding the dimensions that minimize the cost to construct a box, and finding how fast the profit is changing at different production levels. These applications and others can be found in Chapters 5, 7, 10, 11, and 12. Integral calculus and its applications are introduced in the last three chapters.

I hope you find this book easy to use and that you come to appreciate the beauty of this powerful subject.

Rhonda Huettenmueller

BUSINESS CALCULUS DEMYSTIFIED

Algebra Review

Success in calculus requires a solid algebra background. Although most of the algebra steps (as well as calculus steps) are provided in the book, it is worth reviewing algebra basics. In this chapter, we will briefly review how to factor, simplify fractions, solve equations, find equations of lines, and more.

Factoring

One of the most important properties in mathematics is the distributive property: $a(b+c) = ab+ac$. This property allows us to either add b and c before multiplying by a or to multiply b and c by a before adding. For example, $2(4+5)$ could be computed either as 2×9 or as $8+10$. Factoring is working with the distributive property in reverse. To factor an expression means to write the sum or difference as a product.

EXAMPLES

Factor the expression. The common factor is in bold print.

- $6x + 9xy$

  Each of $6x$ and $9xy$ is divisible by $3x$: $6x = 2 \cdot \mathbf{3x}$ and $9xy = 3 \cdot \mathbf{3x} \cdot y$. When we divide $6x$ by $3x$, we are left with 2. When we divide $9xy$ by $3x$, we are left with $3y$.

  $$6x + 9xy = 2 \cdot \mathbf{3x} + 3 \cdot \mathbf{3x} \cdot y = \mathbf{3x}(2 + 3y)$$

- $x^2y + 4xy^2 + 5x = \boldsymbol{x} \cdot xy + 4 \cdot \boldsymbol{x} \cdot y^2 + 5 \cdot \boldsymbol{x} = \boldsymbol{x}(xy + 4y^2 + 5)$
- $8xh - 2xyh + 7yh^2 + h = 8x \cdot \boldsymbol{h} - 2xy \cdot \boldsymbol{h} + 7yh \cdot \boldsymbol{h} + 1 \cdot \boldsymbol{h} = \boldsymbol{h}(8x - 2xy + 7yh + 1)$

Using the distributive property on such quantities as $(x + 2)(y - 3)$ requires several steps. In this book, we will use the FOIL method. The letters in "FOIL" help us to keep track of which quantities are multiplied and which are added. The "F" stands for "first times first." We multiply the first two quantities. In $(x + 2)(y - 3)$, this means we multiply x and y. "O" stands for "outer times outer." We multiply the outside quantities: x and -3. "I" stands for "inner times inner." We multiply the inside quantities: 2 and y. "L" stands for "last times last." We multiply the last quantities: 2 and -3.

$$(x + 2)(y - 3) = \overbrace{xy}^{F\times F} + \overbrace{(-3)x}^{O\times O} + \overbrace{2y}^{I\times I} + \overbrace{2(-3)}^{L\times L} = xy - 3x + 2y - 6$$

EXAMPLES

- $(2x - 5)(x + 3) = 2x \cdot x + 2x \cdot 3 + (-5)x + (-5)3$

  $$= 2x^2 + 6x - 5x - 15 = 2x^2 + x - 15$$

- $(4x - 3)(4x + 3) = 4x \cdot 4x + 4x \cdot 3 + (-3)4x + (-3)3$

  $$= 16x^2 + 12x - 12x - 9 = 16x^2 - 9$$

- $(x^2 + 7)(x - 2) = x^2 \cdot x + x^2(-2) + 7 \cdot x + 7(-2)$

  $$= x^3 - 2x^2 + 7x - 14$$

Expressions in the form $ax^2 + bx + c$ are *quadratic* expressions. The letters a, b, and c stand for fixed numbers.

EXAMPLES

- $x^2 - x - 6$ $\quad a = 1 \quad b = -1 \quad c = -6$
- $x^2 + 7x + 10$ $\quad a = 1 \quad b = 7 \quad c = 10$
- $3x^2 + 10x - 8$ $\quad a = 3 \quad b = 10 \quad c = -8$
- $9x^2 - 4$ $\quad a = 9 \quad b = 0 \quad c = -4$
- $2x^2 + x$ $\quad a = 2 \quad b = 1 \quad c = 0$

Many quadratic expressions can be factored with little trouble. We will begin with expressions of the form $x^2 + bx + c$. The first step is to write $(x \quad)(x \quad)$ so that when we use the FOIL method, the first term is $x \cdot x = x^2$. Next, we will choose two numbers whose product is c. For example, if we factor $x^2 + 6x + 5$, we would try 5 and 1: $(x \quad 5)(x \quad 1)$. Finally, we will decide if we need to use two plus signs, two minus signs, or one of each. The second sign in $x^2 + 6x + 5$ tells us whether or not the signs are the same. If the second sign is plus, then both signs are the same. The second sign is plus, so both signs in $(x \quad 5)(x \quad 1)$ are the same. If the signs are the same, then they will be the first sign. In $x^2 + 6x + 5$ the first sign is plus, so we need to plus signs in $(x \quad 5)(x \quad 1)$: $(x + 5)(x + 1)$. We will use the FOIL method on $(x + 5)(x + 1)$ to see if our factorization is correct.

$$(x + 5)(x + 1) = x \cdot x + x \cdot 1 + 5 \cdot x + 5 \cdot 1 = x^2 + 6x + 5 \checkmark$$

EXAMPLE

- Factor $x^2 - 2x - 15$.

  We have several choices for $(x \quad)(x \quad)$. Beginning with the factors of 15, we need to choose between 1 and 15 or 3 and 5. That is, we either want $(x \quad 1)(x \quad 15)$ or $(x \quad 3)(x \quad 5)$. Because the second sign in $x^2 - 2x - 15$ is a minus sign, the signs in the factors are different. We have four possibilities.

  $(x - 1)(x + 15) \qquad (x + 1)(x - 15) \qquad (x - 3)(x + 5) \qquad (x + 3)(x - 5)$

  The last possibility is correct: $(x + 3)(x - 5) = x^2 - 5x + 3x - 15 = x^2 - 2x - 15$.

If both signs in the factors are the same, b is the sum of the factors of c. If the signs in the factors are different, the difference of the factors of c is b. In the first example, the sum of 1 and 5 is 6. In the second example, the difference of 5 and 3 is 2.

EXAMPLES

- $x^2 - x - 6 = (x-\quad)(x+\quad)$

  The signs are different, so the difference of the factors of 6 is 1. We will choose 2 and 3 (instead of 6 and 1, whose difference is 5). The first sign in $x^2 - x - 6$ is a minus sign, so the larger factor has the minus sign. The factorization is $(x-3)(x+2)$.

- $x^2 + 7x + 10 = (x+\quad)(x+\quad)$

  Both signs are plus, so the sum of the factors of 10 is 7. The sum of 5 and 2 is 7. The factorization is $x^2 + 7x + 10 = (x+2)(x+5)$.

- $3x^2 + 10x - 8$

  When a is not 1 (here a is 3), factoring is a little more work. We always begin factoring by deciding what two factors give us ax^2. Here we need two factors that give us $3x^2$. We will try $3x$ and x. Because the signs in $3x^2+10x-8$ are different, one of $(3x\quad)$ and $(x\quad)$ has a plus sign and the other has a minus sign. Now we have $(3x+\quad)(x-\quad)$ and $(3x-\quad)(x+\quad)$. We have two pairs of factors of 8 to try: 1 and 8 and 2 and 4. There are eight possibilities.

$$(3x+1)(x-8) \qquad (3x-1)(x+8) \qquad (3x-2)(x+4) \qquad (3x+2)(x-4)$$

$$(3x+8)(x-1) \qquad (3x-8)(x+1) \qquad (3x-4)(x+2) \qquad (3x+4)(x-2)$$

  The correct factorization is $(3x-2)(x+4) = 3x^2 + 12x - 2x - 8 = 3x^2 + 10x - 8$.

- $9x^2 - 4$

  We factor quadratic expressions of the form $(ax)^2 - c^2$ with the formula $A^2 - B^2 = (A-B)(A+B)$. In this example, $9x^2$ is $(3x)^2$ and 4 is 2^2.

$$9x^2 - 4 = \overbrace{(3x)^2}^{A^2} - \overbrace{2^2}^{B^2} = \overbrace{(3x-2)}^{A-B}\overbrace{(3x+2)}^{A+B}$$

  When the FOIL method is used on expressions of the form $(A-B)(A+B)$, the middle terms always cancel.

$$(3x-2)(3x+2) = 9x^2 + 6x - 6x - 4 = 9x^2 - 4$$

Some quadratic expressions do not factor easily. For example, $x^2 + x + 1$ cannot be factored using the techniques we have learned so far.

PRACTICE

Use the FOIL method for problems 1–4. Factor the expression in problems 5–10.

1. $(x - 8)(x + 3)$
2. $(5x - 2)(x + 4)$
3. $(x - 3)(x + 3)$
4. $(4x - 5)^2 = (4x - 5)(4x - 5)$
5. $x^2 - 3x + 2$
6. $x^2 - 3x - 4$
7. $x^2 + 5x - 6$
8. $x^2 - 16$
9. $25x^2 - 9$
10. $4x^2 + 11x - 3$

SOLUTIONS

1. $(x - 8)(x + 3) = x^2 + 3x - 8x - 24 = x^2 - 5x - 24$
2. $(5x - 2)(x + 4) = 5x^2 + 20x - 2x - 8 = 5x^2 + 18x - 8$
3. $(x - 3)(x + 3) = x^2 + 3x - 3x - 9 = x^2 - 9$
4. $(4x - 5)^2 = (4x - 5)(4x - 5) = 16x^2 - 20x - 20x + 25 = 16x^2 - 40x + 25$
5. $x^2 - 3x + 2 = (x - 1)(x - 2)$
6. $x^2 - 3x - 4 = (x - 4)(x + 1)$
7. $x^2 + 5x - 6 = (x + 6)(x - 1)$
8. $x^2 - 16 = (x - 4)(x + 4)$
9. $25x^2 - 9 = (5x - 3)(5x + 3)$
10. $4x^2 + 11x - 3 = (4x - 1)(x + 3)$

Fractions

A fraction is reduced to its lowest terms, or simplified, when the numerator and denominator have no common factors. The fraction $\frac{2x}{6}$ is not reduced to its lowest terms because the numerator, $2x$, and denominator, 6, are each divisible by 2. We simplify fractions by factoring the numerator and denominator, using their common factors, and canceling.

EXAMPLES

Reduce the fraction to its lowest terms.

- $\dfrac{2x}{6} = \dfrac{2 \cdot x}{2 \cdot 3} = \dfrac{x}{3}$
- $\dfrac{4x^2y}{6xy} = \dfrac{2xy \cdot 2x}{2xy \cdot 3} = \dfrac{2x}{3}$
- $\dfrac{10xy^2 - 8xy}{12x^2y^2} = \dfrac{2xy(5y-4)}{2xy \cdot 6xy} = \dfrac{5y-4}{6xy}$
- $\dfrac{3xh - h^2 + h}{4h} = \dfrac{h(3x - h + 1)}{h \cdot 4} = \dfrac{3x - h + 1}{4}$
- $\dfrac{x^2 + 3x - 18}{x - 3} = \dfrac{(x-3)(x+6)}{(x-3) \cdot 1} = x + 6$
- $\dfrac{x^2 + 3x + 2}{3x + 6} = \dfrac{(x+2)(x+1)}{(x+2) \cdot 3} = \dfrac{x+1}{3}$
- $\dfrac{x^2 + 5x + 4}{x^2 - 1} = \dfrac{(x+1)(x+4)}{(x+1)(x-1)} = \dfrac{x+4}{x-1}$

Only *factors* can be canceled in a fraction. For example, $\frac{2+x}{2}$ cannot be reduced. It is incorrect to "cancel" the 2 from the numerator and denominator, $\frac{2+x}{2}$ is not the same as x nor as $1 + x$. We can rewrite the expression as the sum of two fractions and reduce one of them.

$$\frac{2+x}{2} = \frac{2}{2} + \frac{x}{2} = 1 + \frac{x}{2}$$

PRACTICE

Reduce the fraction to lowest terms.

1.

$$\frac{15xy^2}{20x^2y}$$

2.

$$\frac{4h^2}{h}$$

3.

$$\frac{12x^2y + 6xy^2}{6xy - 18xy^2}$$

4.

$$\frac{x^2 - x - 12}{x^2 + 4x + 3}$$

5.

$$\frac{x^2 - 9x + 20}{x^2 - 25}$$

SOLUTIONS

1.

$$\frac{15xy^2}{20x^2y} = \frac{5xy \cdot 3y}{5xy \cdot 4x} = \frac{3y}{4x}$$

2.

$$\frac{4h^2}{h} = \frac{h \cdot 4h}{h \cdot 1} = 4h$$

3.

$$\frac{12x^2y + 6xy^2}{6xy - 18xy^2} = \frac{6xy(2x + y)}{6xy(1 - 3y)} = \frac{2x + y}{1 - 3y}$$

4.

$$\frac{x^2 - x - 12}{x^2 + 4x + 3} = \frac{(x + 3)(x - 4)}{(x + 3)(x + 1)} = \frac{x - 4}{x + 1}$$

5.

$$\frac{x^2-9x+20}{x^2-25}=\frac{(x-5)(x-4)}{(x-5)(x+5)}=\frac{x-4}{x+5}$$

Compound fractions have a fraction in the numerator, denominator, or both. Often these fractions can be simplified by writing the compound fraction as a product of two separate fractions. Remember that the fraction $\frac{a}{b}$ is another way of writing $a \div b$ and that $\frac{a}{b} \div \frac{c}{d}$ is the same as $\frac{a}{b} \cdot \frac{d}{c}$.

EXAMPLES

Simplify the fraction.

- $\frac{\frac{2}{3}}{\frac{1}{2}}=\frac{2}{3}\div\frac{1}{2}=\frac{2}{3}\cdot\frac{2}{1}=\frac{4}{3}$

- $\frac{\frac{5x}{3}}{15}=\frac{5x}{3}\div 15=\frac{5x}{3}\cdot\frac{1}{15}=\frac{5x}{45}=\frac{5\cdot x}{5\cdot 9}=\frac{x}{9}$

- $\frac{\frac{x}{x^2-9}}{\frac{1}{x-3}}=\frac{x}{x^2-9}\div\frac{1}{x-3}=\frac{x}{x^2-9}\cdot\frac{x-3}{1}=\frac{(x-3)\cdot x}{(x-3)(x+3)}=\frac{x}{x+3}$

- $\frac{\frac{5h}{x+h}}{h}=\frac{5h}{x+h}\div h=\frac{5h}{x+h}\cdot\frac{1}{h}=\frac{5}{x+h}$

- $\frac{\frac{1}{x+h}-\frac{1}{x}}{h}$

We will begin by writing $\frac{1}{x+h}-\frac{1}{x}$ as one fraction.

$$\begin{aligned}\frac{1}{x+h}-\frac{1}{x}&=\frac{1}{x+h}\cdot\frac{x}{x}-\frac{1}{x}\cdot\frac{x+h}{x+h}\\&=\frac{x}{x(x+h)}-\frac{x+h}{x(x+h)}=\frac{x-(x+h)}{x(x+h)}\\&=\frac{x-x-h}{x(x+h)}=\frac{-h}{x(x+h)}\end{aligned}$$

We will replace $\frac{1}{x+h} - \frac{1}{x}$ with $\frac{-h}{x(x+h)}$.

$$\frac{\frac{1}{x+h} - \frac{1}{x}}{h} = \frac{\frac{-h}{x(x+h)}}{h} = \frac{-h}{x(x+h)} \div h = \frac{-h}{x(x+h)} \cdot \frac{1}{h} = \frac{-1}{x(x+h)}$$

PRACTICE

Simplify the fraction.

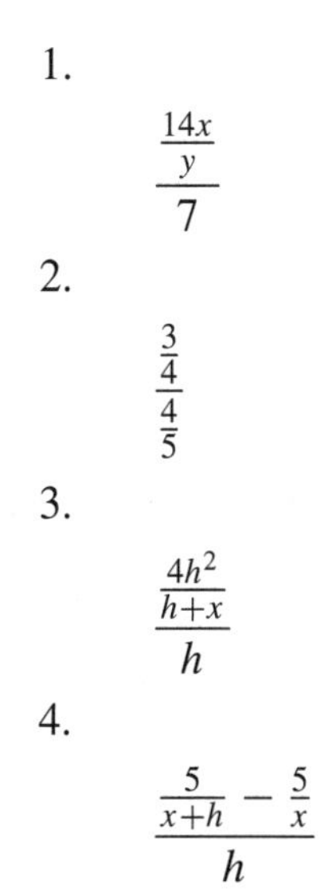

1. $$\frac{\frac{14x}{y}}{7}$$

2. $$\frac{\frac{3}{4}}{\frac{4}{5}}$$

3. $$\frac{\frac{4h^2}{h+x}}{h}$$

4. $$\frac{\frac{5}{x+h} - \frac{5}{x}}{h}$$

SOLUTIONS

1. $$\frac{\frac{14x}{y}}{7} = \frac{14x}{y} \div 7 = \frac{14x}{y} \cdot \frac{1}{7} = \frac{2x}{y}$$

2. $$\frac{\frac{3}{4}}{\frac{4}{5}} = \frac{3}{4} \div \frac{4}{5} = \frac{3}{4} \cdot \frac{5}{4} = \frac{15}{16}$$

3. $$\frac{\frac{4h^2}{h+x}}{h} = \frac{4h^2}{h+x} \div h = \frac{4h^2}{h+x} \cdot \frac{1}{h} = \frac{4h}{h+x}$$

4.

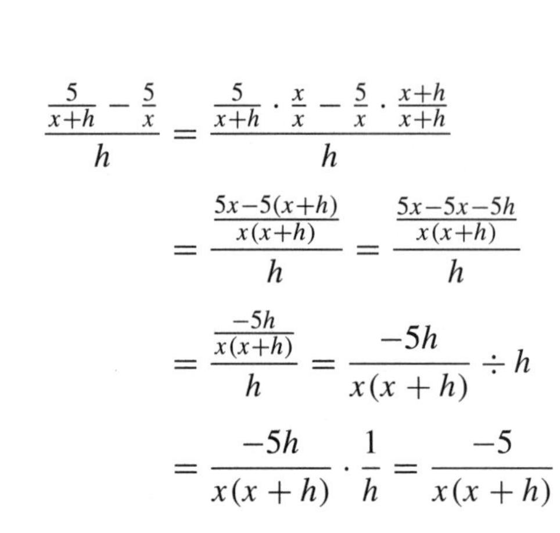

$$\frac{\frac{5}{x+h}-\frac{5}{x}}{h}=\frac{\frac{5}{x+h}\cdot\frac{x}{x}-\frac{5}{x}\cdot\frac{x+h}{x+h}}{h}$$

$$=\frac{\frac{5x-5(x+h)}{x(x+h)}}{h}=\frac{\frac{5x-5x-5h}{x(x+h)}}{h}$$

$$=\frac{\frac{-5h}{x(x+h)}}{h}=\frac{-5h}{x(x+h)}\div h$$

$$=\frac{-5h}{x(x+h)}\cdot\frac{1}{h}=\frac{-5}{x(x+h)}$$

Exponents and Roots

In order to use two important formulas in calculus, we need exponent and root properties to rewrite expressions as quantities raised to a power. Properties 5–7 below are the most important.

1. $a^m \cdot a^n = a^{m+n}$
2. $\dfrac{a^m}{a^n} = a^{m-n}$
3. $(a^m)^n = a^{mn}$
4. $a^0 = 1$
5. $\dfrac{1}{a^n} = a^{-n}$
6. $\sqrt[n]{a} = a^{1/n}$
7. $\sqrt[n]{a^m} = a^{m/n}$

EXAMPLES

Use Properties 5–7 to rewrite the original expression as a quantity to a power.

- $\sqrt[3]{x} = x^{1/3}$ — Property 6
- $\dfrac{1}{x^6} = x^{-6}$ — Property 5
- $\sqrt{x} = \sqrt[2]{x} = x^{1/2}$ — Property 6
- $\sqrt{x^3} = \sqrt[2]{x^3} = x^{3/2}$ — Property 7

- $\frac{1}{\sqrt{x}} = \frac{1}{x^{1/2}} = x^{-1/2}$ Properties 6 and 5
- $\frac{1}{\sqrt[3]{x-8}} = \frac{1}{(x-8)^{1/3}} = (x-8)^{-1/3}$ Properties 6 and 5

PRACTICE

Use Properties 5–7 to rewrite the original expression as a quantity to a power.

1.

$$\frac{1}{x}$$

2.

$$\frac{1}{x^2}$$

3.

$$\sqrt[4]{x^3}$$

4.

$$\frac{1}{\sqrt[4]{x^3}}$$

5.

$$\frac{1}{(3x^2+4)^2}$$

6.

$$\frac{1}{\sqrt{x+4}}$$

SOLUTIONS

1.

$$\frac{1}{x} = \frac{1}{x^1} = x^{-1}$$

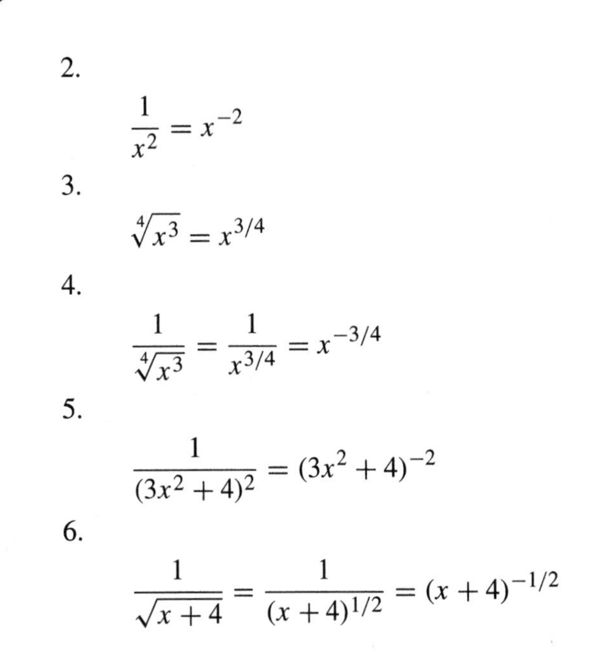

2.
$$\frac{1}{x^2} = x^{-2}$$

3.
$$\sqrt[4]{x^3} = x^{3/4}$$

4.
$$\frac{1}{\sqrt[4]{x^3}} = \frac{1}{x^{3/4}} = x^{-3/4}$$

5.
$$\frac{1}{(3x^2+4)^2} = (3x^2+4)^{-2}$$

6.
$$\frac{1}{\sqrt{x+4}} = \frac{1}{(x+4)^{1/2}} = (x+4)^{-1/2}$$

Miscellaneous Notation

Interval notation is used to describe regions on the number line. The infinity symbols, ∞ and $-\infty$, are used for unbounded intervals. A parenthesis around a number means that the number is not included in the interval. A bracket around a number means that the number is included in the interval (see Figure R.1).

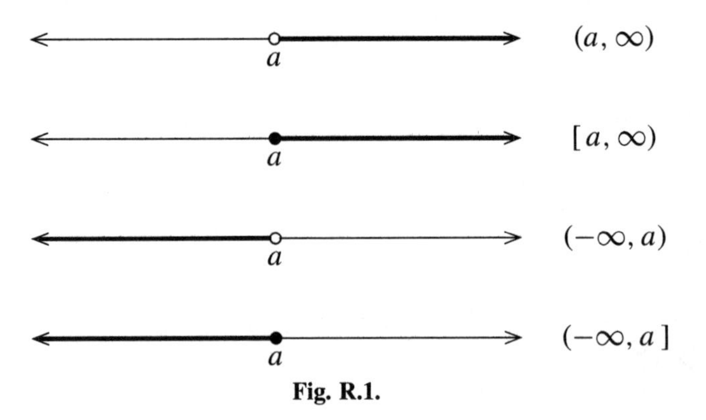

Fig. R.1.

The region between two numbers $x = a$ and $x = b$ (with a smaller than b), is one of (a, b), $(a, b]$, $[a, b)$ or $[a, b]$, depending on whether a and/or b is included in the interval (see Figure R.2).

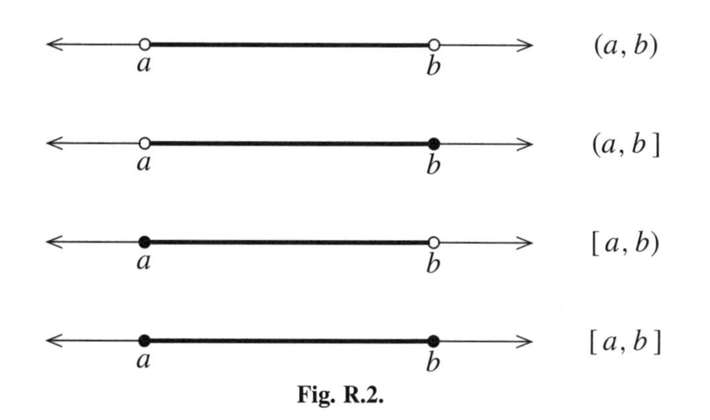

Fig. R.2.

EXAMPLE

Match the shaded regions in Figure R.3 with the interval.

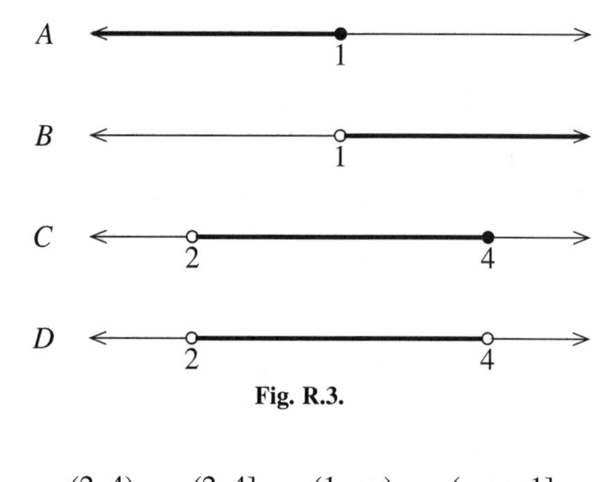

Fig. R.3.

$(2, 4)$ $(2, 4]$ $(1, \infty)$ $(-\infty, 1]$

$(2, 4)$ describes Graph D. $(2, 4]$ describes Graph C. $(1, \infty)$ describes Graph B. $(-\infty, 1]$ describes Graph A.

The union symbol, "$\cup$," is used to describe two or more regions. For example, $(-\infty, 3) \cup (5, \infty)$ describes all numbers smaller than 3 or all numbers larger than 5 (see Figure R.4).

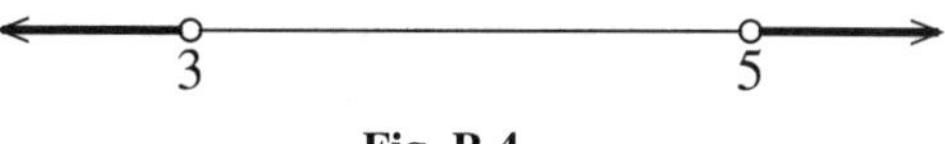

Fig. R.4.

EXAMPLE

Match the shaded regions in Figure R.5 with the intervals.

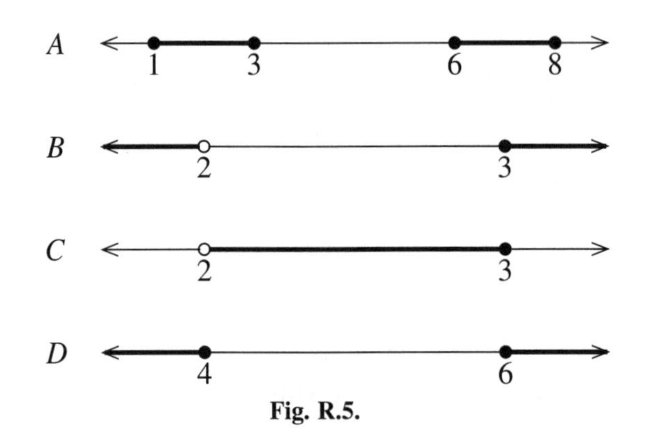

Fig. R.5.

$(2, 3]$ $(-\infty, 2) \cup [3, \infty)$ $[1, 3] \cup [6, 8]$ $(-\infty, 4] \cup [6, \infty)$

$(2, 3]$ describes Graph C. $(-\infty, 2) \cup [3, \infty)$ describes Graph B. $[1, 3] \cup [6, 8]$ describes graph A. $(-\infty, 4] \cup [6, \infty)$ describes Graph D.

PRACTICE

Match the shaded regions in Figures R.6 and R.7 with the intervals.

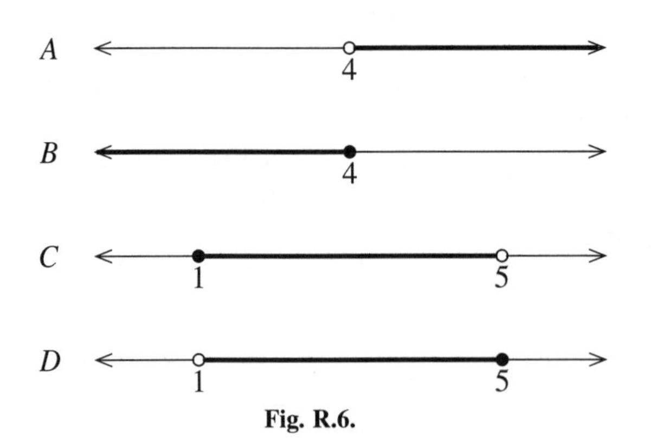

Fig. R.6.

1. $(-\infty, 4]$
2. $(1, 5]$
3. $[1, 5)$
4. $(4, \infty)$

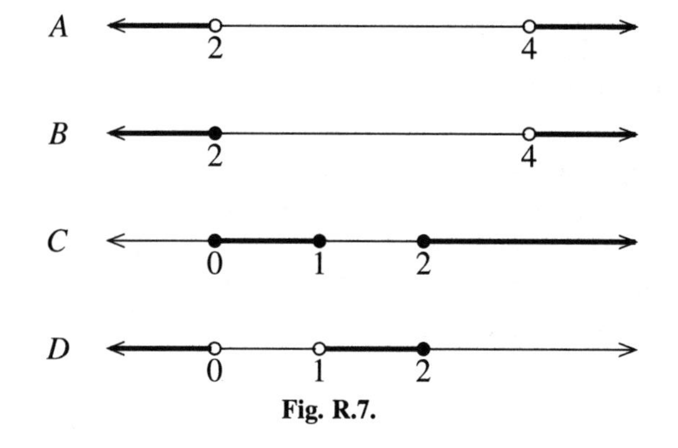

Fig. R.7.

5. $(-\infty, 2] \cup (4, \infty)$
6. $[0, 1] \cup [2, \infty)$
7. $(-\infty, 0) \cup (1, 2]$
8. $(-\infty, 2) \cup (4, \infty)$

SOLUTIONS

1. B
2. D
3. C
4. A
5. B
6. C
7. D
8. A

The Greek letter sigma, "Σ," is used in Chapter 14 to describe a sum. There is usually a subscript and a superscript on Σ. The subscript tells us where the sum begins, and the superscript tells us where the sum ends.

- $\sum_{i=1}^{4} 3i$

This sum begins at $3 \cdot 1\ (i = 1)$ and ends at $3 \cdot 4\ (i = 4)$.

$$\sum_{i=1}^{4} 3i = \overbrace{3(1)}^{i=1} + \overbrace{3(2)}^{i=2} + \overbrace{3(3)}^{i=3} + \overbrace{3(4)}^{i=4} = 3 + 6 + 9 + 12 = 30$$

The absolute value of a number is its distance from 0. The absolute value of -5 is 5 because it is 5 units away from 0. The absolute value of a positive number is the number itself. The absolute value of a quantity is denoted with absolute value bars, "$|\ |$." The absolute value of -5 is denoted "$|-5|$." This notation is used on occasion beginning in Chapter 13.

Solving Equations

We will solve many equations in this book, most of them linear equations or quadratic equations. Linear equations are the easiest to solve. Our goal is to write the equation so that the term(s) having an x is (are) on one side of the equation and terms without an x are on the other side. Once this is done, we will divide both sides of the equation by the coefficient of x (the number multiplying x).

EXAMPLES

- $4x + 10 = 0$

$$4x + 10 = 0$$

$$-10 \quad -10 \qquad \text{Move the non-}x\text{ term to the right side.}$$

$$4x = -10$$

$$x = \frac{-10}{4} \qquad \text{Divide both sides by 4, the coefficient of } x.$$

$$x = -\frac{5}{2}$$

- $\frac{2}{3}x - 8 = 0$

$$\frac{2}{3}x - 8 = 0$$

$$\frac{2}{3}x = 8 \qquad \text{Add 8 to both sides.}$$

$$x = \frac{3}{2} \cdot 8 \qquad \text{Dividing by } \frac{2}{3} \text{ is the same as multiplying by } \frac{3}{2}.$$

$$x = 12$$

A quadratic equation is an equation that can be put in the form $ax^2+bx+c = 0$. Most quadratic equations in this book can be solved by factoring. We will use the quadratic formula on others. No matter which method we use, we need to have a zero on one side of the equation. Once this is done, we will try to factor the quadratic expression $ax^2 + bx + c$. If it factors easily, we will set each factor equal to zero and will solve for x. If it does not factor easily, we will use the quadratic formula. Most of these equations have two solutions.

EXAMPLES

- $x^2 - 2x - 3 = 0$

  $x^2 - 2x - 3$ factors as $(x - 3)(x + 1)$. We will set each of $x - 3$ and $x + 1$ equal to zero.

$$\begin{aligned} x^2 - 2x - 3 &= 0 \\ (x-3)(x+1) &= 0 \\ x - 3 &= 0 \qquad x + 1 = 0 \\ x &= 3 \qquad\quad x = -1 \end{aligned}$$

- $3x^2 + x - 2 = 0$

$$\begin{aligned} 3x^2 + x - 2 &= 0 \\ (3x-2)(x+1) &= 0 \\ 3x - 2 &= 0 \qquad x + 1 = 0 \\ 3x &= 2 \qquad\quad x = -1 \\ x &= \frac{2}{3} \end{aligned}$$

The quadratic formula can solve any quadratic equation. If $ax^2+bx+c = 0$, then

$$x = \frac{-b \pm \sqrt{b^2 - 4ac}}{2a}.$$

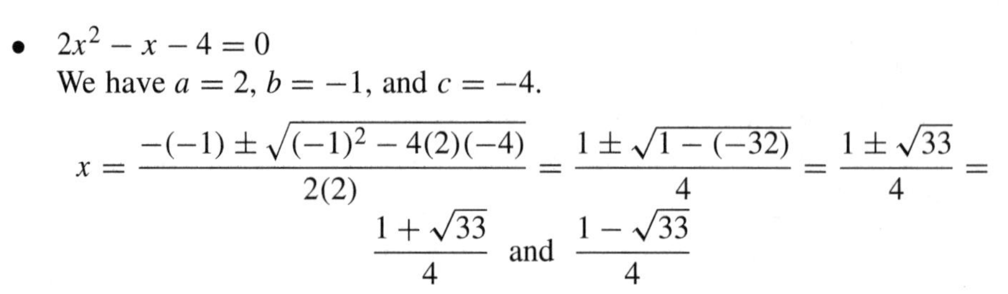

- $2x^2 - x - 4 = 0$
  We have $a = 2$, $b = -1$, and $c = -4$.

$$x = \frac{-(-1) \pm \sqrt{(-1)^2 - 4(2)(-4)}}{2(2)} = \frac{1 \pm \sqrt{1 - (-32)}}{4} = \frac{1 \pm \sqrt{33}}{4} = \frac{1 + \sqrt{33}}{4} \text{ and } \frac{1 - \sqrt{33}}{4}$$

When an equation is in the form "fraction = fraction," we will cross-multiply to solve for x. That is, we will multiply the numerator of each fraction by the denominator of the other fraction.

$$\frac{a}{b} = \frac{c}{d}$$ Multiply a by d and c by b.

$$ad = bc$$

- $\frac{4x}{5} = \frac{1}{x}$

$$4x \cdot x = 5 \cdot 1$$

$$4x^2 = 5$$

$4x^2 = 5$ is a quadratic equation. We could use the quadratic formula for $4x^2 - 5 = 0$, but we can solve it more quickly by dividing both sides of the equation by 4 and then taking the square root of each side.

$$x^2 = \frac{5}{4}$$

$$x = \pm\sqrt{\frac{5}{4}} = -\sqrt{\frac{5}{4}}, \sqrt{\frac{5}{4}}$$

PRACTICE

Solve the equation.

1. $\frac{4}{5}x + 8 = 0$
2. $x^2 - 2x - 8 = 0$
3. $5x^2 - 7x - 6 = 0$
4. $x^2 - 3x - 6 = 0$ (Hint: use the quadratic formula.)
5. $2x^2 + 7x + 1 = 0$

6.

$$\frac{2x}{3} = \frac{5}{7x}$$

SOLUTIONS

1.

$$\frac{4}{5}x + 8 = 0$$

$$\frac{4}{5}x = -8$$

$$x = \frac{5}{4} \cdot -8 = -10$$

2.

$$x^2 - 2x - 8 = 0$$

$$(x - 4)(x + 2) = 0$$

$$x - 4 = 0 \qquad x + 2 = 0$$

$$x = 4 \qquad x = -2$$

3.

$$5x^2 - 7x - 6 = 0$$

$$(5x + 3)(x - 2) = 0$$

$$5x + 3 = 0 \qquad x - 2 = 0$$

$$5x = -3 \qquad x = 2$$

$$x = -\frac{3}{5}$$

4. $a = 1$, $b = -3$, and $c = -6$

$$x = \frac{-(-3) \pm \sqrt{(-3)^2 - 4(1)(-6)}}{2(1)} = \frac{3 \pm \sqrt{9 + 24}}{2} = \frac{-3 \pm \sqrt{33}}{2}$$

$$= \frac{-3 + \sqrt{33}}{2} \text{ and } \frac{-3 - \sqrt{33}}{2}$$

5. $a = 2$, $b = 7$, and $c = 1$

$$x = \frac{-7 \pm \sqrt{7^2 - 4(2)(1)}}{2(2)} = \frac{-7 \pm \sqrt{49 - 8}}{4} = \frac{-7 \pm \sqrt{41}}{4}$$

$$= \frac{-7 + \sqrt{41}}{4} \text{ and } \frac{-7 - \sqrt{41}}{4}$$

6.

$$\frac{2x}{3} = \frac{5}{7x}$$

$$2x \cdot 7x = 3 \cdot 5$$

$$14x^2 = 15$$

$$x^2 = \frac{15}{14}$$

$$x = \pm\sqrt{\frac{15}{14}} = -\sqrt{\frac{15}{14}}, \sqrt{\frac{15}{14}}$$

The Equation of a Line

Throughout much of the book, we find equations of lines. Although there are several forms for the equation of a line, we will use the form $y = mx + b$. We will be given an x-value, a y-value, and m. (Later, we will use a formula to find m.) Having values for x, y, and m, gives us enough information to find b.

EXAMPLES

Find an equation of the line with the given values.

- $x = 2$, $y = 8$, and $m = 3$

  We will substitute 2 for x, 8 for y, and 3 for m in $y = mx + b$ to find b.

$$8 = 3(2) + b$$

$$8 = 6 + b$$

$$2 = b$$

  The equation is $y = 3x + 2$.

- $x = -1$, $y = 5$, and $m = \frac{1}{2}$

$$5 = \frac{1}{2}(-1) + b$$

$$5 + \frac{1}{2} = b$$

$$\frac{11}{2} = b$$

The equation is $y = \frac{1}{2}x + \frac{11}{2}$.

PRACTICE

Find an equation of the line with the given values.

1. $x = 4$, $y = -3$, and $m = 2$
2. $x = -2$, $y = 5$, and $m = -1$

SOLUTIONS

1.

$$-3 = 2(4) + b$$

$$-11 = b$$

The line is $y = 2x - 11$.

2.

$$5 = -1(-2) + b$$

$$3 = b$$

The line is $y = -1 \cdot x + 3 = -x + 3$.

Functions and Their Graphs

The definition of a function is a relation between two sets A and B such that every element in set A is paired with exactly one element in set B. Functions in this book are equations, usually with the variables x and y. At times, we will use the name of the function such as $f(x)$, $C(x)$, $R(x)$, etc., instead of y. Here are some examples of functions.

- $y = 2x - 1$
- $f(x) = x^2 + x - 2$
- $R(x) = \sqrt{x}$

To *evaluate* a function at a number or expression means to substitute the number or expression for x. This gives us the y-value, also called the *functional value*, for a particular x-value. The notation "$f(6)$" means that 6 has been substituted in the equation for x.

EXAMPLES

Evaluate the function at the given value of x.

- $f(x) = x^2 + 3x - 4$; 1, 5

$$f(1) = (1)^2 + 3(1) - 4 = 0 \qquad f(5) = (5)^2 + 3(5) - 4 = 36$$

- $f(x) = \sqrt{x-5}$; 5, 14

$$f(5) = \sqrt{5-5} = \sqrt{0} = 0 \qquad f(14) = \sqrt{14-5} = \sqrt{9} = 3$$

- $C(x) = \dfrac{20}{x+15}$, 10, 40

$$C(10) = \frac{20}{10+15} = \frac{20}{25} = \frac{4}{5} \qquad C(40) = \frac{20}{40+15} = \frac{20}{55} = \frac{4}{11}$$

- $f(x) = 100$; 6, 28

  $f(x) = 100$ is a linear function whose slope is 0. No matter what value x has, the functional value (the y-value) is always 100.

$$f(6) = 100 \qquad f(28) = 100$$

In Chapter 3, we evaluate functions at algebraic expressions. Again, we will substitute the given quantity for x.

EXAMPLES

Evaluate the function at the given quantity.

- $f(x) = 4x + 3$; $a + 2b$ and $5l$

$$f(a+2b) = 4(a+2b) + 3 = 4a + 8b + 3$$

$$f(5l) = 4(5l) + 3 = 20l + 3$$

- $f(x) = x^2 + 8x - 10$; $5w$ and $l + 3$

$$f(5w) = (5w)^2 + 8(5w) - 10 = 25w^2 + 40w - 10$$

$$f(l+3) = (l+3)^2 + 8(l+3) - 10 = (l+3)(l+3) + 8(l+3) - 10$$
$$= l^2 + 6l + 9 + 8l + 24 - 10 = l^2 + 14l + 23$$

- $f(x) = x^2 + 3$; $x + h$

$$f(x+h) = (x+h)^2 + 3 = (x+h)(x+h) + 3 = x^2 + 2xh + h^2 + 3$$

- $f(x) = \dfrac{x}{2x+1}$; $x + h$

$$f(x+h) = \frac{x+h}{2(x+h)+1} = \frac{x+h}{2x+2h+1}$$

- $f(x) = \dfrac{7}{x-3}$; $x + h$

$$f(x+h) = \frac{7}{x+h-3}$$

PRACTICE

Evaluate the function at the given quantity.

1. $f(x) = -3x + 10$; 0, 4
2. $g(x) = x^3 - x^2 + x - 1$; 1, 3
3. $f(x) = \sqrt{x^2 + 1}$; -4, 5
4. $f(x) = \sqrt{x + 9}$; $l - 4$, $3l$
5. $f(x) = -6x + 2$; $x + h$
6. $g(x) = x^2 + 4x + 1$; $x + h$
7. $f(x) = \sqrt{4x - 8}$; $x + h$
8. $R(x) = \frac{1}{x+2}$; $x + h$

SOLUTIONS

1.

$$f(0) = -3(0) + 10 = 10$$

$$f(4) = -3(4) + 10 = -2$$

2.

$$g(1) = (1)^3 - (1)^2 + (1) - 1 = 0$$

$$g(3) = (3)^3 - (3)^2 + (3) - 1 = 20$$

3.

$$f(-4) = \sqrt{(-4)^2 + 1} = \sqrt{16 + 1} = \sqrt{17}$$

$$f(5) = \sqrt{5^2 + 1} = \sqrt{25 + 1} = \sqrt{26}$$

4.

$$f(l - 4) = \sqrt{l - 4 + 9} = \sqrt{l + 5} \qquad f(3l) = \sqrt{3l + 9}$$

5.

$$f(x + h) = -6(x + h) + 2 = -6x - 6h + 2$$

6.

$$g(x + h) = (x + h)^2 + 4(x + h) + 1 = (x + h)(x + h) + 4(x + h) + 1$$

$$= x^2 + 2xh + h^2 + 4x + 4h + 1$$

7.

$$f(x + h) = \sqrt{4(x + h) - 8} = \sqrt{4x + 4h - 8}$$

8.

$$R(x + h) = \frac{1}{x + h + 2}$$

The graph of an equation shows all the pairs of x and y that make the equation true. The graph in Figure R.8 shows the graph of $x + y = 4$ (or $y = -x + 4$). For every point (x, y) on the graph, the sum of the x-coordinate and y-coordinate is 4. For example, (3, 1) is on the graph because $3 + 1 = 4$.

The graph of a line having a zero slope is a horizontal line. The graph in Figure R.9 is the graph of $y = 5$ (or $y = 0x + 5$).

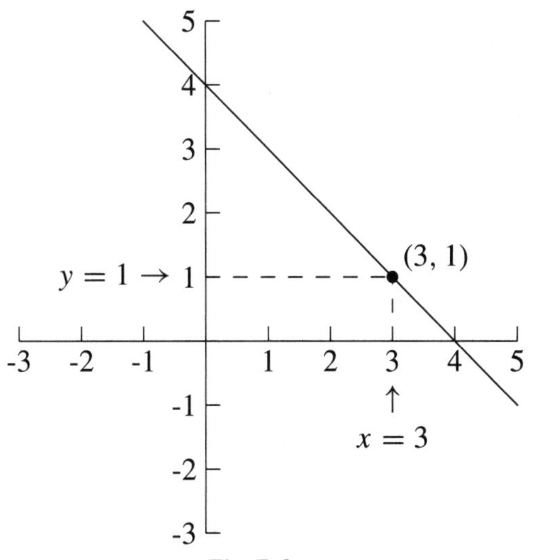

Fig. R.8.

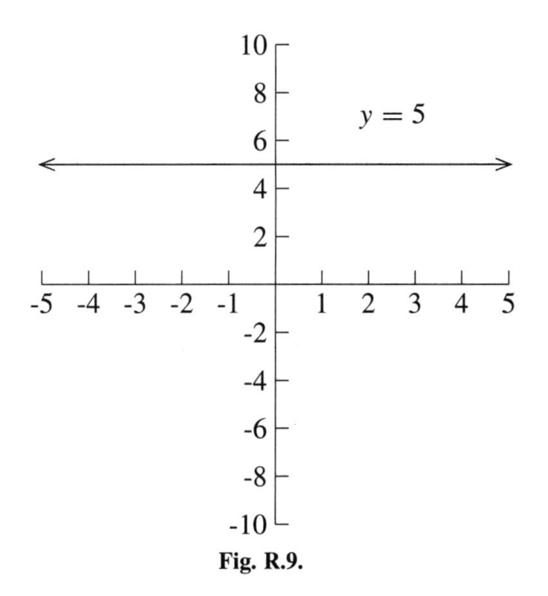

Fig. R.9.

We can look at the graph of a function to find functional values. The y-coordinate of the point (the second number) is the functional value for the x-coordinate of the point (the first number). For example, the point $(2, 5)$ is on the graph of $f(x) = x^2 + 1$ because $f(2) = 2^2 + 1 = 5$.

EXAMPLES

- The graph in Figure R.10 is the graph of $f(x) = x^2 - 3$. Use the graph to find $f(-1)$, $f(0)$, and $f(2)$.

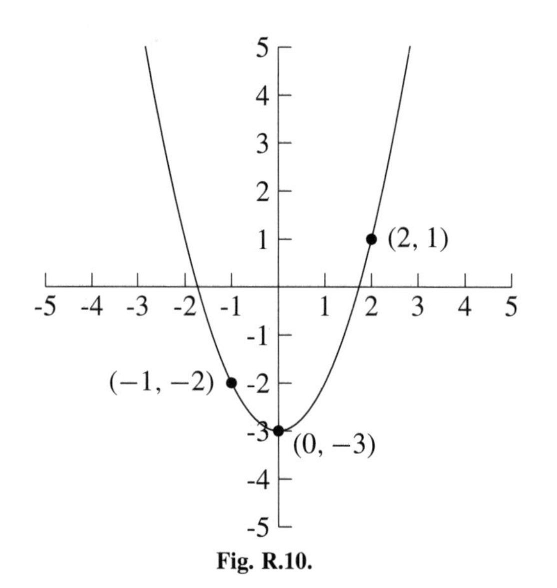

Fig. R.10.

The point $(-1, -2)$ is on the graph, so $f(-1) = -2$. The point $(0, -3)$ is on the graph, so $f(0) = -3$. The point $(2, 1)$ is on the graph, so $f(2) = 1$.
- The graph in Figure R.11 is the graph of a function $f(x)$. Find $f(4)$, $f(-3)$, and $f(-4)$.
The point $(4, 1)$ is on the graph, so $f(4) = 1$. The point $(-3, 1)$ is on the graph, so $f(-3) = 1$. There is a hole in the curve at $x = -4$, so the curve does not give us the functional value at $x = -4$. The dot at $(-4, -2)$ indicates that the function is defined there for $x = -4$. The dot is the point $(-4, -2)$, so $f(-4) = -2$.

PRACTICE

1. The graph of $f(x) = \sqrt{x}$ is given in Figure R.12. Find $f(4)$ and $f(9)$.
2. The graph of $f(x)$ is given in Figure R.13. Find $f(2)$, $f(4)$, and $f(1)$.

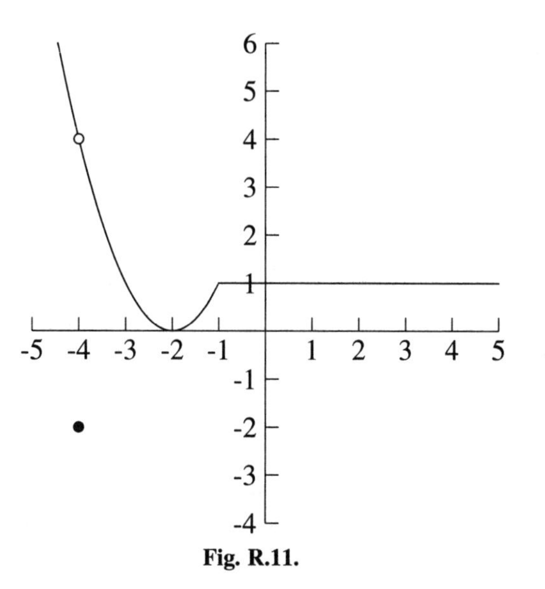

Fig. R.11.

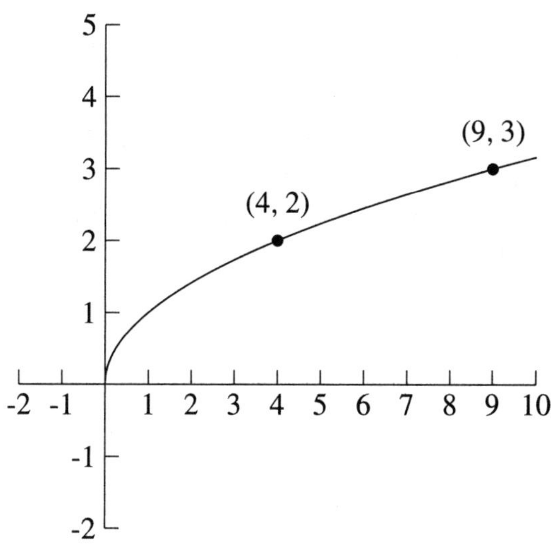

Fig. R.12.

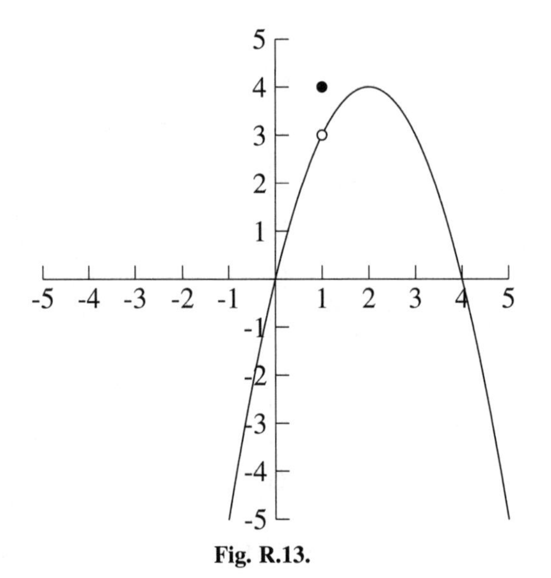

Fig. R.13.

SOLUTIONS

1. The point $(4, 2)$ is on the graph, so $f(4) = 2$. The point $(9, 3)$ is on the graph, so $f(9) = 3$.

2. The point $(2, 4)$ is on the graph, so $f(2) = 4$. The point $(4, 0)$ is on the graph, so $f(4) = 0$. There is a hole in the curve at $x = 1$, so $f(1)$ is not on the curve. There is a dot at $(1, 4)$, indicating that this point is part of the function, so $f(1) = 4$.

The Slope of a Line and the Average Rate of Change

Calculus is the study of the rate of change. We use the slope of a line to describe the rate of change of a function. Instead of thinking of the slope of a line as a simple "rise over run," we need to think of it as a number that measures how one variable changes compared to a change in the other variable. The numerator of the slope describes the change in y, and the denominator describes the change in x. For example, a slope of $\frac{3}{5}$ says that as x increases by 5, y increases by 3. A slope of $\frac{-3}{5}$ says that as x increases by 5, y decreases by 3.

EXAMPLES

Interpret the slope.

- $y = \frac{2}{3}x + 15$

  As x increases by 3, y increases by 2.
- $y = \frac{-20}{9}x + 4$

  As x increases by 9, y decreases by 20.
- $y = \frac{1}{8}x - 6$

  As x increases by 8, y increases by 1.
- $y = -0.03x + 10 = \frac{-0.03}{1}x + 10$

  As x increases by 1, y decreases by 0.03. If we view -0.03 as $\frac{-3}{100}$ instead, we see that as x increases by 100, y decreases by 3.
- $y = x = \frac{1}{1}x$

  As x increases by 1, y increases by 1.
- $y = 7$ (This is the same as $y = 0x + 7$.)

  The slope of this line is 0. If we think of 0 as $\frac{0}{1}$, then we see that as x increases by 1, y does not increase nor does it decrease. In other words, the y-value does not change. In fact, x can change by any amount and y does not change.
- The daily cost of producing x units of a product is given by the equation $y = 3.52x + 490$.

  The cost is y, and x is the number of units produced. The slope of $3.52 = \frac{3.52}{1}$ tells us that as x increases by 1, y increases by 3.52. In other words, each unit costs \$3.52 to produce.
- The property tax for a property valued at x dollars is $y = \frac{0.5981}{100}x$.

  As the value of property increases by \$100, the tax increases by \$0.5981.
- The demand function for a product is given by $y = -\frac{4}{5}x + 300$, where y units are demanded when x is the price per unit.

  As the demand increases by 5 units, the price decreases by \$4. (We could also interpret this slope to mean that as the price decreases by \$4, demand increases by 5 units.)
- The monthly salary for an office manager is given by the equation $y = 3800$.

  The slope of the line for this equation is 0, which means that no matter what happens to x, the y-value is always \$3800. No matter how much (or how little) the manager works, her monthly salary stays the same.

PRACTICE

Interpret the slope.

1. $y = \frac{7}{3}x - 8$
2. $y = -2x + 1$
3. $y = -x$
4. $y = 10$
5. The sales tax on purchases costing x dollars is $y = 0.08x$.
6. The pressure on a certain object submerged in the ocean is approximated by $y = 170x + 6000$, where x is the depth of the object, in feet, and y is the pressure, in pounds.
7. The nonfarm average weekly pay from 1997 to 2002 can be approximated by the equation $y = 15.09x - 29708$. (This equation is based on data from *The Statistical Abstract of the United States*, 123rd edition.)

SOLUTIONS

1. As x increases by 3, y increases by 7.
2. As x increases by 1, y decreases by 2.
3. As x increases by 1, y decreases by 1.
4. No matter how x changes, y does not change.
5. As the amount spent on purchases increases by \$1, the sales tax increases by \$0.08.
6. As the depth increases by 1 foot, the pressure on the object increases by 170 lbs.
7. The average weekly nonfarm wage increased by \$15.09 each year from 1997 to 2002.

The rate of change for most functions is not the same for all x-values as it is with linear functions. For example, if a cup of hot coffee sits on a table for ten minutes, it will cool down faster in the third minute than in the eighth minute. So, the rate of temperature change varies for different periods of time. For most of the functions in this book, the y-values will increase or decrease at different rates for different values of x. In fact, for some values of x, the y-values can increase and for other values of x, the y-values can decrease. We will look at the average rate of change of a function between two x-values. The average rate of change of the function between two values of x is the slope of the line containing the two points on the graph of the function.

EXAMPLES

- Find the average rate of change for $f(x) = x^2 - x - 2$ between $(1, -2)$ and $(3, 4)$ and between $(-2, 4)$ and $(0, -2)$.

  The average rate of change between $(1, -2)$ and $(3, 4)$ is the slope of the line between these two points.

$$\text{Average rate of change} = m = \frac{y_2 - y_1}{x_2 - x_1} = \frac{4 - (-2)}{3 - 1} = \frac{6}{2} = \frac{3}{1} = 3$$

  Between $x = 1$ and $x = 3$, the average increase of the function is 3 as x increases by 1. See Figure 1.1.

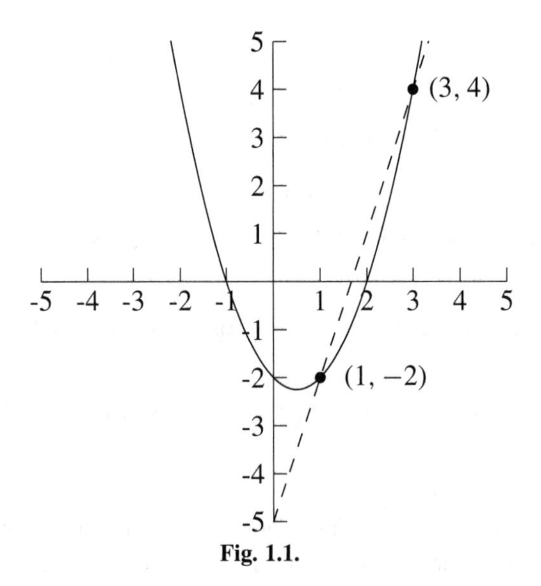

Fig. 1.1.

  The rate of change between $(-2, 4)$ and $(0, -2)$ is

$$\text{Average rate of change} = m = \frac{y_2 - y_1}{x_2 - x_1} = \frac{-2 - 4}{0 - (-2)} = \frac{-6}{2} = \frac{-3}{1} = -3.$$

Between $x = -2$ and $x = 0$, the average decrease of the function is 3 as x increases by 1. See Figure 1.2.

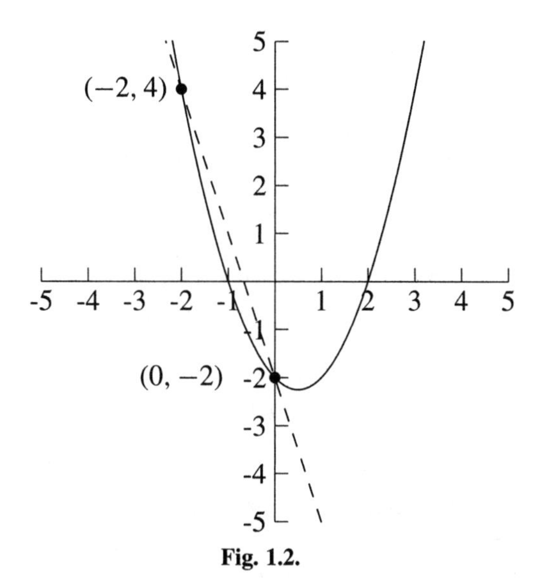

Fig. 1.2.

- Find the average rate of change for $f(x) = \sqrt{x+1}$ between $x = 3$ and $x = 8$.

  Once we have computed the y-values for $x = 3$ and $x = 8$, we will put the points into the slope formula.

  $y_1 = f(x_1) = f(3) = \sqrt{3+1} = 2 \qquad y_2 = f(x_2) = f(8) = \sqrt{8+1} = 3$

  The points are (3, 2) and (8, 3). The average rate of change is

$$\text{Average rate of change} = m = \frac{y_2 - y_1}{x_2 - x_1} = \frac{3-2}{8-3} = \frac{1}{5}.$$

  Between $x = 3$ and $x = 8$, the function increases by 1 on average as x increases by 5.

- Find the average rate of change for $f(x) = \frac{1}{2}x^4 - \frac{5}{2}x^2 - 3$ between $x = -2$ and $x = 2$. See Figure 1.3.

$$\text{Average rate of change} = m = \frac{y_2 - y_1}{x_2 - x_1} = \frac{-5-(-5)}{2-(-2)} = \frac{0}{4} = 0$$

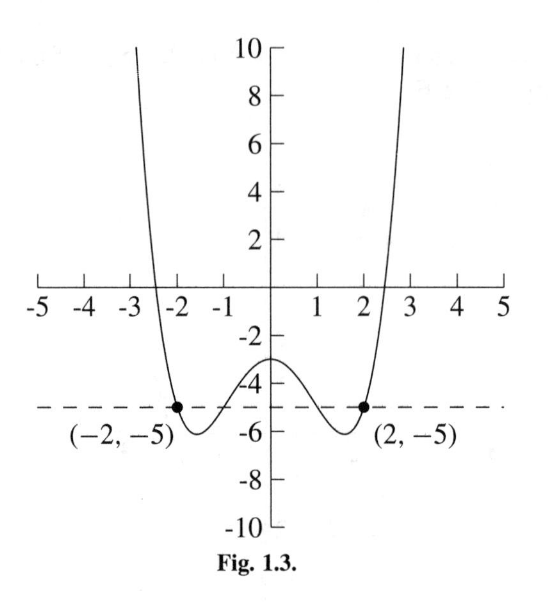

Fig. 1.3.

Because the slope of the line is 0, the average rate of change of the function is zero between $x = -2$ and $x = 2$. The function obviously changes in value but the changes negate each other.

- Find the average rate of change for $f(x) = \frac{2}{3}x - 4$ between $x = -3$ and $x = 0$ and between $x = 6$ and $x = 12$.

We will first find the y-values for $x = -3$ and $x = 0$.

$$y_1 = f(x_1) = f(-3) = \frac{2}{3}(-3) - 4 = -6$$

$$y_2 = f(x_2) = f(0) = \frac{2}{3}(0) - 4 = -4$$

$$\text{Average rate of change} = m = \frac{y_2 - y_1}{x_2 - x_1} = \frac{-4 - (-6)}{0 - (-3)} = \frac{2}{3}$$

Between $x = -3$ and $x = 0$, the function increases, on average, by 2 as x increases by 3. The average rate of change is the same between $x = 6$ and $x = 12$.

$$y_1 = f(x_1) = f(6) = \frac{2}{3}(6) - 4 = 0$$

$$y_2 = f(x_2) = f(12) = \frac{2}{3}(12) - 4 = 4$$

$$\text{Average rate of change} = m = \frac{y_2 - y_1}{x_2 - x_1} = \frac{4-0}{12-6} = \frac{4}{6} = \frac{2}{3}$$

The average rate of change for a linear function is the same between any two points on its graph.

PRACTICE

Find the average rate of change.

1. $f(x) = x^3 + x^2 - 4$ between the points $(-1, -4)$ and $(2, 8)$.
2. $f(x) = x^4 - 4x^2$, between $x = -3$ and $x = 0$.
3. $f(x) = \frac{x-2}{x+3}$ between $x = 0$ and $x = 2$.
4. See Figure 1.4.

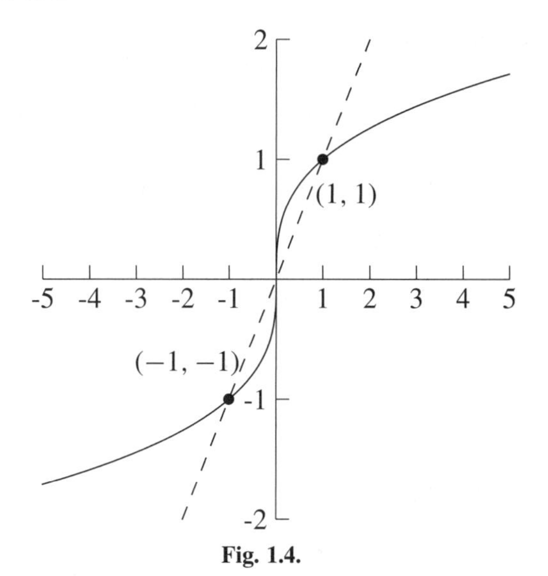

Fig. 1.4.

SOLUTIONS

1.

$$\text{Average rate of change} = \frac{8-(-4)}{2-(-1)} = \frac{12}{3} = \frac{4}{1} = 4$$

Between $x = -1$ and $x = 2$, the function increases by 4, on average, as x increases by 1.

2.

$$y_1 = f(x_1) = f(-3) = (-3)^4 - 4(-3)^2 = 45$$

$$y_2 = f(x_2) = f(0) = 0^4 - 4(0)^2 = 0$$

$$\text{Average rate of change} = \frac{0-45}{0-(-3)} = \frac{-45}{3} = \frac{-15}{1} = -15$$

Between $x = 0$ and $x = -3$, the function decreases, on average, by 15 as x increases by 1.

3.

$$f(x_1) = f(0) = \frac{0-2}{0+3} = \frac{-2}{3} \qquad f(x_2) = f(2) = \frac{2-2}{2+3} = \frac{0}{5} = 0$$

$$\text{Average rate of change} = \frac{0-(-\frac{2}{3})}{2-0} = \frac{\frac{2}{3}}{2} = \frac{2}{3} \div 2 = \frac{2}{3} \cdot \frac{1}{2} = \frac{1}{3}$$

Between $x = 0$ and $x = 2$, the function increases, on average, by 1 as x increases by 3.

4. We need to find the average rate of change of the function between $(-1, -1)$ and $(1, 1)$.

$$\text{Average rate of change} = \frac{1-(-1)}{1-(-1)} = \frac{2}{2} = \frac{1}{1} = 1$$

Between $x = -1$ and $x = 1$, the function increases, on average, by 1 as x increases by 1.

CHAPTER 1 REVIEW

1. The value of a certain car can be approximated by $y = -1500x + 9000$, x years after the car's purchase. What does the slope mean?
   (a) The car decreases in value $150 per year.
   (b) The car decreases in value $1500 per year.
   (c) The car decreases in value $900 per year.
   (d) The car decreases in value $9000 per year.

2. The monthly bill for a family's electricity usage is $y = 0.05x + 18$, when x kilowatt hours are used. Which of the following is true?
   (a) Each kilowatt of electricity costs $0.05.
   (b) Each kilowatt of electricity costs $0.50.

(c) Each kilowatt of electricity costs \$1.80.

(d) Each kilowatt of electricity costs \$0.18.

3. What is the average rate of change of the function $f(x) = x^3 - 2x^2 + x - 5$ between $x = -1$ and $x = 2$?
   (a) $-\frac{8}{5}$
   (b) $\frac{5}{8}$
   (c) 2
   (d) -2

4. What is the average rate of change of the function $f(x) = 25$ between $x = 3$ and $x = 8$?
   (a) 3
   (b) 8
   (c) $\frac{3}{8}$
   (d) 0

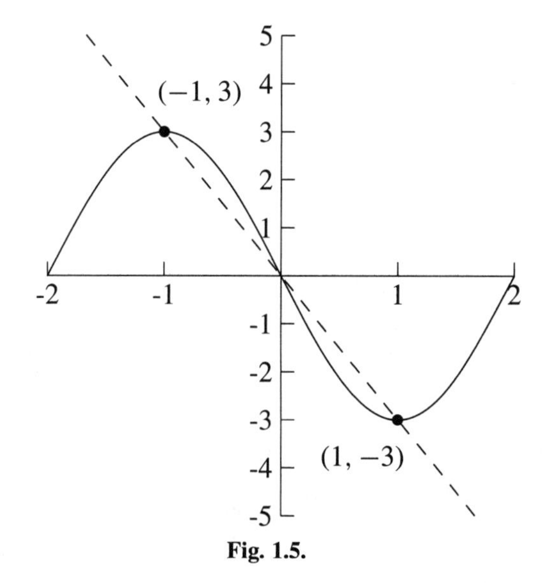

Fig. 1.5.

5. What is the average rate of change of the function whose graph is in Figure 1.5 (see page 37) between the indicated points?
   (a) 3
   (b) -3
   (c) $\frac{2}{3}$
   (d) $-\frac{2}{3}$

SOLUTIONS

1. b 2. a 3. c 4. d 5. b

CHAPTER 2

The Limit and Continuity

The Limit

An important concept in calculus is that of the limit of a function. The ancient Greeks used the notion of a limit to approximate the area inside a curve (like a circle) by using the area of a polygon because they could easily find the area of a polygon. Take, for example, using the area of polygons to approximate the area of a circle. The more sides the polygon has, the better the approximation. The area of the square in Figure 2.1 is not a good approximation of the area of the circle. The area of the hexagon in Figure 2.2 is a better approximation, and the area of the 12-sided polygon in Figure 2.3 is even better. The Greeks called this the method of exhaustion. In modern language, we say that as the number of sides of the polygon increases, the area of the polygon approaches the area of the circle.

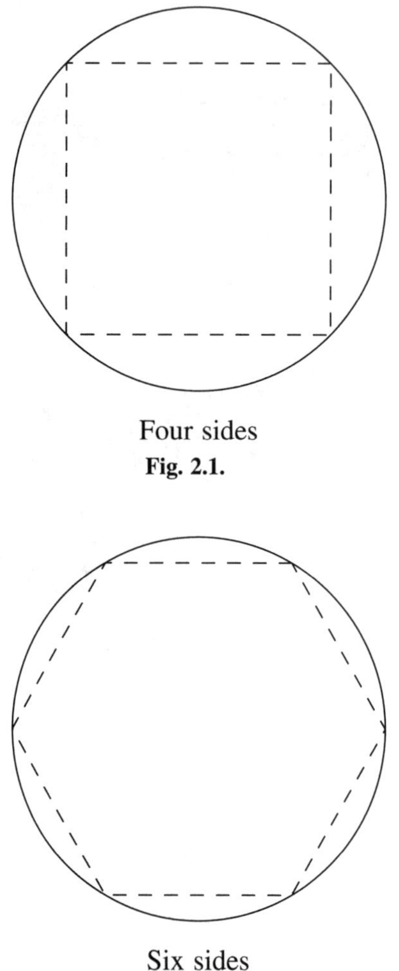

Four sides
Fig. 2.1.

Six sides
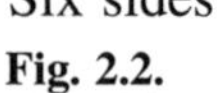
Fig. 2.2.

Another example of a limit involves the irrational number e (you probably have an e or e^x key on your calculator). The value of e can be approximated by rational numbers of the form $(1 + 1/n)^n$. The decimal approximation for e is 2.718281828.... As you can see from Table 2.1, the larger n is, the better the approximation for e. Using mathematical terms, we say that e is the limit of $(1 + 1/n)^n$ as n gets large without bound.

We will work with the limits of functions. Usually, x will get close to a fixed number. As x is getting closer to the fixed number, we want to know what y is getting close to (if anything). In Tables 2.2 and 2.3 x is "approaching" the

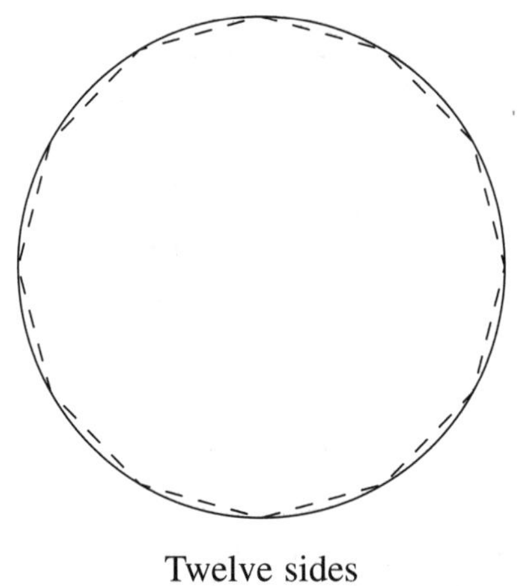

Twelve sides

Fig. 2.3.

Table 2.1

n	$\left(1+\frac{1}{n}\right)^n$
$n = 5$	$\left(1+\frac{1}{5}\right)^5 = 2.48832$
$n = 10$	$\left(1+\frac{1}{10}\right)^{10} = 2.59374246$
$n = 100$	$\left(1+\frac{1}{100}\right)^{100} = 2.704813829$
$n = 1000$	$\left(1+\frac{1}{1000}\right)^{1000} = 2.716923932$
$n = 10{,}000$	$\left(1+\frac{1}{10{,}000}\right)^{10{,}000} = 2.718145927$

Table 2.2

x	y
4.5	2.9155
4.6	2.9326
4.8	2.9665
4.9	2.9833
4.99	2.9983
4.999	2.9983
5	?

Table 2.3

x	y
4.5	19.25
4.6	20.16
4.8	22.04
4.9	23.01
4.99	23.90
4.999	23.99
5	?

number 5, and we will observe what y is approaching. The y-values in Table 2.2 appear to be getting closer to 3. We say that the limit of y as x approaches 5 is 3. The y-values in Table 2.3 appear to be getting closer to 24. We say the limit of y as x approaches 5 is 24.

PRACTICE

1. See Table 2.4. As x approaches __________ y approaches __________ .

Table 2.4

x	y
3.5	5.5
3.8	6.4
3.9	6.7
3.99	6.97
3.999	6.997
3.9999	6.9997
4	?

2. See Table 2.5. As x approaches __________ y approaches __________ .

Table 2.5

x	y
7.5	2.9574
7.8	2.9832
7.9	2.9916
7.99	2.9992
7.999	2.9999
8	?

SOLUTIONS

1. As x approaches 4, y approaches 7.
2. As x approaches 8, y approaches 3.

We can find the limit of a function (the y-values) by looking at the function's graph. Consider the function whose graph is in Figure 2.4. Suppose we want to find the limit of the function as x approaches 4. Look at the region of the graph near $x = 4$, what are the y-values of this region close to? The y-values (for example 1.73, 1.87, 1.97) are approaching 2: as x approaches 4, the limit of the function is 2.

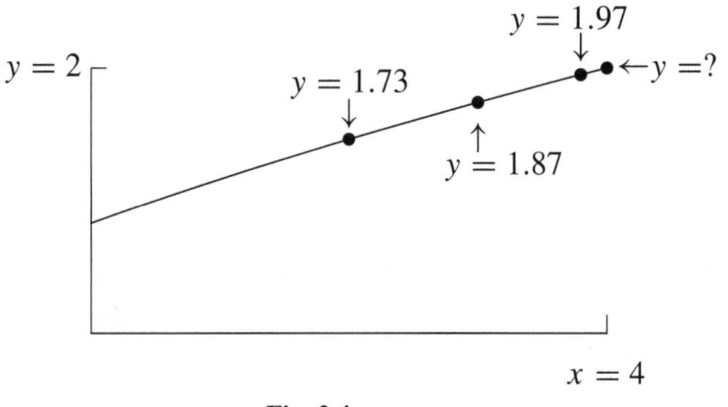

Fig. 2.4.

EXAMPLES

Find the limit.

- The graph of a function is given in Figure 2.5. Find the limit of y as x approaches 1.

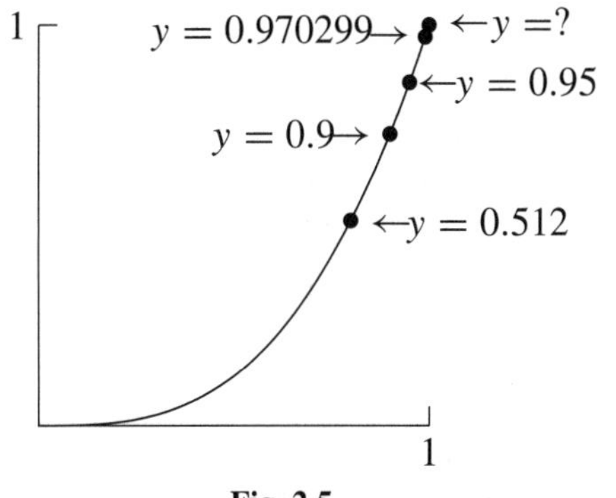

Fig. 2.5.

The y-values approach 1 as x approaches 1, so the limit of y is also 1.

- The graph of a function is given in Figure 2.6. Find the limit of y as x approaches -1.

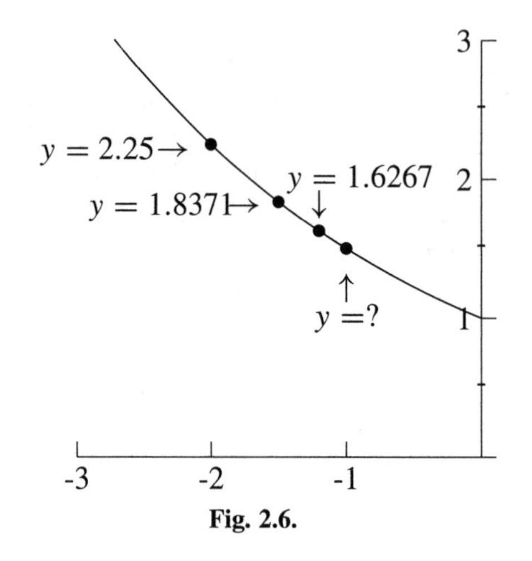

Fig. 2.6.

The limit of y as x approaches -1 is 1.5.

- The graph of $y = -(x+1)(x-1)^2$ is given in Figure 2.7. What is the limit of y as x approaches 2?

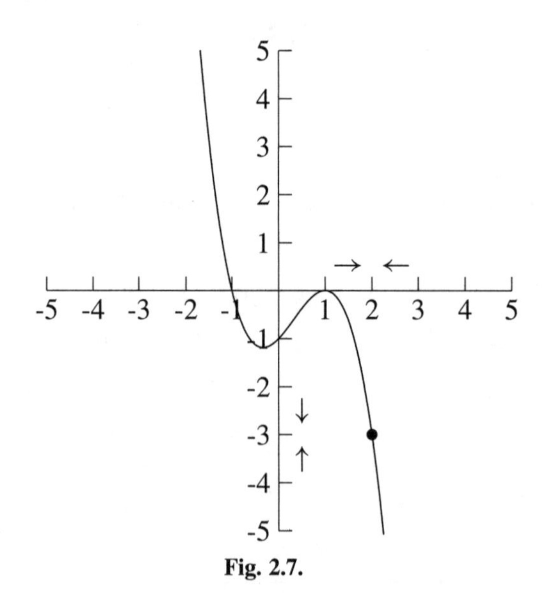

Fig. 2.7.

If we are on the graph near the point $x = 2$ and move toward the point at $x = 2$, the y-values move close to -3 (Figure 2.8). As x approaches 2, the limit of y is -3.

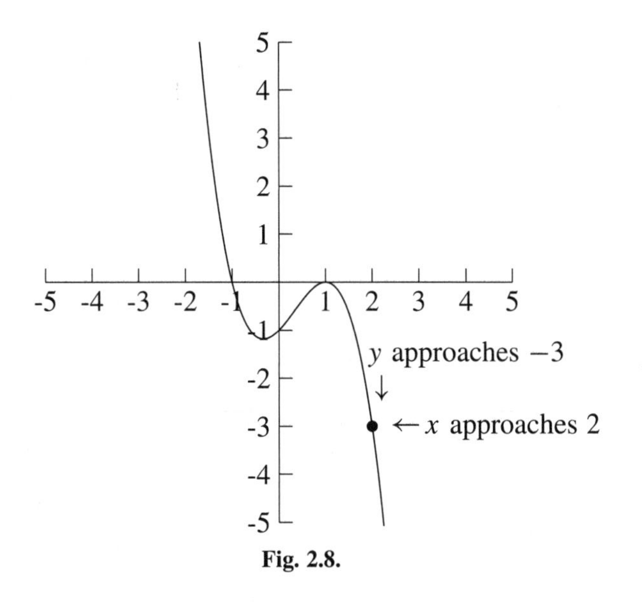

Fig. 2.8.

- The graph of a function is given in Figure 2.9. What is the limit of the function as x approaches 1?

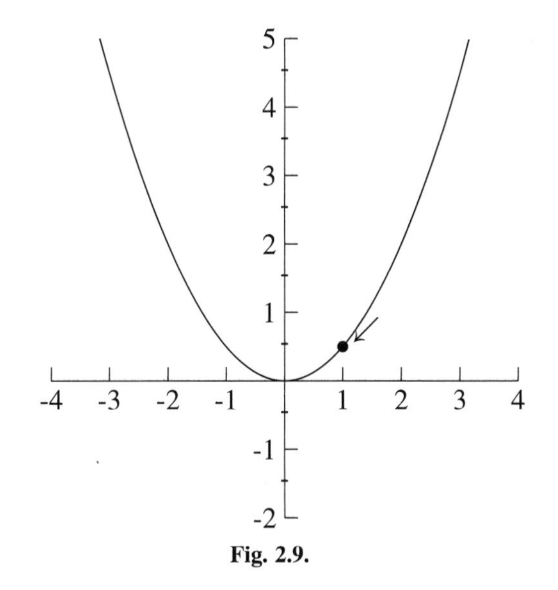
Fig. 2.9.

If we are on the graph near the point whose x-coordinate is 1 and move toward this point. What is the y-coordinate of this point? The y-values approach 0.5, so as x approaches 1, the limit of the function is 0.5.

PRACTICE

1. For the graph in Figure 2.10, what is the limit of y as x approaches 1.5?

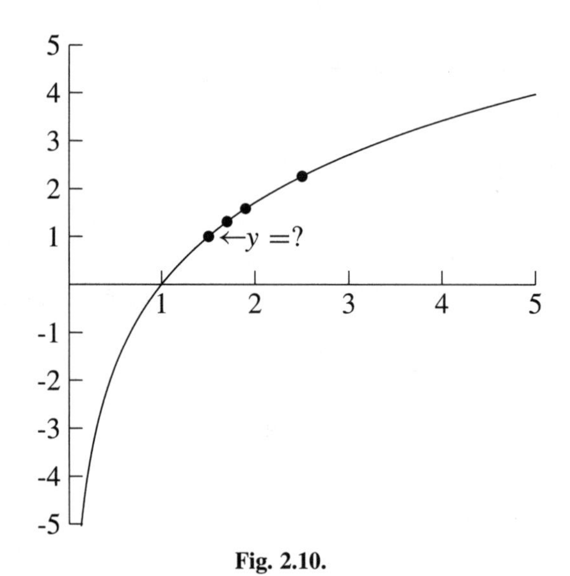

Fig. 2.10.

2. For the graph in Figure 2.11, what is the limit of y as x approaches $\frac{1}{2}$?

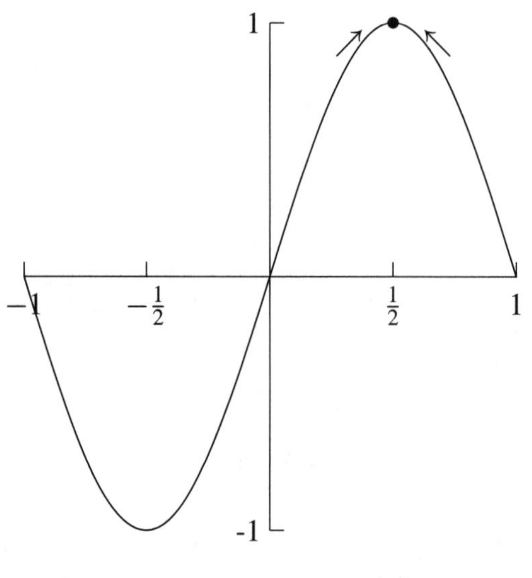

Fig. 2.11.

SOLUTIONS

1. The limit of y as x approaches 1.5 is 1.
2. The limit of y as x approaches $\frac{1}{2}$ is 1.

Strictly speaking, in order to say "the limit of y as x approaches 2 is 6," the y-values must approach 6 when x approaches 2 from both the left (such as 1.9, 1.99, 1.999, ...) and the right (such as 2.1, 2.01, 2.001, ...). Refer to Table 2.6. The y-values approach 12 as x approaches 4 from both the left and the right.

Table 2.6

	x	y
Approaching	3.9	11.31
4 from	3.99	11.93
the left	3.999	11.993
	4	?
Approaching	4.001	12.007
4 from the	4.01	12.07
right	4.1	12.71

For the numbers in Table 2.7, we would say the limit of y as x approaches 4 does not exist. The reason the limit does not exist is that the y-values approach 12 as x approaches 4 from the *left*, but the y-values approach 20 as x approaches 4 from the *right*. In order for the limit to exist, the y-values need to approach the same number on both sides of x.

Table 2.7

x	y
3.9	11.31
3.99	11.93
3.999	11.993
4	?
4.001	20.009
4.01	20.09
4.1	20.91

We can tell from the graph of a function if a limit exists. If there is a big gap in the graph, then the limit will not exist at the gap. Consider the graph in Figure 2.12. The limit of y as x approaches 3 does not exist.

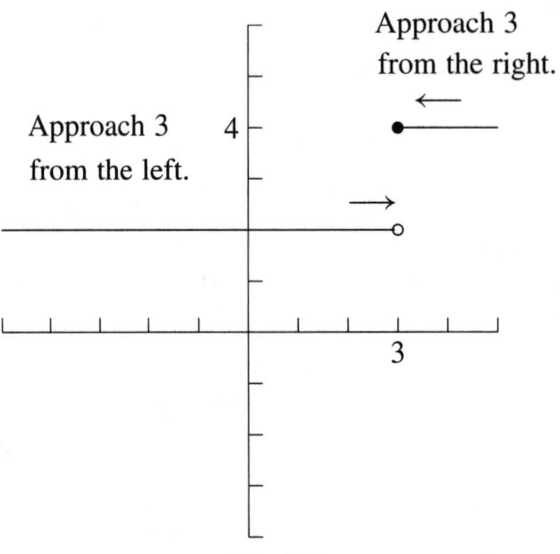

Fig. 2.12.

The y-values approach 4 as x approaches 3 from the right, but the y-values approach 2 as x approaches 3 from the left.

EXAMPLE

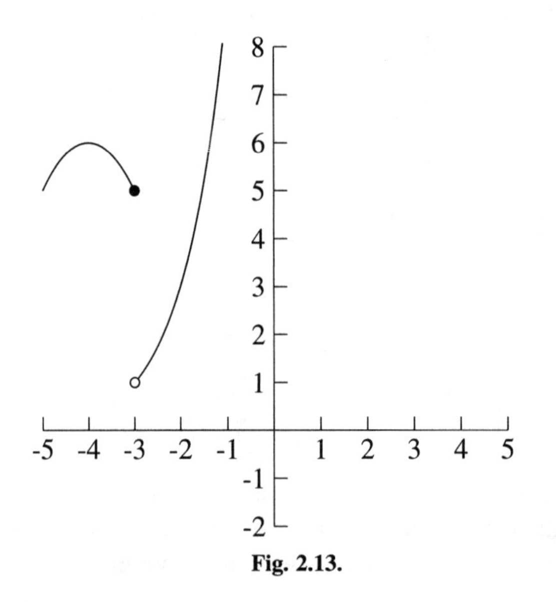

Fig. 2.13.

- There is a gap in the graph shown in Figure 2.13 at $x = -3$. When x approaches -3 from the right, the y-values approach 1. When x approaches -3 from the left, the y-values approach 5. The limit of y as x approaches -3 does not exist. Although the limit of y as x approaches -3 does not exist for the graph above, both *one-sided* limits do exist. The limit of y as x approaches -3 from the left is 5. The limit as x approaches -3 from the right is 1.

  The notation $x \to a$ means "x approaches a." The notation

$$\lim_{x \to a} f(x) = b$$

is saying, "the limit of y as x approaches a is b," where we use $f(x)$ for y.

EXAMPLES

- $\lim_{x \to 7} f(x) = 12$

  The limit of $f(x)$ as x approaches 7 is 12.
- $\lim_{x \to 1} (x^2 - 2x) = -1$

  The limit of y (or of $x^2 - 2x$) as x approaches 1 is -1.
- The graph in Figure 2.14 is the graph of a function $f(x)$. The limit as x approaches -1 is -3.

$$\lim_{x \to -1} f(x) = -3$$

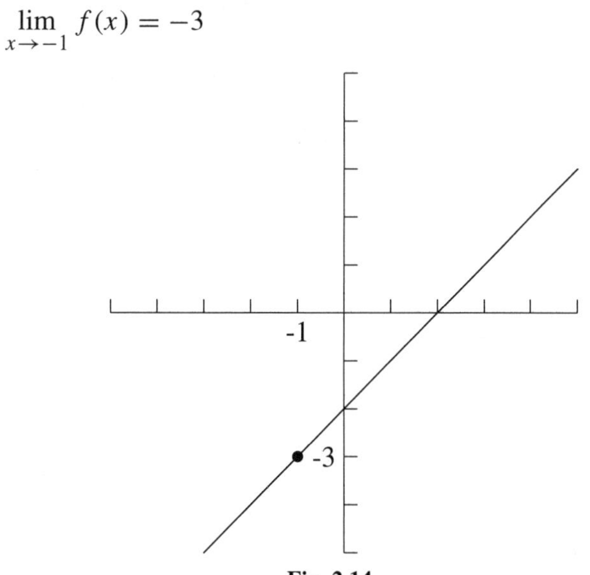

Fig. 2.14.

The notation for a one-sided limit uses a plus or minus sign as a superscript to the right of the number. $\lim_{x\to a^-} f(x)$ means "the limit of $f(x)$ as x approaches a from the left." $\lim_{x\to a^+} f(x)$ means "the limit of $f(x)$ as x approaches a from the right."

- The graph of $f(x)$ is given in Figure 2.15.

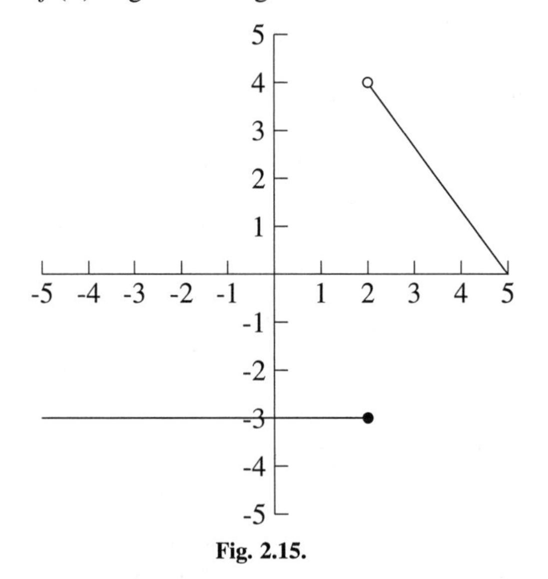

Fig. 2.15.

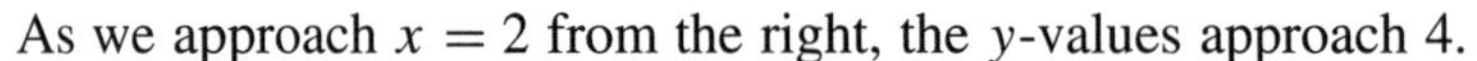

As we approach $x = 2$ from the right, the y-values approach 4.

$$\lim_{x\to 2^+} f(x) = 4$$

As we approach $x = 2$ from the left, the y-values approach -3.

$$\lim_{x\to 2^-} f(x) = -3$$

PRACTICE

1. Refer to Table 2.8.

   (a)

   $$\lim_{x\to 2^-} f(x) =$$

   (b)

   $$\lim_{x\to 2^+} f(x) =$$

Table 2.8

x	$f(x)$
1.9	4.959
1.999	5.8906
1.999	5.9892
2	?
2.001	6.011
2.01	6.1106
2.1	7.161

(c) Does $\lim_{x\to 2} f(x)$ exist? If so, what is it?

2. Refer to Table 2.9

Table 2.9

x	$f(x)$
−4.1	73.02
−4.01	68.49
−4.001	68.05
−4	?
−3.999	11.993
−3.99	11.93
−3.9	19.11

(a)

$$\lim_{x\to -4^-} f(x) =$$

(b)

$$\lim_{x\to -4^+} f(x) =$$

(c) Does $\lim_{x\to -4} f(x)$ exist? If so, what is it?

3. The graph in Figure 2.16 is the graph of a function $f(x)$.

(a)

$$\lim_{x\to 1^-} f(x) =$$

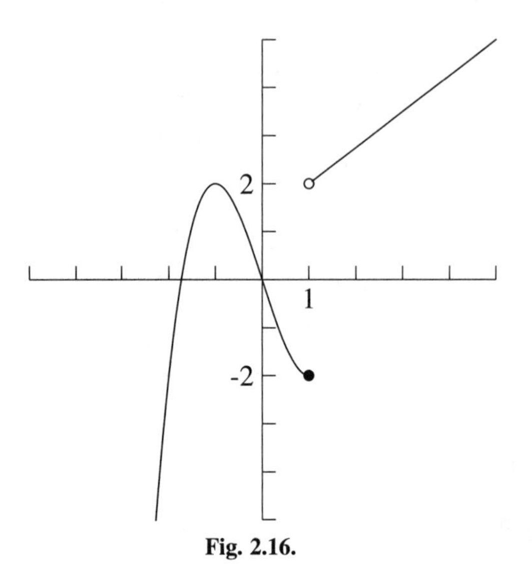

Fig. 2.16.

(b)

$$\lim_{x\to 1^+} f(x) =$$

(c) Does $\lim_{x\to 1} f(x)$ exist? If so, what is it?

4. The graph in Figure 2.17 is the graph of a function $g(x)$.

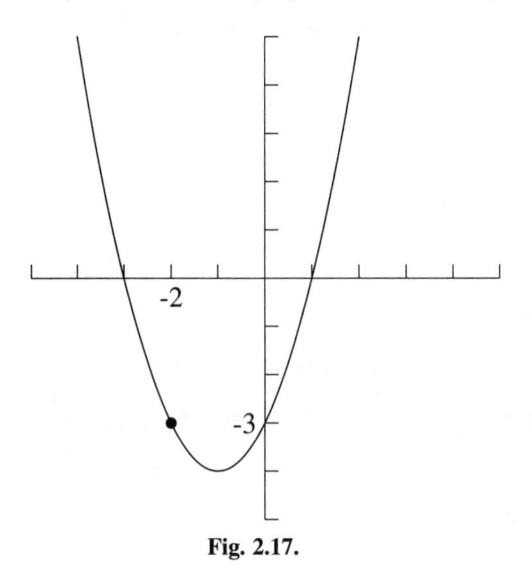

Fig. 2.17.

(a)

$$\lim_{x \to -2^-} g(x) =$$

(b)

$$\lim_{x \to -2^+} g(x) =$$

(c) Does $\lim_{x \to -2} g(x)$ exist? If so, what is it?

SOLUTIONS

1. (a)

$$\lim_{x \to 2^-} f(x) = 6$$

(b)

$$\lim_{x \to 2^+} f(x) = 6$$

(c) Yes, both one-sided limits are the same number, so $\lim_{x \to 2} f(x) = 6$.

2. (a)

$$\lim_{x \to -4^-} f(x) = 68$$

(b)

$$\lim_{x \to -4^+} f(x) = 12$$

(c) $\lim_{x \to -4} f(x)$ does not exist because the left limit, 68, is not the same as the right limit, 12.

3. (a)

$$\lim_{x \to 1^-} f(x) = -2$$

(b)

$$\lim_{x \to 1^+} f(x) = 2$$

(c) $\lim_{x \to 1} f(x)$ does not exist because the left limit, -2, is not the same as the right limit, 2.

4. (a)

$$\lim_{x \to -2^-} g(x) = -3$$

(b)

$$\lim_{x \to -2^+} g(x) = -3$$

(c) $\lim_{x \to -2} g(x)$ exists because both one-sided limits are the same number. The limit is -3.

We are ready to evaluate limits directly from the function, without having to look at a graph or a table. We will begin with some important limit properties.

1. $\lim_{x \to a} x = a$
2. For a constant number, c, $\lim_{x \to a} c = c$
   For the rest of the properties, assume $\lim_{x \to a} f(x) = L$ and $\lim_{x \to a} g(x) = M$.
3. $\lim_{x \to a}[f(x) \pm g(x)] = \lim_{x \to a} f(x) \pm \lim_{x \to a} g(x) = L \pm M$
   We can find the limit of the sum (or difference) of two functions by first finding their individual limits, then adding (or subtracting).
4. $\lim_{x \to a} f(x) \cdot g(x) = [\lim_{x \to a} f(x)] \cdot [\lim_{x \to a} g(x)] = L \cdot M$
   We can find the limit of the product of two functions by first finding their individual limits, then multiplying.
5.

$$\lim_{x \to a} \frac{f(x)}{g(x)} = \frac{\lim_{x \to a} f(x)}{\lim_{x \to a} g(x)} = \frac{L}{M} \quad \text{(provided } M \neq 0\text{)}$$

   We can find the limit of the quotient of two functions by first finding their individual limits, then dividing.
6. For any real number n, $\lim_{x \to a}(f(x))^n = [\lim_{x \to a} f(x)]^n = L^n$
   We can find the limit of a function to a power by first finding the limit of the function, then raising the limit to the power.
7. For any positive integer n, $\lim_{x \to a} \sqrt[n]{f(x)} = \sqrt[n]{\lim_{x \to a} f(x)} = \sqrt[n]{L}$, when n is even, we must have $L \geq 0$.
   We can find the limit of the nth root of a function by first finding the limit of the function, then by taking the nth root of the limit.
8. For a constant number c, $\lim_{x \to a} c \cdot f(x) = c \cdot \lim_{x \to a} f(x) = c \cdot L$
   We can find the limit of a constant times a function by first finding the limit of the function, and then by multiplying the limit by the constant.

EXAMPLES

We will use examples to see why these properties work.

- $\lim_{x \to 6} x = ?$
  A table of values is given in Table 2.10 and its graph is given in Figure 2.18.

Table 2.10

x	$y = x$
5.9	5.9
5.99	5.99
5.999	5.999
6	?
6.001	6.001
6.01	6.01
6.1	6.1

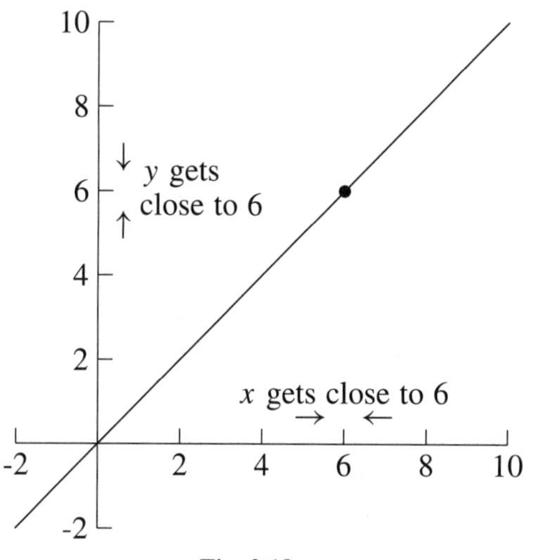

Fig. 2.18.

Because y approaches 6 as x approaches 6, $\lim_{x \to 6} x = 6$.

- $\lim_{x \to -2} 3 = ?$
  A table of values is given in Table 2.11 and its graph is given in Figure 2.19

The y-values are 3 no matter what x is, so as x approaches -2, y approaches 3. Now we can see that $\lim_{x \to -2} 3 = 3$.

Table 2.11

x	$y = 3$
-1.9	3
-1.99	3
-1.999	3
-2	?
-2.001	3
-2.01	3
-2.1	3

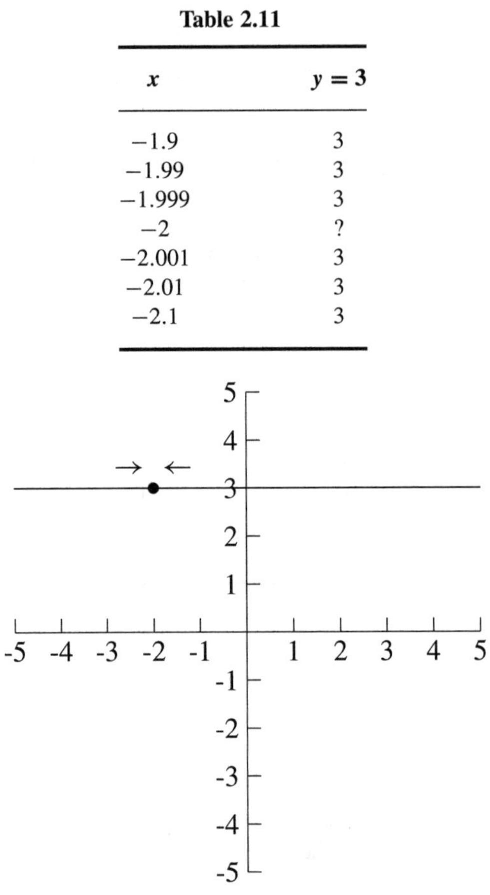

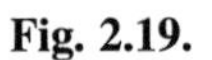
Fig. 2.19.

These properties allow us to evaluate many limits by simply substituting a for x.

EXAMPLES

Evaluate the limit.

- $\lim_{x \to 1}(3x^2 - x + 2)$
  By Properties 1, 2, 3, 6, and 8, all we have to do is to substitute 1 for x.

$$\lim_{x \to 1} 3x^2 - x + 2 = \lim_{x \to 1} 3x^2 - \lim_{x \to 1} x + \lim_{x \to 1} 2$$

$$= 3(\lim_{x \to 1} x)^2 - \lim_{x \to 1} x + \lim_{x \to 1} 2$$

$$= 3(1)^2 - (1) + 2 = 4$$

- $\lim_{x\to 0}(5x-6) = 5(0) - 6 = -6$
- $\lim_{x\to -2}(7 - x + x^2) = 7 - (-2) + (-2)^2 = 13$
- $\lim_{x\to 8} \dfrac{x+1}{x-1} = \dfrac{8+1}{8-1} = \dfrac{9}{7}$
- $\lim_{x\to 4}(x-2)^3 = (4-2)^3 = 8$
- $\lim_{x\to 3} \sqrt{x+6} = \sqrt{3+6} = 3$

What happens to $\lim_{x\to a} \frac{f(x)}{g(x)}$ if letting $x = a$ causes a zero in the denominator? Sometimes the limit exists, sometimes it does not. If letting $x = a$ gets us 0/0, then very often the limit does exist. When we get 0/0, we will reduce the fraction to lowest terms, then let $x = a$. If we get a number, then the limit is this number. If we get $\frac{\text{nonzero number}}{0}$, then the limit will not exist, or might be infinite (more about this later).

EXAMPLE

- $\lim\limits_{x\to 2} \dfrac{x^2-4}{x-2}$

  If we let $x = 2$, we have $\frac{2^2-4}{2-2} = \frac{0}{0}$. Of course, $\frac{0}{0}$ is not a number, but this limit might exist. We will look at both a table of values and the graph. As we can see from the Table 2.12 and Figure 2.20, $\lim_{x\to 2} \frac{x^2-4}{x-2} = 4$. We can find this limit algebraically. Factor the numerator and denominator and reduce the fraction to lowest terms. Then try letting $x = 2$.

$$\lim_{x\to 2} \frac{x^2-4}{x-2} = \lim_{x\to 2} \frac{(x-2)(x+2)}{x-2} = \lim_{x\to 2}(x+2) = 2+2 = 4$$

Table 2.12

x	$\frac{x^2-4}{x-2}$
1.9	3.9
1.99	3.99
1.999	3.999
2	?
2.001	4.001
2.01	4.01
2.1	4.1

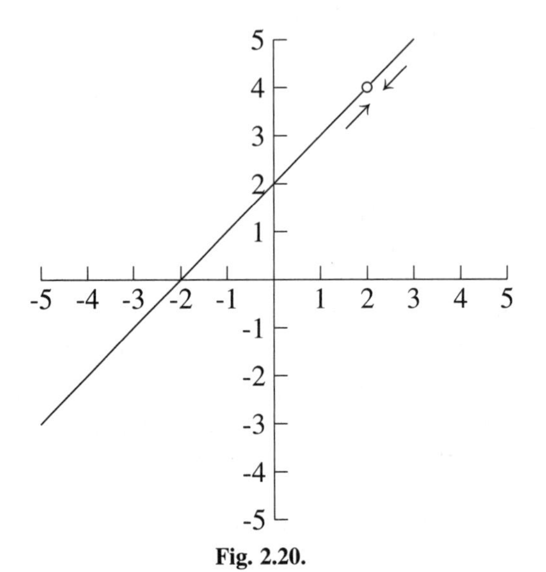

Fig. 2.20.

- $\lim\limits_{x \to -3} \dfrac{x^2 + x - 6}{x^2 + 4x + 3}$

  Letting $x = -3$ gives us 0/0, so we will factor the numerator and denominator and reduce the fraction to lowest terms.

$$\lim_{x \to -3} \frac{x^2 + x - 6}{x^2 + 4x + 3} = \lim_{x \to -3} \frac{(x+3)(x-2)}{(x+3)(x+1)} = \lim_{x \to -3} \frac{x-2}{x+1} = \frac{-3-2}{-3+1} = \frac{5}{2}$$

Sometimes 0/0 means the limit does not exist (or is infinite).

- $\lim\limits_{x \to -1} \dfrac{x^2 + 5x + 4}{x^2 + 2x + 1}$

  Once we reduce the fraction to lowest terms and let $x = -1$, we will have $\frac{\text{nonzero number}}{0}$, which is not a number. This means the limit will not exist (or is infinite).

$$\lim_{x \to -1} \frac{x^2 + 5x + 4}{x^2 + 2x + 1} = \lim_{x \to -1} \frac{(x+1)(x+4)}{(x+1)(x+1)}$$

$$= \lim_{x \to -1} \frac{x+4}{x+1} = \frac{3}{0} \qquad \text{This is not a number.}$$

The limit does not exist (or is infinite).

PRACTICE

Evaluate the limit, if it exists.

1. $\lim_{x\to 0}(4x^2 - 6) =$
2. $$\lim_{x\to 2}\frac{x^2+1}{x^2-1} =$$
3. $\lim_{x\to 7}\sqrt{x+2} =$
4. $\lim_{x\to -3}(x^2 - x + 5) =$
5. $$\lim_{x\to 0}\frac{3x^2+2x}{x} =$$
6. $$\lim_{x\to 4}\frac{2x^2-7x-4}{x^2-2x-8} =$$
7. $$\lim_{x\to 5}\frac{x^2-6x+5}{x^2-10x+25} =$$

SOLUTIONS

1. $\lim_{x\to 0}(4x^2 - 6) = 4(0)^2 - 6 = -6$
2. $$\lim_{x\to 2}\frac{x^2+1}{x^2-1} = \frac{2^2+1}{2^2-1} = \frac{5}{3}$$
3. $\lim_{x\to 7}\sqrt{x+2} = \sqrt{7+2} = 3$
4. $\lim_{x\to -3}(x^2 - x + 5) = (-3)^2 - (-3) + 5 = 17$
5. $$\lim_{x\to 0}\frac{3x^2+2x}{x} = \lim_{x\to 0}\frac{x(3x+2)}{x} = \lim_{x\to 0}(3x+2) = 3(0)+2 = 2$$
6. $$\lim_{x\to 4}\frac{2x^2-7x-4}{x^2-2x-8} = \lim_{x\to 4}\frac{(x-4)(2x+1)}{(x-4)(x+2)} = \lim_{x\to 4}\frac{2x+1}{x+2}$$
$$= \frac{2(4)+1}{4+2} = \frac{9}{6} = \frac{3}{2}$$

7.

$$\lim_{x\to 5}\frac{x^2-6x+5}{x^2-10x+25}=\lim_{x\to 5}\frac{(x-5)(x-1)}{(x-5)(x-5)}=\lim_{x\to 5}\frac{x-1}{x-5}$$

We cannot let $x=5$ in $(x-1)/(x-5)$ because we would have 4/0, so this limit does not exist (or is infinite).

Sometimes as x approaches a zero in the denominator, the y-values get large. For example, as x approaches 3 in $f(x)=\frac{1}{(x-3)^2}$, the y's get larger and larger (see Table 2.13 and Figure 2.21). When this happens, we say the limit is infinite:

$$\lim_{x\to 3}\frac{1}{(x-3)^2}=\infty.$$

Table 2.13

x	$\frac{1}{(x-3)^2}$
2.9	100
2.99	10,000
2.999	1,000,000
3	?
3.001	1,000,000
3.01	10,000
3.1	100

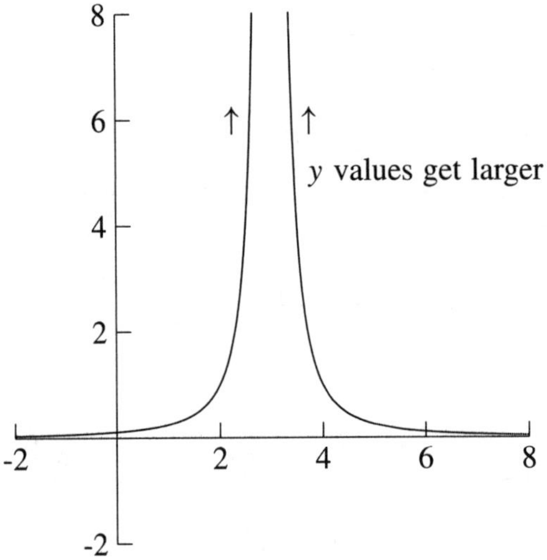

Fig. 2.21.

The limit does not exist, however, when the y-values get larger in different directions. For example, the limit does not exist for $\lim_{x\to 2} \frac{1}{x-2}$ because the y-values get large in the *positive* direction on the right of $x = 2$, but they get large in the *negative* direction to the left of $x = 2$ (see Figure 2.22).

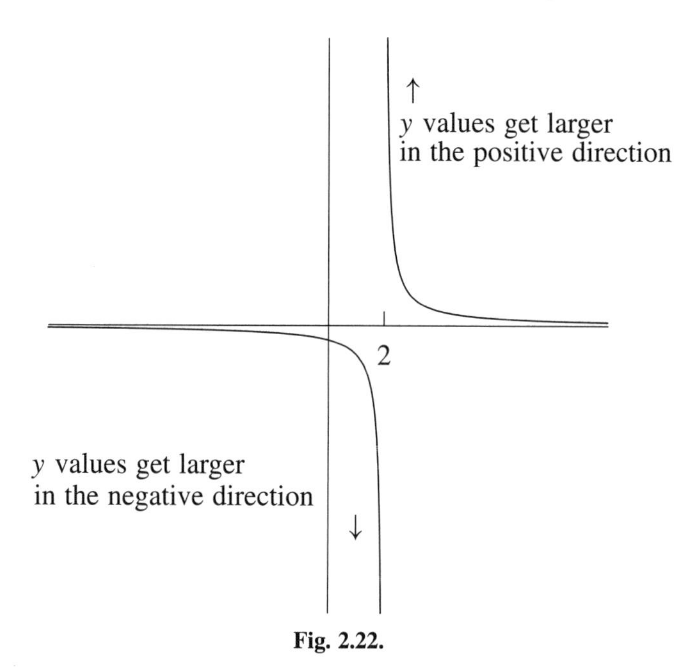

Fig. 2.22.

How can we tell if a limit is a number, is infinite, or does not exist? First, we need to make sure that the fraction is reduced to lowest terms. If letting $x = a$ in the fraction gives a number, this number is the limit. Otherwise, the limit will either be infinite or will not exist. We can determine which is the case by letting x be a number a little to the left of $x = a$ and again with x a number a little to the right of $x = a$. If we get both numbers to be large positive numbers, the limit will be positive infinity (∞). If we get both numbers to be large negative numbers, the limit will be negative infinity ($-\infty$). If one number is a large positive number and the other a large negative number, the limit will not exist.

EXAMPLES

Evaluate the limit.

- $\lim\limits_{x\to -5} \dfrac{x+6}{x+5}$

The fraction is reduced to lowest terms. We will let $x = -4.99$ as our number that is a little to the right of $x = -5$ and $x = -5.01$ as our number that is a little to the left of $x = -5$.

$$\frac{-4.99+6}{-4.99+5} = 101 \qquad \frac{-5.01+6}{-5.01+5} = -99$$

These large numbers have different signs, so the limit does not exist.

- $\lim_{x \to 1} \frac{x-1}{x^2-1}$

This fraction can be reduced.

$$\lim_{x \to 1} \frac{x-1}{x^2-1} = \lim_{x \to 1} \frac{x-1}{(x-1)(x+1)} = \lim_{x \to 1} \frac{1}{x+1} = \frac{1}{1+1} = \frac{1}{2}$$

- $\lim_{x \to 0} \frac{x+5}{x^2}$

This fraction is reduced to lowest terms. We will let $x = -0.1$ and $x = 0.1$ to see if the y-values are both large positive numbers, both negative numbers, or one positive and one negative number.

$$\frac{-0.1+5}{(-0.1)^2} = 490 \qquad \frac{0.1+5}{(0.1)^2} = 510$$

Both y-values are large positive numbers, so the limit is positive infinity: $\lim_{x \to 0} \frac{x+5}{x^2} = \infty$.

PRACTICE

Evaluate the limit.

1. $$\lim_{x \to -4} \frac{x+1}{2x+8} =$$

2. $$\lim_{x \to 9} \frac{2x}{(9-x)^2} =$$

3. $$\lim_{x \to 4} \frac{x-4}{x^2-16} =$$

4.

$$\lim_{x \to 0} \frac{-1}{x^2} =$$

SOLUTIONS

1. The fraction is reduced to lowest terms.

   For $x = -3.99$, $\dfrac{-3.99 + 1}{2(-3.99) + 8} = -149.5$,

   for $x = -4.01$, $\dfrac{-4.01 + 1}{2(-4.01) + 8} = 150.5$

   These large y-values have different signs, so the limit does not exist.

2. The fraction is reduced to lowest terms.

   For $x = 9.1$, $\dfrac{2(9.1)}{(9 - 9.1)^2} = 1820$, for $x = 8.9$, $\dfrac{2(8.9)}{(9 - 8.9)^2} = 1780$

   These large y-values are both positive, so the limit is infinite.

   $\lim_{x \to 9} \dfrac{2x}{(9 - x)^2} = \infty$.

3. We need to reduce the fraction to lowest terms, and then try to let $x = 4$.

$$\lim_{x \to 4} \frac{x - 4}{x^2 - 16} = \lim_{x \to 4} \frac{x - 4}{(x - 4)(x + 4)} = \lim_{x \to 4} \frac{1}{x + 4} = \frac{1}{4 + 4} = \frac{1}{8}$$

4. The fraction is reduced to lowest terms.

   For $x = -0.1$, $\dfrac{-1}{(-0.1)^2} = -100$, for $x = 0.1$, $\dfrac{-1}{(0.1)^2} = -100$

   These large y-values are both negative, so the limit is infinite.

   $\lim_{x \to 0} \dfrac{-1}{x^2} = -\infty$.

When we take the limit of an algebraic expression that has more than one variable, the limit is usually another algebraic expression instead of a number. The variables that do not "move" are treated as if they were fixed numbers when the limit is taken. In the limit $\lim_{x \to 1}(x^2 + xy + y^2)$, only x is changing. We let x "go to 1" but leave y and y^2 as they are.

$$\lim_{x \to 1} (x^2 + xy + y^2) = 1^2 + 1(y) + y^2 = 1 + y + y^2$$

In the next chapter, we will work with limits of functions having an x (or t) as well as h as variables. We will take the limit as h goes to 0.

EXAMPLES

Evaluate the limit.

- $\lim_{x \to 2} (y - 6xy^2 + 5x)$

  We only need to replace x with 2.

  $$\lim_{x \to 2} (y - 6xy^2 + 5x) = y - 6(2)y^2 + 5(2) = y - 12y^2 + 10$$

- $\lim_{h \to 0} (x^2 - 2xh + 4) = x^2 - 2x(0) + 4 = x^2 + 4$

- $$\lim_{h \to 0} \frac{10 + h}{(x + h + 1)(x + 1)} = \frac{10 + 0}{(x + 0 + 1)(x + 1)} = \frac{10}{(x + 1)(x + 1)} = \frac{10}{(x + 1)^2}$$

- $\lim_{h \to 0} \frac{h^2}{2xh}$

  If we let $h = 0$, we would get $0/0$. We must reduce the fraction before attempting to let $h = 0$.

  $$\lim_{h \to 0} \frac{h^2}{2xh} = \lim_{h \to 0} \frac{h}{2x} = \frac{0}{2x} = 0$$

- $\lim_{h \to 0} \frac{2xh + h^2 - 2h}{h}$

  Again, if we let $h = 0$, we would get $0/0$. First we will factor h from each term in the numerator. And then we can reduce the fraction and take the limit.

  $$\lim_{h \to 0} \frac{2xh + h^2 - 2h}{h} = \lim_{h \to 0} \frac{h(2x + h - 2)}{h} = \lim_{h \to 0} (2x + h - 2)$$
  $$= 2x + 0 - 2 = 2x - 2$$

PRACTICE

Evaluate the limit.

1. $\lim_{x \to -3} (x^2 + xy - 2x + 4y) =$

2.
$$\lim_{x \to 4} \frac{x + y}{x - y^2} =$$
3.
$$\lim_{h \to 0} \frac{7xh - 3}{h^2 + 2} =$$
4. $\lim_{h \to 0}(3x - 8h) =$
5.
$$\lim_{h \to 0} \frac{2xh}{h} =$$
6.
$$\lim_{h \to 0} \frac{4xh^2 - 2h}{h} =$$
7.
$$\lim_{h \to 0} \frac{2xh - 3h^2}{h} =$$
8.
$$\lim_{h \to 0} \frac{5xh - 3x^2h + h^2}{h} =$$

SOLUTIONS

1.
$$\lim_{x \to -3} (x^2 + xy - 2x + 4y) = (-3)^2 + (-3)y - 2(-3) + 4y$$
$$= 9 - 3y + 6 + 4y = 15 + y$$
2.
$$\lim_{x \to 4} \frac{x + y}{x - y^2} = \frac{4 + y}{4 - y^2}$$
3.
$$\lim_{h \to 0} \frac{7xh - 3}{h^2 + 2} = \frac{7x(0) - 3}{0^2 + 2} = -\frac{3}{2}$$
4. $\lim_{h \to 0}(3x - 8h) = 3x - 8(0) = 3x$
5.
$$\lim_{h \to 0} \frac{2xh}{h} = \lim_{h \to 0} 2x = 2x$$

6.

$$\lim_{h\to 0} \frac{4xh^2 - 2h}{h} = \lim_{h\to 0} \frac{h(4xh - 2)}{h} = \lim_{h\to 0} (4xh - 2) = 4x(0) - 2 = -2$$

7.

$$\lim_{h\to 0} \frac{2xh - 3h^2}{h} = \lim_{h\to 0} \frac{h(2x - 3h)}{h} = \lim_{h\to 0} (2x - 3h) = 2x - 3(0) = 2x$$

8.

$$\lim_{h\to 0} \frac{5xh - 3x^2h + h^2}{h} = \lim_{h\to 0} \frac{h(5x - 3x^2 + h)}{h}$$

$$= \lim_{h\to 0} (5x - 3x^2 + h) = 5x - 3x^2 + 0 = 5x - 3x^2$$

Continuity

A function is *continuous* at an x-value if its graph can be drawn through the point. A graph is not continuous at an x-value if there is a break at the x-value. The graph in Figure 2.23 is not continuous at both $x = -2$ and $x = 1$. The graph in Figure 2.24 is not continuous at $x = 1$.

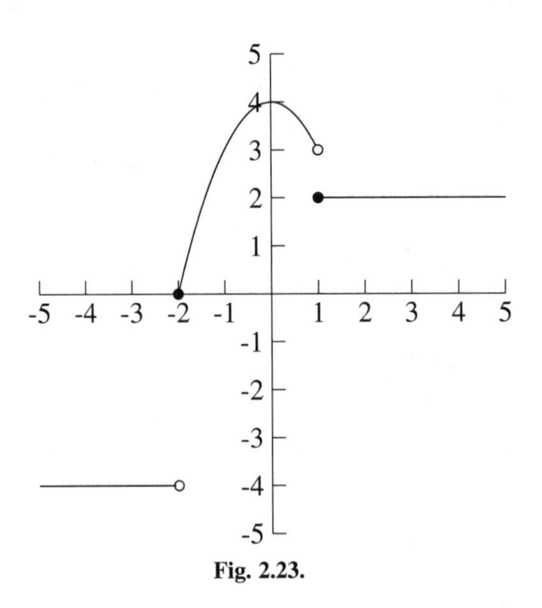

Fig. 2.23.

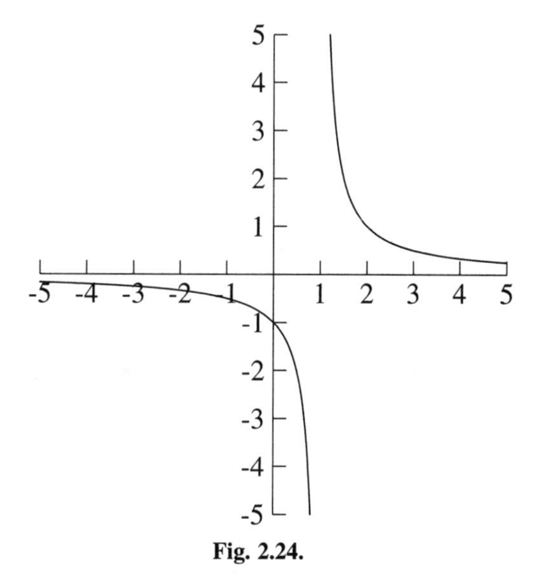

Fig. 2.24.

EXAMPLES

Determine where the functions are not continuous.

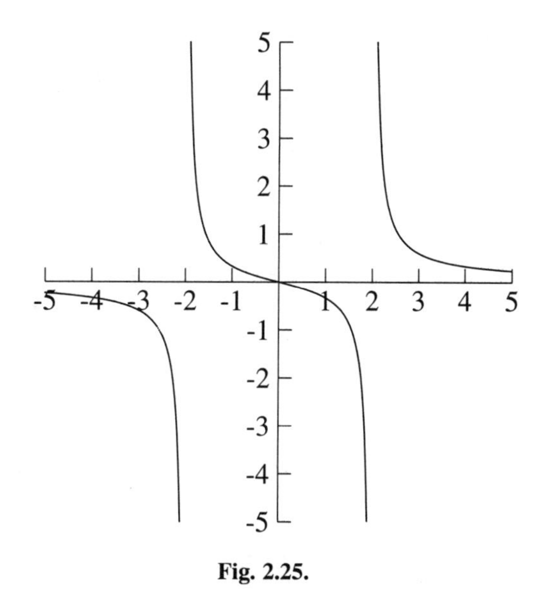

Fig. 2.25.

- The function shown in Figure 2.25 is not continuous at $x = -2$ and at $x = 2$.

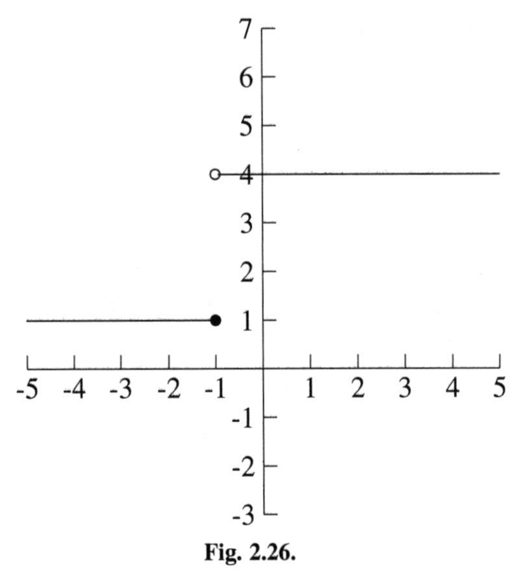

Fig. 2.26.

- The function shown in Figure 2.26 is not continuous at $x = -1$.

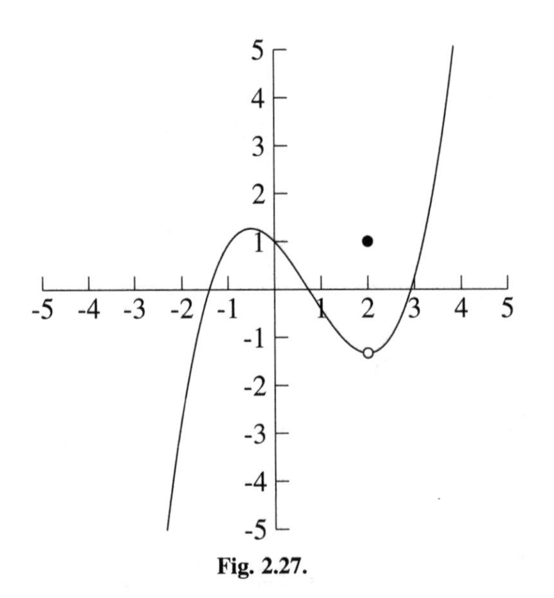

Fig. 2.27.

- The function shown in Figure 2.27 is not continuous at $x = 2$.

Each of the functions above failed to be continuous for different reasons. The function whose graph is in Figure 2.25 is $f(x) = \frac{x}{x^2-4}$. It is not defined for $x = -2$ and $x = 2$. These x-values cause a zero in the denominator. The function whose graph is in Figure 2.26 is not continuous at $x = -1$ because the left limit is different from the right limit, so the limit does not exist. The function whose graph is in Figure 2.27 is not continuous at $x = 2$ because there is a hole at $x = 2$, even though the function is defined at $x = 2$. The limit (as x approaches 2) exists but is different from the value of the function there.

In short, a function $f(x)$ is continuous at $x = a$ if all three of the following are true.

1. $f(a)$ exists. (There is a point on the graph for $x = a$.)
2. $\lim_{x\to a} f(x)$ exists. (The left limit is the same number as the right limit.)
3. $\lim_{x\to a} f(x) = f(a)$. (The limit exists and is the same number as the y-value for $x = a$.)

PRACTICE

Determine where the function is not continuous and which of the three conditions it fails.

1. The function of Figure 2.28.

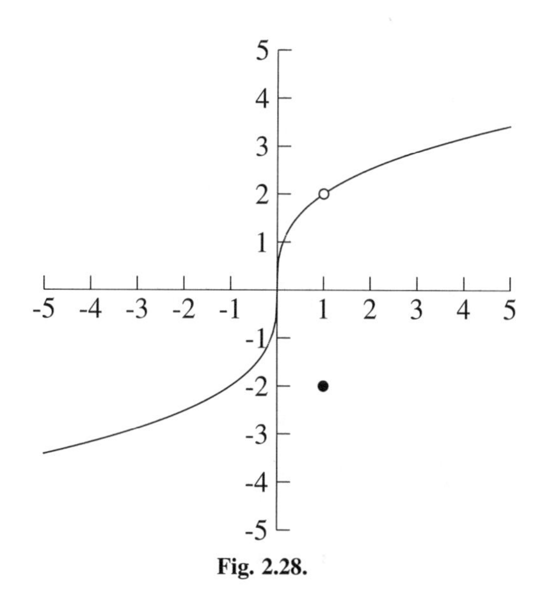

Fig. 2.28.

2. The function of Figure 2.29.

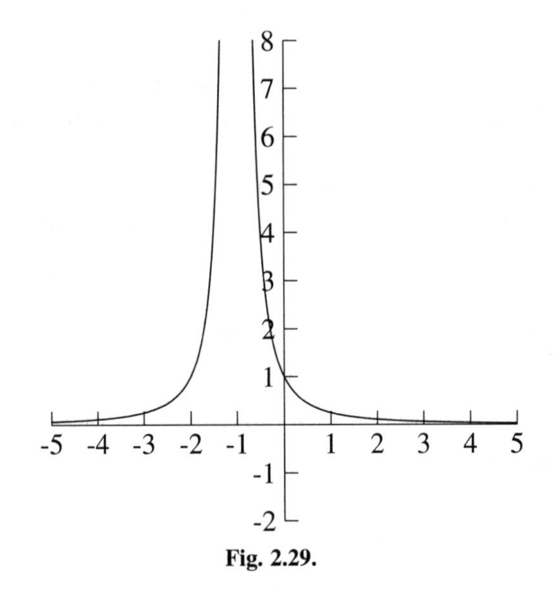

Fig. 2.29.

3. The function of Figure 2.30.

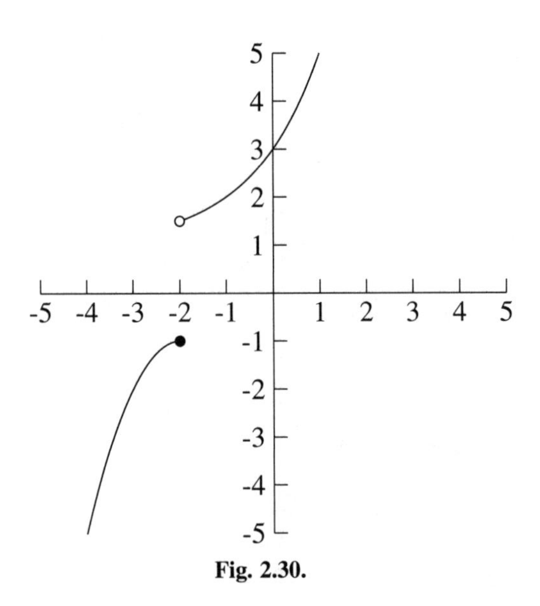

Fig. 2.30.

SOLUTIONS

1. The function is not continuous at $x = 1$. Even though the function exists at $x = 1$ (its y-value is -2) and the limit exists (it is 2), these two numbers are not the same, so the function is not continuous at $x = 1$. It fails the third condition, $\lim_{x\to 1} \neq f(1)$.
2. The function is not continuous at $x = -1$ because there is no point on the graph for $x = -1$. The function is not continuous at $x = -1$ because it fails the first condition, $f(-1)$ does not exist.
3. The function is not continuous at $x = -2$ because the limit does not exist (the left limit is -1 and the right limit is 1.5).

CHAPTER 2 REVIEW

1. Estimate the $\lim_{x\to 10} y$ from the numbers in Table 2.14.

Table 2.14

x	y
9.5	−14
9.9	−14.8
9.99	−14.98
10	?
10.01	−15.02
10.1	−15.2
10.5	−16

a) 10 b) −15 c) 25 d) The limit does not exist.

2. Estimate $\lim_{x\to 10} y$ from the numbers in Table 2.15.

Table 2.15

x	y
9.5	−14
9.9	−14.8
9.99	−14.98
10	?
10.01	25.02
10.1	25.2
10.5	26

a) 10 b) −15 c) 25 d) The limit does not exist.

3. Use the graph of $f(x)$ in Figure 2.31 to find $\lim_{x\to 1} f(x)$.

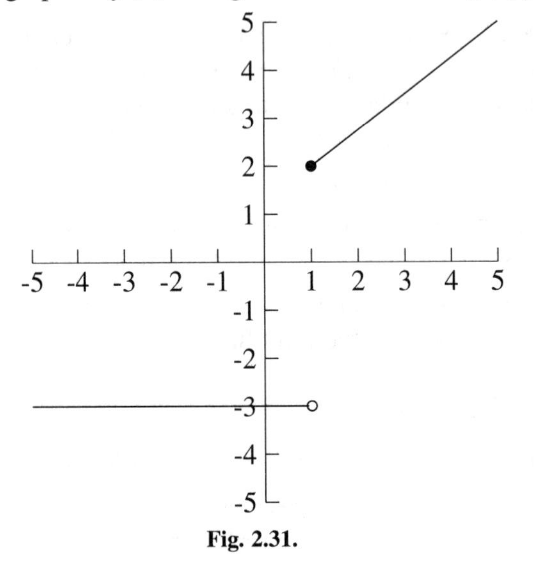

Fig. 2.31.

a) 2 b) -3 c) 1 d) The limit does not exist.

4. Use the graph of $f(x)$ in Figure 2.31 to find $\lim_{x\to 1^-} f(x)$.
a) 2 b) -3 c) 1 d) The limit does not exist.

5. Use the graph of $f(x)$ in Figure 2.31 to determine where $f(x)$ is not continuous.
a) $x = 2$ b) $x = -3$ c) $x = 1$ d) The function is continuous everywhere.

6. Use the graph of $g(x)$ in Figure 2.32 to find $\lim_{x\to -2} g(x)$.
a) 3 b) 2 c) -5 d) The limit does not exist.

7. Use the graph of $f(x)$ in Figure 2.33 to find $\lim_{x\to 3} f(x)$.
a) 10 b) $-\infty$ c) ∞ d) The limit does not exist.

8.
$$\lim_{x\to 0} \frac{x^2 - 2}{x + 4} =$$

a) -2 b) $-\frac{1}{2}$ c) 2 d) The limit does not exist.

9.
$$\lim_{x\to 1} \frac{x^2 - 1}{x - 1} =$$

a) 2 b) 0/0 c) 0 d) The limit does not exist.

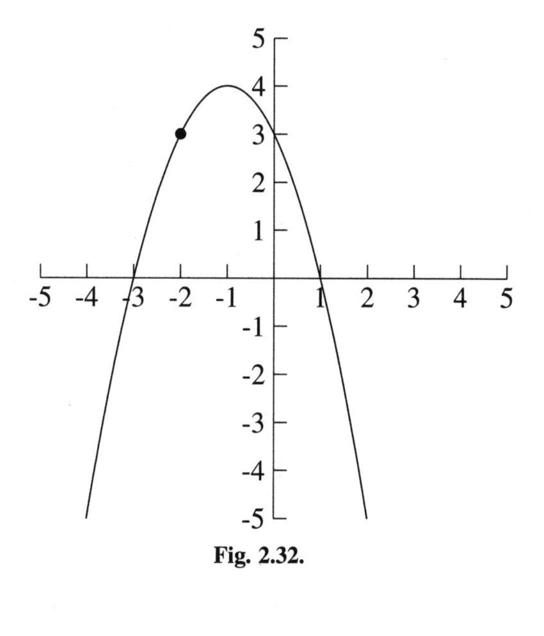

Fig. 2.32.

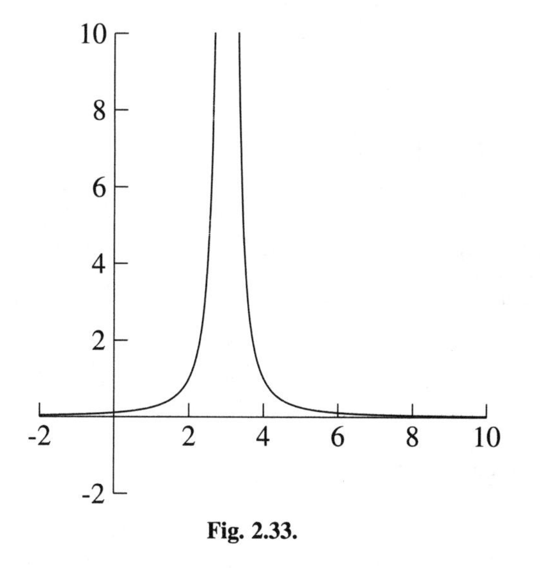

Fig. 2.33.

10. The graph of $f(x)$ is in Figure 2.34. Why is the function not continuous at $x = 1$?
a) $f(1)$ does not exist b) $\lim_{x\to 1} f(x)$ does not exist
c) $\lim_{x\to 1} f(x) \neq f(1)$ d) None of the above

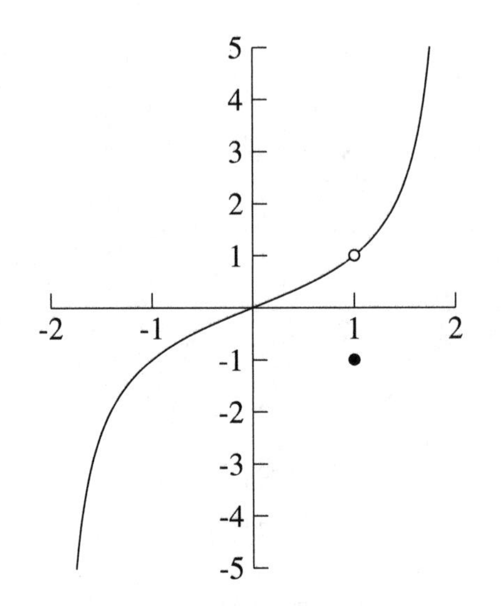

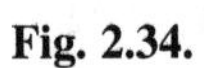
Fig. 2.34.

11. $\lim_{h\to 0}(4x^2 - 2x + h) =$
    (a) $4x^2 - 2x$
    (b) 0
    (c) $4x^2$
    (d) $4x^2 - 2x + 1$

12.

$$\lim_{h\to 0} \frac{6xh - 3h}{h} =$$

    (a) 3
    (b) -3
    (c) $6x$
    (d) $6x - 3$

SOLUTIONS

1. b	2. d	3. d	4. b	5. c	6. a
7. c	8. b	9. a	10. c	11. a	12. d

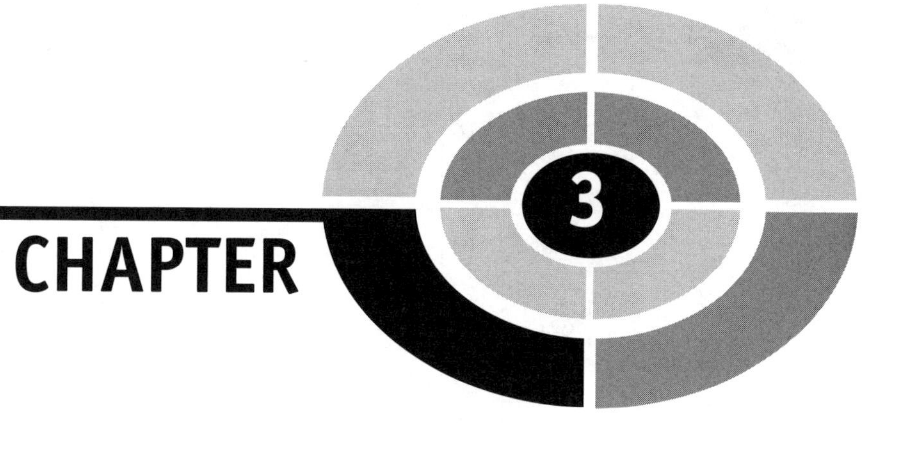

The Derivative

Suppose the graph in Figure 3.1 represents the relationship between the weekly sales budget and the number of cars sold for a small car dealership. As the budget increases from \$1000 to \$2000 (from $x = 1$ to $x = 2$ on the graph), the number of cars sold increases from 2 to 4 ($y = 2$ to $y = 4$ on the graph). As the budget increases from \$3000 to \$4000, the number of cars sold increases from 8 to 16. This shows that an extra \$1000 in the sales budget from \$3000 to \$4000 results in an extra 8 cars sold; whereas, an extra \$1000 in the sales budget from \$1000 to \$2000 results in only an extra 2 cars sold.

The graph in Figure 3.2 shows the annual revenue for a product during the years 1990 to 2005. From 1991 to 1992, the revenue increased about half a million dollars. From 2004 to 2005, though, sales hardly increased at all.

These examples show us that the rate of change of a function can be different for different values of x. Calculus gives us a way to find and describe the rate of change at different x-values. The slope of the *tangent line* describes the rate of change at different x-values, and the slope of the tangent line is found using the *derivative*.

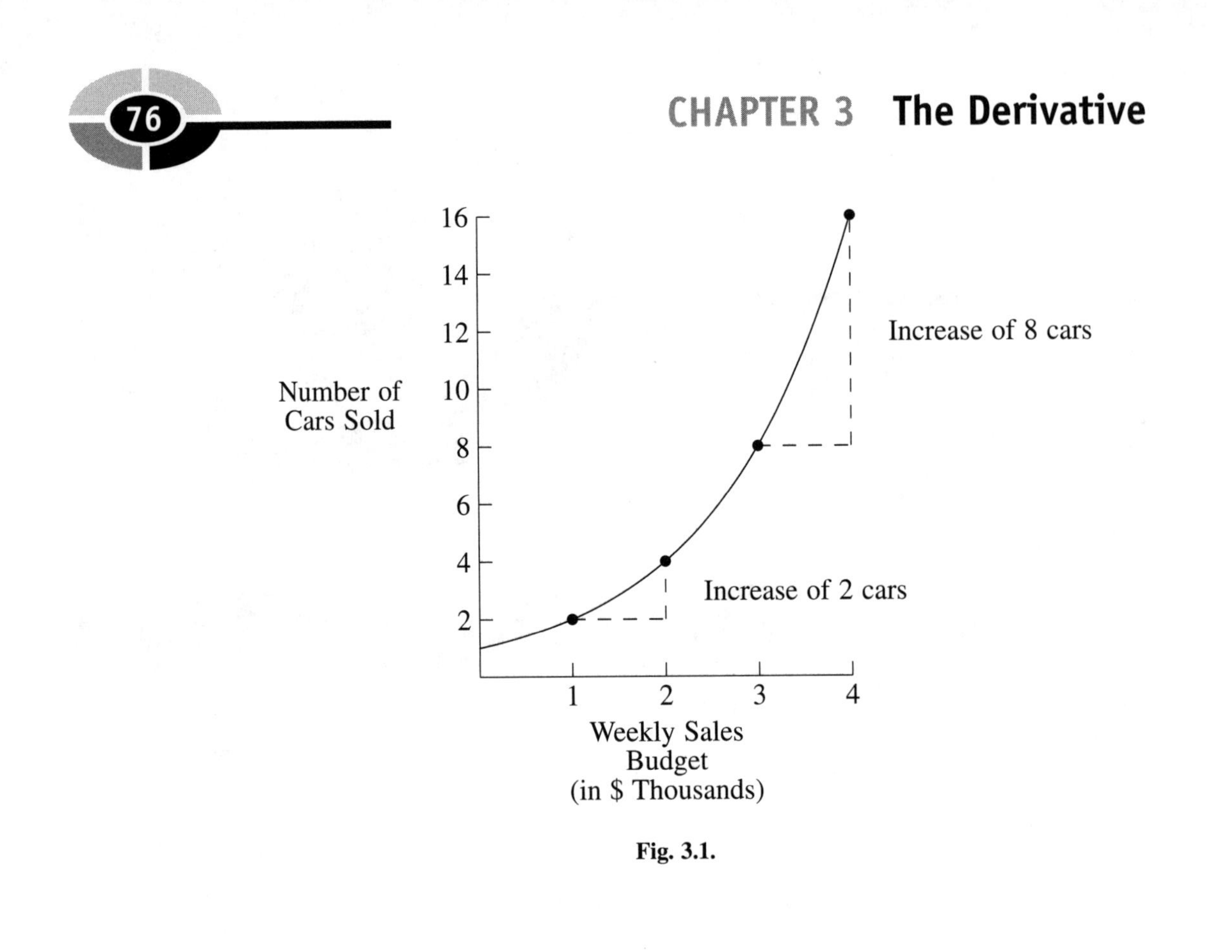
16
14
12
10
8
6
4
2
Number of
Cars Sold
Increase of 8 cars
Increase of 2 cars
1
2
3
4
Weekly Sales
Budget
(in $ Thousands)

Fig. 3.1.

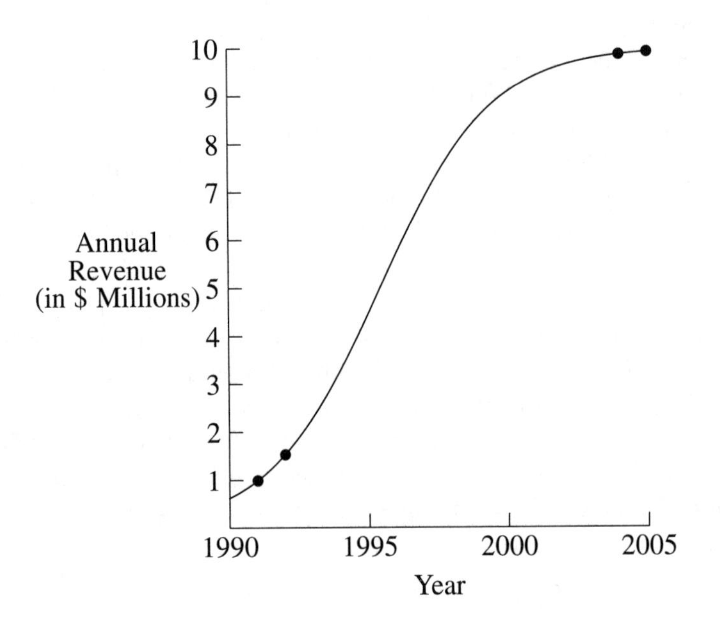
10
9
8
7
6
5
4
3
2
1
Annual
Revenue
(in $ Millions)
1990
1995
2000
2005
Year

Fig. 3.2.

Before learning about the tangent line and the derivative, we need to learn about *secant lines*. A secant line on the graph of a function is formed by drawing a line through two points on the graph (see Figure 3.3).

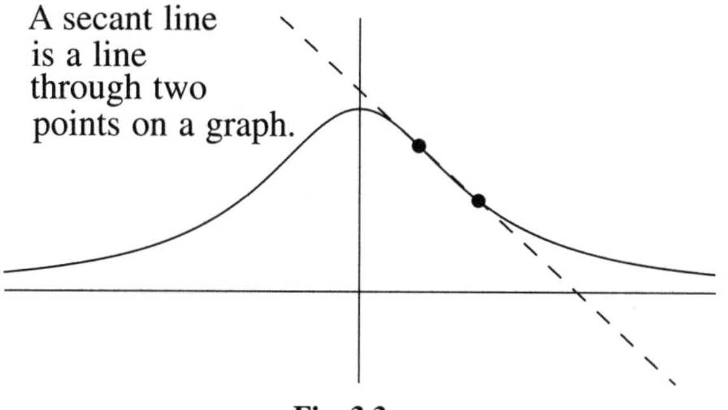

Fig. 3.3.

Suppose we have a series of secant lines that all go through a fixed point on the graph. Each of its other points gets closer and closer to the fixed point. For the graphs in Figures 3.4–3.6, one point is fixed and the other point of the graph is moving closer to the fixed point. We are interested in the slope of these lines. Consider the distance between the x-values of these points. This distance is called h (see Figure 3.7). What is the slope of the secant line between the point $(x, f(x))$ (the fixed point) and $(x+h, f(x+h))$ (the point moving closer to the fixed point)? Using the slope formula with $x_1 = x$, $y_1 = f(x)$, $x_2 = x+h$, and $y_2 = f(x+h)$, the slope of a secant line is

$$m = \frac{y_2 - y_1}{x_2 - x_1} = \frac{f(x+h) - f(x)}{x + h - x} = \frac{f(x+h) - f(x)}{h}.$$

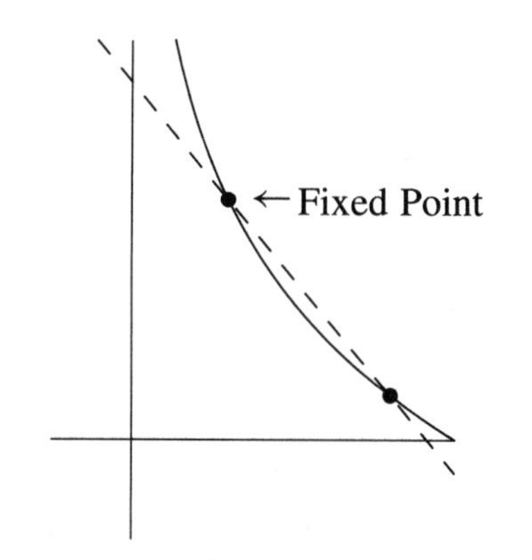

Fig. 3.4.

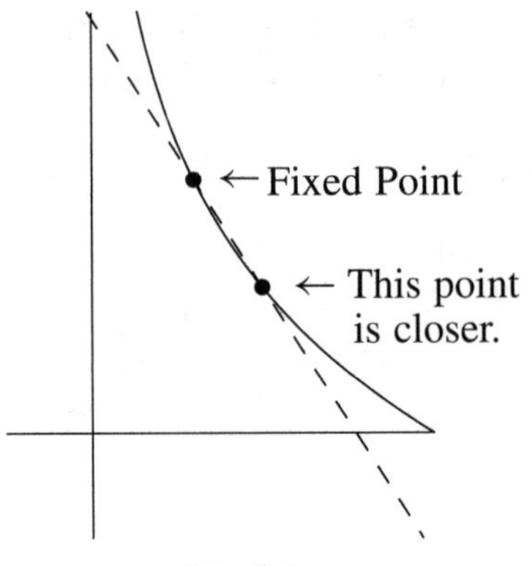

Fig. 3.5.

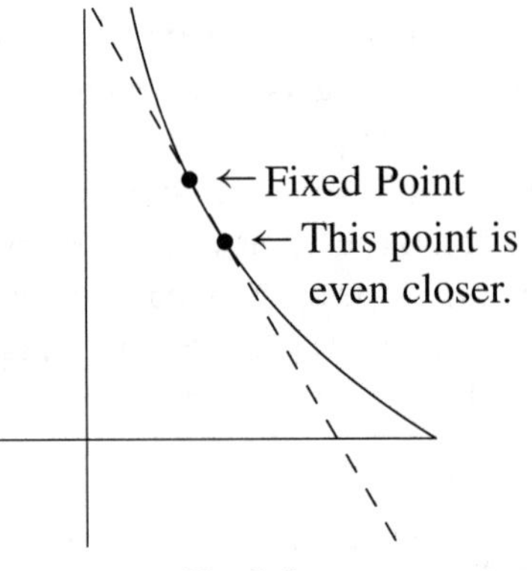

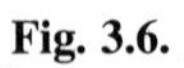
Fig. 3.6.

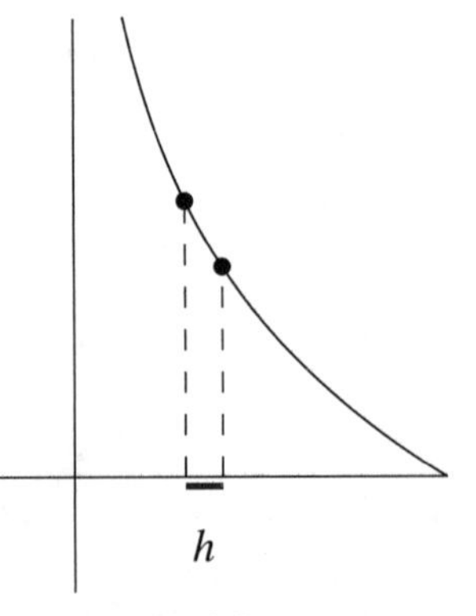

Fig. 3.7.

EXAMPLE

- $f(x) = x^2 - x$
  Let $x = 3$, $f(x) = 6$ be the fixed point. We will find the slope of the secant line between $x = 3$ and $x = 3.5$, between $x = 3$ and $x = 3.1$, and between $x = 3$ and $x = 3.01$ (see Table 3.1).

Table 3.1

$x+h$	h	$y=(x+h)^2-(x+h)$	$m = \frac{y_2-y_1}{x_2-x_1} = \frac{y-6}{h}$
3.5	$3.5 - 3 = 0.5$	$3.5^2 - 3.5 = 8.75$	$\frac{8.75-6}{3.5-3} = 5.5$
3.1	$3.1 - 3 = 0.1$	$3.1^2 - 3.1 = 6.51$	$\frac{6.51-6}{3.1-3} = 5.1$
3.01	$3.01 - 3 = 0.01$	$3.01^2 - 3.01 = 6.0501$	$\frac{6.0501-6}{3.01-3} = 5.01$

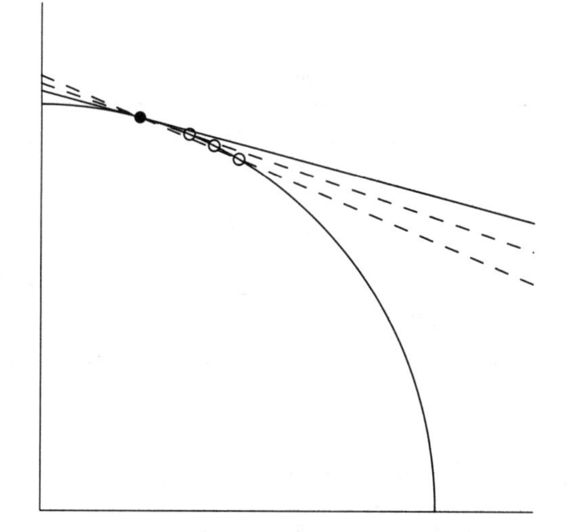

Fig. 3.8.

The slope of the tangent line is the limit of the slope of the secant lines as the moving point approaches the fixed point. In the above example, the slope of the tangent line is 5. For the graph in Figure 3.8, the tangent line (the solid line) is the limit of the secant lines (the dashed lines).

PRACTICE

1. Fill in Table 3.2 for $f(x) = x^2 + 5x$ and the fixed point (3, 24).

Table 3.2

$x + h$	h	$y = (x + h)^2 + 5(x + h)$	$m = \frac{y - 24}{h}$ (Slope of the secant line)
3.5			
3.1			
3.01			

2. What appears to be the slope of the tangent line?

SOLUTIONS

1. See Table 3.3.

Table 3.3

$x + h$	h	$y = (x + h)^2 + 5(x+h)$	$m = \frac{y - 24}{h}$ (Slope of the secant line)
3.5	0.5	$3.5^2 + 5(3.5) = 29.75$	$\frac{29.75-24}{0.5} = 11.5$
3.1	0.1	$3.1^2 + 5(3.1) = 25.11$	$\frac{25.11-24}{0.1} = 11.1$
3.01	0.01	$3.01^2 + 5(3.01) = 24.1101$	$\frac{24.1101-24}{0.01} = 11.01$

2. The slope of the tangent line appears to be 11.

The slope of the tangent line is the slope of the secant lines as h approaches 0.

$$\text{Tangent slope} = \lim_{h \to 0} \frac{f(x + h) - f(x)}{h}$$

Once this limit is simplified, we have a formula for the slope of the tangent line for any x. This formula is called the *derivative*. It has many notations, among them are y' (pronounced "y-prime"), $f'(x)$ (pronounced "f prime of x"), and

$\frac{dy}{dx}$ (pronounced "dee-y, dee-x"). We will practice evaluating this limit for various kinds of functions. In the next chapter, we will see that there are formulas that can eliminate most of the messy algebra.

EXAMPLES

- Find $f'(x)$ for $f(x) = x^2$.

  We will find and simplify $\lim_{h \to 0} \frac{f(x+h)-f(x)}{h}$. We will begin by finding and simplifying $f(x+h)$. Then we will put this as well as x^2 in the formula. Finally, we will simplify the fraction and take the limit.

$$f(x+h) = (x+h)^2 = (x+h)(x+h) = x^2 + 2xh + h^2$$

$$f'(x) = \lim_{h \to 0} \frac{f(x+h) - f(x)}{h} = \lim_{h \to 0} \frac{\overbrace{x^2 + 2xh + h^2}^{f(x+h)} - \overbrace{x^2}^{f(x)}}{h}$$

$$= \lim_{h \to 0} \frac{2xh + h^2}{h} \qquad \text{Factor } h \text{ from } 2xh + h^2\text{, leaving } 2x + h.$$

$$= \lim_{h \to 0} \frac{h(2x+h)}{h} \qquad h \text{ in the numerator and denominator cancels.}$$

$$= \lim_{h \to 0} (2x + h) = 2x + 0 = 2x$$

- Find y' for $y = 3x - 4$.

  The "$f(x)$" part of the formula is "$3x - 4$" and the "$f(x+h)$" part of the formula is "$3(x+h) - 4$."

$$y' = \lim_{h \to 0} \frac{f(x+h) - f(x)}{h} = \lim_{h \to 0} \frac{\overbrace{3(x+h) - 4}^{f(x+h)} - \overbrace{(3x-4)}^{f(x)}}{h}$$

$$= \lim_{h \to 0} \frac{3x + 3h - 4 - 3x + 4}{h} = \lim_{h \to 0} \frac{3h}{h} = 3$$

  Any time the function is linear (in the form $f(x) = mx + b$), the slope of the tangent line for any point on the graph is the same as the slope of the line.

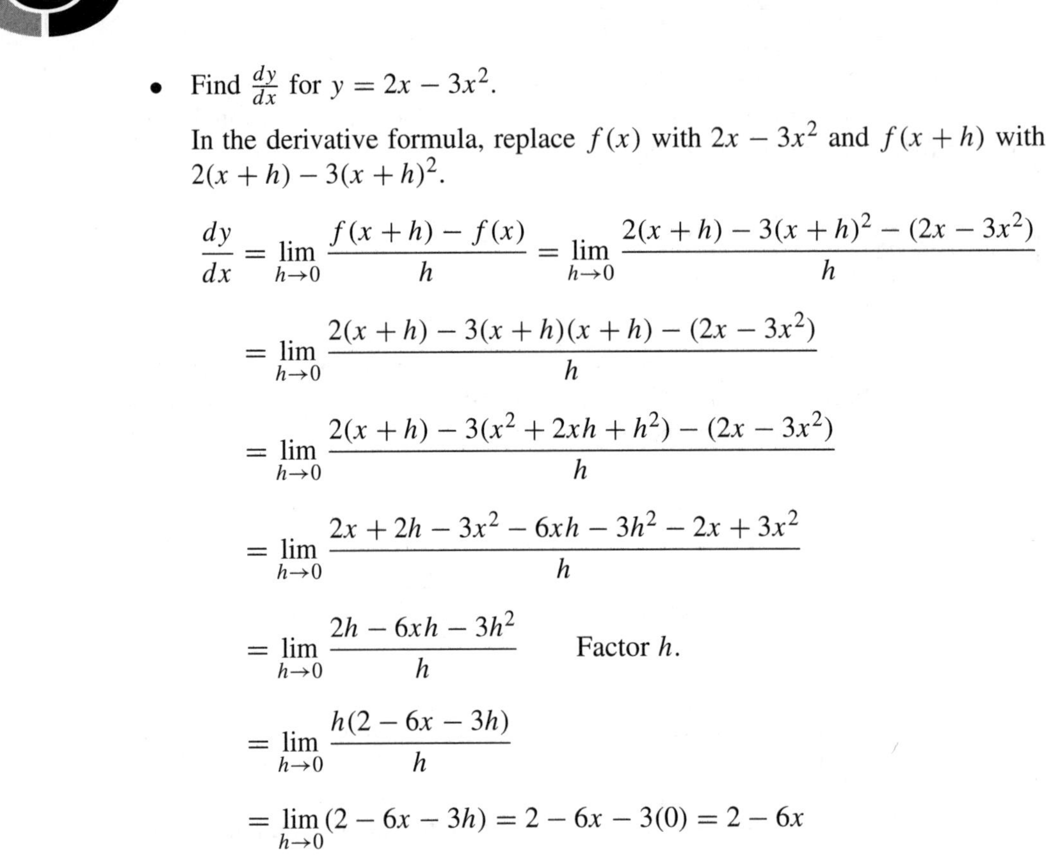

- Find $\frac{dy}{dx}$ for $y = 2x - 3x^2$.

  In the derivative formula, replace $f(x)$ with $2x - 3x^2$ and $f(x+h)$ with $2(x+h) - 3(x+h)^2$.

$$\frac{dy}{dx} = \lim_{h\to 0} \frac{f(x+h) - f(x)}{h} = \lim_{h\to 0} \frac{2(x+h) - 3(x+h)^2 - (2x - 3x^2)}{h}$$

$$= \lim_{h\to 0} \frac{2(x+h) - 3(x+h)(x+h) - (2x - 3x^2)}{h}$$

$$= \lim_{h\to 0} \frac{2(x+h) - 3(x^2 + 2xh + h^2) - (2x - 3x^2)}{h}$$

$$= \lim_{h\to 0} \frac{2x + 2h - 3x^2 - 6xh - 3h^2 - 2x + 3x^2}{h}$$

$$= \lim_{h\to 0} \frac{2h - 6xh - 3h^2}{h} \qquad \text{Factor } h.$$

$$= \lim_{h\to 0} \frac{h(2 - 6x - 3h)}{h}$$

$$= \lim_{h\to 0} (2 - 6x - 3h) = 2 - 6x - 3(0) = 2 - 6x$$

- Find $f'(x)$ for $f(x) = 10$.

  For this function the y-value is always 10, so both $f(x)$ and $f(x+h)$ are 10.

$$f'(x) = \lim_{h\to 0} \frac{f(x+h) - f(x)}{h} = \lim_{h\to 0} \frac{10 - 10}{h} = \lim_{h\to 0} \frac{0}{h}$$

$$= \lim_{h\to 0} 0 = 0 \qquad \left(\frac{0}{h} = 0\right)$$

PRACTICE

Find the derivative.

1. $f(x) = x^2 + 4$

$$f'(x) = \lim_{h\to 0} \frac{(\quad) - (\;)}{h} =$$

2. $y = \frac{1}{2}x - 1$

$$y' = \lim_{h \to 0} \frac{(\quad) - (\ )}{h} =$$

3. $y = 4x^2 - x + 6$

$$\frac{dy}{dx} = \lim_{h \to 0} \frac{(\quad) - (\ )}{h} =$$

4. $f(x) = 8$

$$f'(x) = \lim_{h \to 0} \frac{(\quad) - (\ )}{h} =$$

SOLUTIONS

1.

$$f'(x) = \lim_{h \to 0} \frac{(x+h)^2 + 4 - (x^2+4)}{h} = \lim_{h \to 0} \frac{(x+h)(x+h) + 4 - (x^2+4)}{h}$$

$$= \lim_{h \to 0} \frac{x^2 + 2xh + h^2 + 4 - x^2 - 4}{h} = \lim_{h \to 0} \frac{2xh + h^2}{h}$$

$$= \lim_{h \to 0} \frac{h(2x+h)}{h} = \lim_{h \to 0} (2x+h) = 2x + 0 = 2x$$

2.

$$y' = \lim_{h \to 0} \frac{\frac{1}{2}(x+h) - 1 - (\frac{1}{2}x - 1)}{h} = \lim_{h \to 0} \frac{\frac{1}{2}x + \frac{1}{2}h - 1 - \frac{1}{2}x + 1}{h}$$

$$= \lim_{h \to 0} \frac{\frac{1}{2}h}{h} = \lim_{h \to 0} \frac{1}{2} = \frac{1}{2}$$

3.

$$\frac{dy}{dx} = \lim_{h \to 0} \frac{4(x+h)^2 - (x+h) + 6 - (4x^2 - x + 6)}{h}$$

$$= \lim_{h \to 0} \frac{4(x+h)(x+h) - (x+h) + 6 - (4x^2 - x + 6)}{h}$$

$$= \lim_{h \to 0} \frac{4(x^2 + 2xh + h^2) - (x+h) + 6 - (4x^2 - x + 6)}{h}$$

$$= \lim_{h\to 0} \frac{4x^2 + 8xh + 4h^2 - x - h + 6 - 4x^2 + x - 6}{h}$$

$$= \lim_{h\to 0} \frac{8xh + 4h^2 - h}{h} \quad \text{Factor } h \text{ in the numerator.}$$

$$= \lim_{h\to 0} \frac{h(8x + 4h - 1)}{h} = \lim_{h\to 0} (8x + 4h - 1) = 8x + 4(0) - 1$$

$$= 8x - 1$$

4. Both $f(x)$ and $f(x+h)$ are 8.

$$f'(x) = \lim_{h\to 0} \frac{8-8}{h} = \lim_{h\to 0} \frac{0}{h} = \lim_{h\to 0} 0 = 0$$

The functions in the next set of problems will contain fractions. These fractions make the algebra of the limit more tedious. To make matters a little easier, we will use the following shortcut.

$$\frac{(ab)/c}{a} = \frac{b}{c}$$

Here is why the shortcut works.

$$\frac{(ab)/c}{a} = \frac{ab}{c} \div a = \frac{ab}{c} \cdot \frac{1}{a} = \frac{ab}{ac} = \frac{b}{c}$$

Now we can simplify expressions such as

$$\frac{\frac{h(x^2+2x-1)}{x-1}}{h} = \frac{x^2 + 2x - 1}{x - 1}$$

without having to go through these steps:

$$\frac{\frac{h(x^2+2x-1)}{x-1}}{h} = \frac{h(x^2 + 2x - 1)}{x - 1} \div h = \frac{h(x^2 + 2x - 1)}{x - 1} \cdot \frac{1}{h} = \frac{x^2 + 2x - 1}{x - 1}.$$

Once we have factored h from the numerator (of the main numerator), it will cancel the h in the main denominator.

EXAMPLES

- Find $f'(x)$ for $f(x) = \frac{1}{x}$.

$$f'(x) = \lim_{h \to 0} \frac{\frac{1}{x+h} - \frac{1}{x}}{h}$$

We will take a few steps to simplify $\dfrac{1}{x+h} - \dfrac{1}{x}$.

$$= \lim_{h \to 0} \frac{\frac{1}{x+h} \cdot \frac{x}{x} - \frac{1}{x} \cdot \frac{x+h}{x+h}}{h}$$

$$= \lim_{h \to 0} \frac{\frac{x}{x(x+h)} - \frac{x+h}{x(x+h)}}{h}$$

$$= \lim_{h \to 0} \frac{\frac{x-(x+h)}{x(x+h)}}{h} = \lim_{h \to 0} \frac{\frac{x-x-h}{x(x+h)}}{h}$$

$$= \lim_{h \to 0} \frac{\frac{-h}{x(x+h)}}{h}$$

$$= \lim_{h \to 0} \frac{-1}{x(x+h)}$$ This is the shortcut.

$$= \frac{-1}{x(x+0)}$$ Now it is safe to let $h = 0$.

$$= \frac{-1}{x \cdot x} = \frac{-1}{x^2}$$

- Find $\frac{dy}{dx}$ for $y = \frac{10}{x+1}$.

$$\frac{dy}{dx} = \lim_{h \to 0} \frac{\frac{10}{x+h+1} - \frac{10}{x+1}}{h}$$

$$= \lim_{h \to 0} \frac{\frac{10}{x+h+1} \cdot \frac{x+1}{x+1} - \frac{10}{x+1} \cdot \frac{x+h+1}{x+h+1}}{h}$$

$$= \lim_{h \to 0} \frac{\frac{10(x+1)-10(x+h+1)}{(x+h+1)(x+1)}}{h}$$

$$= \lim_{h \to 0} \frac{\frac{10x+10-10x-10h-10}{(x+h+1)(x+1)}}{h}$$

$$= \lim_{h \to 0} \frac{\frac{-10h}{(x+h+1)(x+1)}}{h}$$

$$= \lim_{h \to 0} \frac{-10}{(x+h+1)(x+1)} \qquad \text{This is the shortcut.}$$

$$= \frac{-10}{(x+0+1)(x+1)} = \frac{-10}{(x+1)(x+1)}$$

$$= \frac{-10}{(x+1)^2}$$

- Find y' for $y = \frac{x}{2x+3}$

$$y' = \lim_{h \to 0} \frac{\frac{x+h}{2(x+h)+3} - \frac{x}{2x+3}}{h}$$

$$= \lim_{h \to 0} \frac{\frac{x+h}{2(x+h)+3} \cdot \frac{2x+3}{2x+3} - \frac{x}{2x+3} \cdot \frac{2(x+h)+3}{2(x+h)+3}}{h}$$

$$= \lim_{h \to 0} \frac{\frac{(x+h)(2x+3)-x[2(x+h)+3]}{[2(x+h)+3](2x+3)}}{h}$$

$$= \lim_{h \to 0} \frac{\frac{2x^2+3x+2xh+3h-x(2x+2h+3)}{(2x+2h+3)(2x+3)}}{h}$$

$$= \lim_{h \to 0} \frac{\frac{2x^2+3x+2xh+3h-2x^2-2xh-3x}{(2x+2h+3)(2x+3)}}{h}$$

$$= \lim_{h \to 0} \frac{\frac{3h}{(2x+2h+3)(2x+3)}}{h}$$

$$= \lim_{h \to 0} \frac{3}{(2x+2h+3)(2x+3)} \qquad \text{This is the shortcut.}$$

$$= \frac{3}{(2x+2 \cdot 0+3)(2x+3)}$$

$$= \frac{3}{(2x+3)(2x+3)} = \frac{3}{(2x+3)^2}$$

PRACTICE

Find the derivative.

1. $y = \frac{2}{x}$
2. $f(x) = \frac{6}{2x+1}$
3. $f(x) = \frac{x}{x-1}$

SOLUTIONS

1.

$$\begin{aligned}
\frac{dy}{dx} &= \lim_{h\to 0} \frac{\frac{2}{x+h} - \frac{2}{x}}{h} \\
&= \lim_{h\to 0} \frac{\frac{2}{x+h}\cdot\frac{x}{x} - \frac{2}{x}\cdot\frac{x+h}{x+h}}{h} \\
&= \lim_{h\to 0} \frac{\frac{2x-2(x+h)}{x(x+h)}}{h} \\
&= \lim_{h\to 0} \frac{\frac{2x-2x-2h}{x(x+h)}}{h} \\
&= \lim_{h\to 0} \frac{\frac{-2h}{x(x+h)}}{h} \\
&= \lim_{h\to 0} \frac{-2}{x(x+h)} \\
&= \frac{-2}{x(x+0)} = \frac{-2}{x\cdot x} = \frac{-2}{x^2}
\end{aligned}$$

2.

$$\begin{aligned}
f'(x) &= \lim_{h\to 0} \frac{\frac{6}{2(x+h)+1} - \frac{6}{2x+1}}{h} \\
&= \lim_{h\to 0} \frac{\frac{6}{2(x+h)+1}\cdot\frac{2x+1}{2x+1} - \frac{6}{2x+1}\cdot\frac{2(x+h)+1}{2(x+h)+1}}{h} \\
&= \lim_{h\to 0} \frac{\frac{6(2x+1)}{(2x+1)[2(x+h)+1]} - \frac{6[2(x+h)+1]}{(2x+1)[2(x+h)+1]}}{h}
\end{aligned}$$

$$= \lim_{h \to 0} \frac{\frac{12x+6}{(2x+1)(2x+2h+1)} - \frac{6(2x+2h+1)}{(2x+1)(2x+2h+1)}}{h}$$

$$= \lim_{h \to 0} \frac{\frac{12x+6-12x-12h-6}{(2x+1)(2x+2h+1)}}{h}$$

$$= \lim_{h \to 0} \frac{\frac{-12h}{(2x+1)(2x+2h+1)}}{h}$$

$$= \lim_{h \to 0} \frac{-12}{(2x+1)(2x+2h+1)}$$

$$= \frac{-12}{(2x+1)(2x+2 \cdot 0+1)} = \frac{-12}{(2x+1)(2x+1)} = \frac{-12}{(2x+1)^2}$$

3.

$$f'(x) = \lim_{h \to 0} \frac{\frac{x+h}{x+h-1} - \frac{x}{x-1}}{h}$$

$$= \lim_{h \to 0} \frac{\frac{x+h}{x+h-1} \cdot \frac{x-1}{x-1} - \frac{x}{x-1} \cdot \frac{x+h-1}{x+h-1}}{h}$$

$$= \lim_{h \to 0} \frac{\frac{(x+h)(x-1)-x(x+h-1)}{(x+h-1)(x-1)}}{h}$$

$$= \lim_{h \to 0} \frac{\frac{x^2-x+xh-h-x^2-xh+x}{(x+h-1)(x-1)}}{h}$$

$$= \lim_{h \to 0} \frac{\frac{-h}{(x+h-1)(x-1)}}{h}$$

$$= \lim_{h \to 0} \frac{-1}{(x+h-1)(x-1)}$$

$$= \frac{-1}{(x+0-1)(x-1)} = \frac{-1}{(x-1)(x-1)} = \frac{-1}{(x-1)^2}$$

The limit property $\lim_{x \to a}[f(x) \pm g(x)] = \lim_{x \to a} f(x) \pm \lim_{x \to a} g(x)$ allows us to differentiate functions one term at a time. The limit property $\lim_{x \to a} cf(x) = c \lim_{x \to a} f(x)$ allows us to take the derivative of a quantity either before or after multiplying the quantity by a constant.

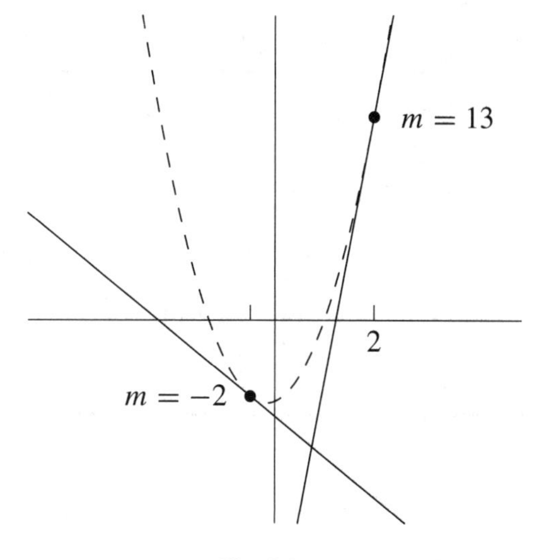

Fig. 3.9.

If $f(x)$ and $g(x)$ are differentiable functions, and if $y = f(x) \pm g(x)$, then $y' = f'(x) \pm g'(x)$.
If $y = cf(x)$, then $y' = cf'(x)$.

The derivative is a formula for the slope of a tangent line for a function. For example, the derivative of $f(x) = 3x^2 + x - 4$ is $f'(x) = 6x + 1$. With this formula, we can find the slope of the tangent line for any x-value. For example, the slope of the tangent line at $x = 2$ is $f'(2) = 6(2) + 1 = 13$. The slope of the tangent line at $x = -\frac{1}{2}$ is $f'(-\frac{1}{2}) = 6(-\frac{1}{2}) + 1 = -2$ (see Figure 3.9).

Later, after we have learned some derivative formulas, we will practice finding equations of tangent lines.

CHAPTER 3 REVIEW

1. What appears to be the slope of the tangent line at $x = 2$ for the numbers in Table 3.4 (see page 90)?
   a) 2 b) 3 c) 4 d) 5

2. Find y' for $y = 4x^2 + 5$.
   a) $8x$ b) $8x + 5$ c) $8x + 5h$ d) $8x - h$

3. Find $f'(x)$ for $f(x) = 7$.
   a) 7 b) 0 c) 49 d) $f'(x)$ does not exist.

Table 3.4

$x+h$	$f(x+h)$	h	$\frac{f(x+h)-f(x)}{h}$ Secant slope
1.5	0.75	$1.5-2=-0.5$	2.5
1.9	1.71	$1.9-2=-0.1$	2.9
1.99	1.9701	$1.99-2=-0.01$	2.99
			?
2.01	2.0301	$2.01-2=0.01$	3.01
2.1	2.31	$2.1-2=0.1$	3.1
2.5	3.75	$2.5-2=0.5$	3.5

4. Find $f'(x)$ for $f(x)=\frac{1}{x+5}$.

   a) $\frac{1}{x+h+5}$ b) $\frac{-1}{x+5}$

   c) $\frac{1}{x+5}$ d) $\frac{-1}{(x+5)^2}$

5. Find $f'(x)$ for $f(x)=6-3x$.

   a) -1 b) -3 c) 6 d) 3

SOLUTIONS

1. b 2. a 3. b 4. d 5. b

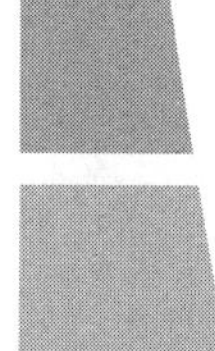

CHAPTER 4

Three Important Formulas

There are fewer than ten differentiation formulas that you will need to know, and three of them are in this chapter. They should be memorized as they will be used extensively throughout most of this book (as well as in any calculus course).

The Power Rule

For a function of the form $f(x) = x^n$, where n is any real number, the derivative is $f'(x) = nx^{n-1}$. The old power moves in front of x and the new power is the old power minus 1.

EXAMPLES

- $f(x) = x^3$
  We will write the old power, 3, in front of x. The new power on x is $3 - 1 = 2$. Now we can see that $f'(x) = 3x^2$.
- $f(x) = x^{-5}$
  $n = -5$, $n - 1 = -5 - 1 = -6$ The derivative is $f'(x) = -5x^{-6}$.
- $y = x$
  Here, $n = 1$, so $n - 1 = 1 - 1 = 0$ and $y' = nx^{n-1}$ becomes $y' = 1 \cdot x^0$. Because $x^0 = 1$, this means that $y' = 1$.
- $f(x) = x^{3/4}$
  $n = \frac{3}{4}$, $n - 1 = \frac{3}{4} - 1 = -\frac{1}{4}$ and $f'(x) = nx^{n-1}$ becomes $f'(x) = \frac{3}{4}x^{-1/4}$.
- $g(t) = t^{-1/2}$
  $n = -\frac{1}{2}$, $n - 1 = -\frac{1}{2} - 1 = -\frac{3}{2}$ and $g'(t) = nt^{n-1}$ becomes $g'(t) = -\frac{1}{2}t^{-3/2}$.

PRACTICE

Use the power rule to find the derivative.

1. $f(x) = x^{10}$
2. $y = x^{-6}$
3. $f(x) = x^7$
4. $g(t) = t^{1/3}$
5. $f(t) = t$
6. $f(x) = x^{-1}$
7. $y = x^{-2/3}$
8. $f(x) = x^{9/4}$
9. $f(x) = x^{-2}$
10. $y = x^{-5/7}$

SOLUTIONS

1. $f'(x) = 10x^{10-1} = 10x^9$
2. $y' = -6x^{-6-1} = -6x^{-7}$
3. $f'(x) = 7x^{7-1} = 7x^6$
4. $g'(t) = \frac{1}{3}t^{1/3-1} = \frac{1}{3}t^{1/3-3/3} = \frac{1}{3}t^{-2/3}$
5. $f'(t) = 1 \cdot t^{1-1} = 1 \cdot t^0 = 1$
6. $f'(x) = -1x^{-1-1} = -x^{-2}$
7. $y' = -\frac{2}{3}x^{-2/3-1} = -\frac{2}{3}x^{-2/3-3/3} = -\frac{2}{3}x^{-5/3}$
8. $f'(x) = \frac{9}{4}x^{9/4-1} = \frac{9}{4}x^{9/4-4/4} = \frac{9}{4}x^{5/4}$
9. $f'(x) = -2x^{-2-1} = -2x^{-3}$
10. $y' = -\frac{5}{7}x^{-5/7-1} = -\frac{5}{7}x^{-5/7-7/7} = -\frac{5}{7}x^{-12/7}$

Sometimes we need to use algebra to put a function into a form that looks like a derivative formula. For example, we do not have formulas for functions such as $f(x) = \sqrt{x}$ and $f(x) = 1/x$, but we do have exponent properties that allow us to write these functions in the form of x to a power. Then we can use the power rule to find their derivatives.

$$\sqrt[n]{x^m} = x^{m/n} \qquad \text{and} \qquad \frac{1}{x^n} = x^{-n}$$

EXAMPLES

- $f(x) = \frac{1}{x^3}$
  Using the fact that $\frac{1}{x^n} = x^{-n}$, we can rewrite the function as $f(x) = x^{-3}$, and use the power rule: $f'(x) = -3x^{-4}$.
- $y = \sqrt{x}$
  Using the fact that $\sqrt[n]{x^m} = x^{m/n}$, we can rewrite the function as $y = x^{1/2}$ and use the power rule: $y' = \frac{1}{2}x^{1/2-1} = \frac{1}{2}x^{-1/2}$.
- $f(x) = \frac{1}{x}$
  We can rewrite this function as $f(x) = x^{-1}$. The derivative is $f'(x) = -1 \cdot x^{-1-1} = -x^{-2}$.
- $y = \frac{1}{\sqrt[3]{x^4}}$
  We need to use both properties for this function.

$$y = \frac{1}{\sqrt[3]{x^4}} = \frac{1}{x^{4/3}} = x^{-4/3}$$

  Now we can use the power rule: $y' = -\frac{4}{3}x^{-4/3-1} = -\frac{4}{3}x^{-7/3}$.

PRACTICE

Use the power rule to find the derivative.

1. $y = \frac{1}{x^5}$
2. $f(x) = \sqrt[3]{x}$
3. $f(x) = \sqrt[3]{x^2}$
4. $f(x) = \frac{1}{x^4}$
5. $g(t) = \frac{1}{t}$
6. $y = \frac{1}{\sqrt[3]{x}}$
7. $f(x) = \frac{1}{\sqrt[3]{x^2}}$
8. $f(x) = \frac{1}{\sqrt[2]{x^5}}$

SOLUTIONS

1. $y = x^{-5}$ and $y' = -5x^{-6}$
2. $f(x) = x^{1/3}$ and $y' = \frac{1}{3}x^{1/3-1} = \frac{1}{3}x^{-2/3}$

3. $f(x) = x^{2/3}$ and $f'(x) = \frac{2}{3}x^{2/3-1} = \frac{2}{3}x^{-1/3}$
4. $f(x) = x^{-4}$ and $f'(x) = -4x^{-5}$
5. $g(t) = t^{-1}$ and $g'(t) = -t^{-2}$
6. $y = x^{-1/3}$ and $y' = -\frac{1}{3}x^{-1/3-1} = -\frac{1}{3}x^{-4/3}$
7. $f(x) = x^{-2/3}$ and $f'(x) = -\frac{2}{3}x^{-2/3-1} = -\frac{2}{3}x^{-5/3}$
8. $f(x) = x^{-5/2}$ and $f'(x) = -\frac{5}{2}x^{-5/2-1} = -\frac{5}{2}x^{-7/2}$

Some instructors want the derivative to look like the original function. For example, if x is in a denominator or under a square root in the original function, then x needs to be in a denominator or under a square root in the derivative. Also, evaluating functions and derivatives is easier when there are no negative exponents and fraction exponents. In this section, we will write the solutions to the previous examples and practice problems in the same form as the original function. This means that we will write the derivatives without using negative exponents and fraction exponents.

EXAMPLES

Write the derivatives without using negative exponents and fraction exponents.

- $f(x) = \frac{1}{x^3}$
  We found $f'(x) = -3x^{-4}$. We can rewrite this as
  $$f'(x) = -3\left(\frac{1}{x^4}\right) = \frac{-3}{x^4}.$$
- $y = \sqrt{x}$
  We found that $y' = \frac{1}{2}x^{-1/2}$. We can rewrite this as
  $$y' = \frac{1}{2}\frac{1}{x^{1/2}} = \frac{1}{2}\frac{1}{\sqrt{x}} = \frac{1}{2\sqrt{x}}.$$
- $y = \frac{1}{\sqrt[3]{x^4}}$
  We found that $y' = -\frac{4}{3}x^{-7/3}$. We can rewrite this as
  $$y' = -\frac{4}{3}\left(\frac{1}{x^{7/3}}\right) = -\frac{4}{3\sqrt[3]{x^7}}.$$

PRACTICE

Write the derivatives without using negative exponents and fraction exponents. These are the same functions as before.

1. $y = \frac{1}{x^5}$
2. $f(x) = \sqrt[3]{x}$
3. $(x) = \sqrt[3]{x^2}$
4. $f(x) = \frac{1}{x^4}$
5. $g(t) = \frac{1}{t}$
6. $y = \frac{1}{\sqrt[3]{x}}$
7. $f(x) = \frac{1}{\sqrt[3]{x^2}}$
8. $f(x) = \frac{1}{\sqrt{x^5}}$

SOLUTIONS

1. $y' = -5x^{-6} = -5\left(\frac{1}{x^6}\right) = \frac{-5}{x^6}$
2. $f'(x) = \frac{1}{3}x^{-2/3} = \frac{1}{3}\left(\frac{1}{x^{2/3}}\right) = \frac{1}{3}\left(\frac{1}{\sqrt[3]{x^2}}\right) = \frac{1}{3\sqrt[3]{x^2}}$
3. $f'(x) = \frac{2}{3}x^{-1/3} = \frac{2}{3}\left(\frac{1}{x^{1/3}}\right) = \frac{2}{3}\frac{1}{\sqrt[3]{x}} = \frac{2}{3\sqrt[3]{x}}$
4. $f'(x) = -4x^{-5} = -4\left(\frac{1}{x^5}\right) = \frac{-4}{x^5}$
5. $g'(t) = -t^{-2} = -\left(\frac{1}{t^2}\right) = -\frac{1}{t^2}$
6. $y' = -\frac{1}{3}x^{-4/3} = -\frac{1}{3}\left(\frac{1}{x^{4/3}}\right) = -\frac{1}{3}\cdot\frac{1}{\sqrt[3]{x^4}} = -\frac{1}{3\sqrt[3]{x^4}}$
7. $f'(x) = -\frac{2}{3}x^{-5/3} = -\frac{2}{3}\left(\frac{1}{x^{5/3}}\right) = -\frac{2}{3}\frac{1}{\sqrt[3]{x^5}} = -\frac{2}{3\sqrt[3]{x^5}}$
8. $f'(x) = -\frac{5}{2}x^{-7/2} = -\frac{5}{2}\left(\frac{1}{x^{7/2}}\right) = -\frac{5}{2}\frac{1}{\sqrt{x^7}} = -\frac{5}{2\sqrt{x^7}}$

We can use the power rule together with the properties in Chapter 3 to find the derivative for a large family of functions. First, we will use the power rule with the property that says if $y = af(x)$, then $y' = af'(x)$. Putting this property together with the power rule, we have a new rule.

> If $y = ax^n$, then $y' = nax^{n-1}$, for any real number n.

EXAMPLES

Find the derivative.

- $f(x) = 4x^5$
  Here, $a = 4$, $n = 5$, so $f'(x) = nax^{n-1}$ becomes $f'(x) = 5(4)x^{5-1} = 20x^4$.

  - $f(x) = -2x^{-3}$ $\quad f'(x) = (-3)(-2)x^{-3-1} = 6x^{-4}$
  - $y = \frac{1}{2}x^6$ $\quad y' = 6\left(\frac{1}{2}\right)x^{6-1} = 3x^5$
  - $y = -4x$ $\quad y' = 1(-4)x^{1-1} = -4x^0 = -4$
  - $y = 3\sqrt{x} = 3x^{1/2}$ $\quad y' = \left(\frac{1}{2}\right)(3)x^{1/2-1} = \frac{3}{2}x^{-1/2}$ or $\frac{3}{2\sqrt{x}}$
  - $f(x) = \frac{10}{\sqrt[4]{x}} = 10x^{-1/4}$ $\quad f'(x) = \left(-\frac{1}{4}\right)(10)x^{-1/4-1}$
    $= -\frac{5}{2}x^{-5/4}$ or $-\frac{5}{2\sqrt[4]{x^5}}$

PRACTICE

Find the derivative.

1. $f(x) = -3x^2$
2. $f(x) = 15x^3$
3. $y = \frac{4}{x^2}$
4. $f(x) = -8\sqrt{x}$
5. $f(x) = 2\sqrt[3]{x^2}$
6. $h(t) = \frac{3}{\sqrt[4]{x^3}}$
7. $f(x) = 17x$
8. $g(t) = \frac{-1}{\sqrt[3]{t^2}}$

SOLUTIONS

1. $f'(x) = 2(-3)x^{2-1} = -6x$
2. $f'(x) = 3(15)x^{3-1} = 45x^2$
3. $y = 4x^{-2}$; $y' = (-2)(4)x^{-2-1} = -8x^{-3}$ or $-\frac{8}{x^3}$

4. $f(x) = -8x^{1/2}$; $f'(x) = \left(\frac{1}{2}\right)(-8)x^{1/2-1} = -4x^{-1/2}$ or $\frac{-4}{\sqrt{x}}$

5. $f(x) = 2x^{2/3}$; $f'(x) = \left(\frac{2}{3}\right)(2)x^{2/3-1} = \frac{4}{3}x^{-1/3}$ or $\frac{4}{3\sqrt[3]{x}}$

6. $h(t) = 3t^{-3/4}$; $h'(t) = \left(-\frac{3}{4}\right)(3)t^{-3/4-1} = -\frac{9}{4}t^{-7/4}$ or $-\frac{9}{4\sqrt[4]{t^7}}$

7. $f'(x) = (1)(17)x^{1-1} = 17x^0 = 17$

8. $g(t) = (-1)t^{-2/3}$; $g'(t) = \left(-\frac{2}{3}\right)(-1)t^{-2/3-1} = \frac{2}{3}t^{-5/3}$ or $\frac{2}{3\sqrt[3]{t^5}}$

Because the derivative of the sum (or difference) is the sum (or difference) of the derivatives, and the fact that if $y = af(x)$, then $y' = af'(x)$, we easily can differentiate functions such as $y = -2x^3 + 5x$. All we need to do is differentiate term by term.

EXAMPLES

Find the derivative.

- $y = -2x^3 + 5x$
  The derivative of $-2x^3$ is $-6x^2$, and the derivative of $5x$ is 5, so the derivative of this function is $y' = -6x^2 + 5$.
- $f(x) = 3x^4 - \frac{1}{2}x^2 + 1$
  (Remember that the derivative of a constant is 0.) Differentiating term by term, we get $f'(x) = (4)(3)x^{4-1} - (\frac{1}{2})(2)x^{2-1} + 0 = 12x^3 - x$.
- $y = -4\sqrt{x} + \frac{3}{x^2} - 8$
  We need to rewrite the function using exponents: $y = -4x^{1/2} + 3x^{-2} - 8$. Differentiating term by term, we have $y' = (\frac{1}{2})(-4)x^{1/2-1} + (-2)(3)x^{-2-1} - 0 = -2x^{-1/2} - 6x^{-3}$ or $-\frac{2}{\sqrt{x}} - \frac{6}{x^3}$.

PRACTICE

Find the derivative.

1. $f(x) = x^3 + 4x^2 - x + 2$
2. $f(x) = -x^2 - x - 10$
3. $f(x) = x + 12$
4. $y = \frac{4}{x^2} + \frac{1}{x} + 3x + 7$
5. $f(x) = 3\sqrt{x} + 2x + 4$
6. $y = 5\sqrt[3]{x^2} - \sqrt{x} + 2$
7. $h(t) = \frac{3}{\sqrt[3]{t}} + \sqrt{t} - 6$
8. $f(x) = x^4 - \frac{1}{x^4}$

SOLUTIONS

1. $f'(x) = 3x^2 + 8x - 1$
2. $f'(x) = -2x - 1$
3. $f'(x) = 1$
4. $y = 4x^{-2} + x^{-1} + 3x + 7; \quad y' = -8x^{-3} - x^{-2} + 3$ or $-\frac{8}{x^3} - \frac{1}{x^2} + 3$
5. $f(x) = 3x^{1/2} + 2x + 4; \quad f'(x) = \frac{3}{2}x^{-1/2} + 2$ or $\frac{3}{2\sqrt{x}} + 2$
6. $y = 5x^{2/3} - x^{1/2} + 2; \quad y' = \frac{10}{3}x^{-1/3} - \frac{1}{2}x^{-1/2}$ or $\frac{10}{3\sqrt[3]{x}} - \frac{1}{2\sqrt{x}}$
7. $h(t) = 3t^{-1/3} + t^{1/2} - 6; \quad h'(t) = -t^{-4/3} + \frac{1}{2}t^{-1/2}$ or $-\frac{1}{\sqrt[3]{t^4}} + \frac{1}{2\sqrt{t}}$
8. $f(x) = x^4 - x^{-4}; \quad f'(x) = 4x^3 + 4x^{-5}$ or $4x^3 + \frac{4}{x^5}$

The Tangent Line

A common problem in calculus is finding an equation for the tangent line. We are given a function and a point and told to find an equation for the tangent line to the graph of the function at the given point. First, we need to find the derivative. The derivative is a formula for the slope of the tangent line. Once we have the slope, we will be ready to find the equation. Suppose we are given a function $f(x)$ and told to find the tangent line at a point (a, b).

Step 1 Find $f'(x)$.
Step 2 Evaluate $f'(x)$ at $x = a$. This number is the slope ($m = f'(a)$).
Step 3 Substitute our numbers $x = a$, $y = b$, and m in the formula $y = mx + b$.
Step 4 Solve the above equation for b and write the equation for the line.

EXAMPLES

Find an equation of the tangent line.

- $f(x) = -2x^3 + x^2 - 4x + 3$ at $(1, -2)$
  We need to find $f'(x)$ so that we can compute m.

$$f'(x) = -6x^2 + 2x - 4 \qquad \text{Step 1}$$

$$m = f'(1) = -6(1)^2 + 2(1) - 4 = -8 \qquad \text{Step 2}$$

Now we can put $x = 1$, $y = -2$, and $m = -8$ into $y = mx + b$ to find b.

$$-2 = -8(1) + b \qquad \text{Step 3}$$

$$6 = b \qquad \text{Step 4}$$

An equation of the tangent line at $(1, -2)$ is $y = -8x + 6$.

- $f(x) = x^2 - 4$ at $(3, 5)$
  $f'(x) = 2x,\ m = f'(3) = 2(3) = 6$
  The slope of the tangent line at $(3, 5)$ is 6, and $y = mx + b$ becomes $5 = 6(3) + b$, making $b = -13$. The tangent line at $(3, 5)$ is $y = 6x - 13$.
- $f(x) = \frac{4}{x^2} - \frac{1}{x} + 1$ at $\left(2, \frac{3}{2}\right)$

$$f(x) = 4x^{-2} - x^{-1} + 1$$

$$f'(x) = -8x^{-3} + x^{-2} = -\frac{8}{x^3} + \frac{1}{x^2}$$

$$m = f'(2) = -\frac{8}{2^3} + \frac{1}{2^2} = -\frac{3}{4}$$

$$\frac{3}{2} = -\frac{3}{4}(2) + b$$

$$3 = b$$

The tangent line at $(2, \frac{3}{2})$ is $y = -\frac{3}{4}x + 3$.

- $y = 3\sqrt[3]{x} - 4\sqrt{x} - 1$ at $(1, -2)$

$$y = 3x^{1/3} - 4x^{1/2} - 1$$

$$y' = x^{-2/3} - 2x^{-1/2} = \frac{1}{\sqrt[3]{x^2}} - \frac{2}{\sqrt{x}}$$

$$m = \frac{1}{\sqrt[3]{1^2}} - \frac{2}{\sqrt{1}} = -1$$

$$-2 = -1(1) + b$$

$$-1 = b$$

The tangent line at $(1, -2)$ is $y = -x - 1$.

- $f(x) = \dfrac{2x^5 - x^3 + 4x^2 + 2x + 4}{x^2}$ at $(-1, 5)$

This function is in the form of a fraction. We will rewrite it in the form of a sum, then we will find the derivative.

$$f(x) = \frac{2x^5}{x^2} - \frac{x^3}{x^2} + \frac{4x^2}{x^2} + \frac{2x}{x^2} + \frac{4}{x^2}$$

$$= 2x^3 - x + 4 + \frac{2}{x} + \frac{4}{x^2} = 2x^3 - x + 4 + 2x^{-1} + 4x^{-2}$$

$$f'(x) = 6x^2 - 1 - 2x^{-2} - 8x^{-3} = 6x^2 - 1 - \frac{2}{x^2} - \frac{8}{x^3}$$

$$m = f'(-1) = 6(-1)^2 - 1 - \frac{2}{(-1)^2} - \frac{8}{(-1)^3} = 11$$

$$5 = 11(-1) + b$$

$$16 = b$$

The tangent line at $(-1, 5)$ is $y = 11x + 16$.

When rewriting the following function as a sum, we will use the exponent fact that $\frac{a^m}{a^n} = a^{m-n}$.

- $y = \dfrac{x^2 + 4x + 6}{\sqrt[3]{x}}$ at $(-8, -19)$

$$y = \frac{x^2}{\sqrt[3]{x}} + \frac{4x}{\sqrt[3]{x}} + \frac{6}{\sqrt[3]{x}} = \frac{x^2}{x^{1/3}} + \frac{4x}{x^{1/3}} + \frac{6}{x^{1/3}}$$

$$= x^{2-1/3} + 4x^{1-1/3} + 6x^{-1/3} = x^{5/3} + 4x^{2/3} + 6x^{-1/3}$$

$$y' = \frac{5}{3}x^{2/3} + \frac{8}{3}x^{-1/3} - 2x^{-4/3} = \frac{5\sqrt[3]{x^2}}{3} + \frac{8}{3\sqrt[3]{x}} - \frac{2}{\sqrt[3]{x^4}}$$

$$m = \frac{5\sqrt[3]{(-8)^2}}{3} + \frac{8}{3\sqrt[3]{-8}} - \frac{2}{\sqrt[3]{(-8)^4}}$$

$$= \frac{5(4)}{3} + \frac{8}{3(-2)} - \frac{2}{16} = \frac{125}{24}$$

Now we will let $x = -8$, $y = -19$, and $m = \frac{125}{24}$ in $y = mx + b$.

$$-19 = \frac{125}{24}(-8) + b$$

$$\frac{68}{3} = b$$

The tangent line at $(-8, -19)$ is $y = \frac{125}{24}x + \frac{68}{3}$.

PRACTICE

Find an equation of the tangent line.

1. $f(x) = -x^3 + 2x + 4$ at $(0, 4)$
2. $f(x) = 2x^4 - 6x^2 + x + 10$ at $(-2, 16)$
3. $y = \frac{4}{x} + 2x - 1$ at $(-1, -7)$
4. $f(x) = -6\sqrt{x} + x$ at $(4, -8)$
5. $h(t) = \frac{1}{\sqrt{t}} + \frac{1}{\sqrt[3]{t}} - 1$ at $(1, 1)$
6. $f(x) = \frac{2x^3 - x^2 - 3}{x^3}$ at $\left(3, \frac{14}{9}\right)$
7. $f(x) = \frac{4x^2 - 3x + 2}{\sqrt{x}}$ at $(4, 27)$

SOLUTIONS

1.

$$f'(x) = -3x^2 + 2$$

$$m = f'(0) = -3(0^2) + 2 = 2$$

$$4 = 2(0) + b$$

$$4 = b$$

$$y = 2x + 4$$

2.

$$f'(x) = 8x^3 - 12x + 1$$

$$m = f'(-2) = 8(-2)^3 - 12(-2) + 1 = -39$$

$$16 = -39(-2) + b$$

$$-62 = b$$

$$y = -39x - 62$$

3.

$$y = 4x^{-1} + 2x - 1$$

$$y' = -4x^{-2} + 2 = -\frac{4}{x^2} + 2$$

$$m = -\frac{4}{(-1)^2} + 2 = -2$$

$$-7 = (-2)(-1) + b$$

$$-9 = b$$

$$y = -2x - 9$$

4.

$$f(x) = -6x^{1/2} + x$$

$$f'(x) = -3x^{-1/2} + 1 = -\frac{3}{\sqrt{x}} + 1$$

$$m = f'(4) = -\frac{3}{\sqrt{4}} + 1 = -\frac{1}{2}$$

$$-8 = -\frac{1}{2}(4) + b$$

$$-6 = b$$

$$y = -\frac{1}{2}x - 6$$

5.

$$h(t) = t^{-1/2} + t^{-1/3} - 1$$

$$h'(t) = -\frac{1}{2}t^{-3/2} - \frac{1}{3}t^{-4/3} = -\frac{1}{2\sqrt{t^3}} - \frac{1}{3\sqrt[3]{t^4}}$$

$$m = h'(1) = -\frac{1}{2\sqrt{1^3}} - \frac{1}{3\sqrt[3]{1^4}} = -\frac{5}{6}$$

$$1 = -\frac{5}{6}(1) + b$$

$$\frac{11}{6} = b$$

$$y = -\frac{5}{6}t + \frac{11}{6}$$

6.

$$f(x) = \frac{2x^3}{x^3} - \frac{x^2}{x^3} - \frac{3}{x^3} = 2 - x^{-1} - 3x^{-3}$$

$$f'(x) = x^{-2} + 9x^{-4} = \frac{1}{x^2} + \frac{9}{x^4}$$

$$m = f'(3) = \frac{1}{3^2} + \frac{9}{3^4} = \frac{2}{9}$$

$$\frac{14}{9} = \frac{2}{9}(3) + b$$

$$\frac{8}{9} = b$$

$$y = \frac{2}{9}x + \frac{8}{9}$$

7.

$$f(x) = \frac{4x^2}{\sqrt{x}} - \frac{3x}{\sqrt{x}} + \frac{2}{\sqrt{x}} = \frac{4x^2}{x^{1/2}} - \frac{3x}{x^{1/2}} + \frac{2}{x^{1/2}}$$

$$= 4x^{2-1/2} - 3x^{1-1/2} + 2x^{-1/2} = 4x^{3/2} - 3x^{1/2} + 2x^{-1/2}$$

$$f'(x) = 6x^{1/2} - \frac{3}{2}x^{-1/2} - x^{-3/2} = 6\sqrt{x} - \frac{3}{2\sqrt{x}} - \frac{1}{\sqrt{x^3}}$$

$$m = f'(4) = 6\sqrt{4} - \frac{3}{2\sqrt{4}} - \frac{1}{\sqrt{4^3}} = \frac{89}{8}$$

$$27 = \frac{89}{8}(4) + b$$

$$-\frac{35}{2} = b$$

$$y = \frac{89}{8}x - \frac{35}{2}$$

The tangent line for a linear function is the linear function itself. For example, suppose we are asked to find the tangent line for $f(x) = 2x + 1$ at (5, 11).

$$f'(x) = 2 \qquad \text{Step 1}$$

$$m = f'(5) = 2 \qquad \text{Step 2}$$

$$11 = 2(5) + b \qquad \text{Step 3}$$

$$1 = b \qquad \text{Step 4}$$

$$y = 2x + 1$$

The tangent line is the same as the function $f(x) = 2x + 1$.

In the next set of problems, we will only be given the x-value of the point. We will use the original function to compute the y-value. All of the other steps will be the same. First we will find the derivative, second we will use the x-value in the derivative to find m, third we will use the x-value in the original function to find the y-value, and fourth we will use x, y, and m in $y = mx + b$ to find b.

EXAMPLE

- Find an equation of the tangent line for $f(x) = 2x^3 - 4x^2 + x + 5$ at $x = -1$.

  $f'(x) = 6x^2 - 8x + 1$, $m = f'(-1) = 6(-1)^2 - 8(-1) + 1 = 15$. We need to find y. We will put $x = -1$ into the *original* equation: $y = f(-1) = 2(-1)^3 - 4(-1)^2 + (-1) + 5 = -2$. Now we can find b.

$$-2 = 15(-1) + b$$

$$13 = b$$

$$y = 15x + 13$$

PRACTICE

Find an equation of the tangent line.

1. $f(x) = x^2 + 3x - 4$ at $x = -3$
2. $f(x) = \frac{6}{x} + 7$ at $x = -2$
3. $h(t) = \frac{4}{\sqrt{t}} + 2$ at $t = 1$

SOLUTIONS

1.

$$f'(x) = 2x + 3$$

$$m = f'(-3) = 2(-3) + 3 = -3$$

$$y = f(-3) = (-3)^2 + 3(-3) - 4 = -4$$

$$-4 = -3(-3) + b$$

$$-13 = b$$

$$y = -3x - 13$$

2.

$$f(x) = 6x^{-1} + 7$$

$$f'(x) = -6x^{-2} = -\frac{6}{x^2}$$

$$m = f'(-2) = -\frac{6}{(-2)^2} = -\frac{3}{2}$$

$$y = f(-2) = \frac{6}{-2} + 7 = 4$$

$$4 = -\frac{3}{2}(-2) + b$$

$$1 = b$$

$$y = -\frac{3}{2}x + 1$$

3.

$$h(t) = 4t^{-1/2} + 2$$

$$h'(t) = -2t^{-3/2} = -\frac{2}{\sqrt{t^3}}$$

$$m = h'(1) = -\frac{2}{\sqrt{1^3}} = -2$$

$$y = h(1) = \frac{4}{\sqrt{1}} + 2 = 6$$

$$6 = -2(1) + b$$

$$8 = b$$

$$y = -2t + 8$$

The Product Rule

Unfortunately, finding the derivative of a product of two functions is not simply a matter of finding the product of their derivatives. That is, if $y = f(x)g(x)$, it is *not* the case that $y' = f'(x)g'(x)$. To see why not, let $y = (x+1)(x-2)$. The derivative of $x + 1$ is 1, and the derivative of $x - 2$ is 1. *If* it were true that $y' = f'(x)g'(x)$, then we would have $y' = 1 \cdot 1 = 1$. But if we use the FOIL method on the function we would have $y = x^2 - x - 2$, and the derivative of this function is $2x - 1$, not 1.

We will begin to find the derivative of a product of two functions by identifying the individual functions and each of their derivatives. The derivative of the product is the derivative of the first function times the second function plus the first function times the derivative of the second function.

If $y = f(x)g(x)$ where $f(x)$ and $g(x)$ are differentiable, then $y' = f'(x)g(x) + f(x)g'(x)$.

Once we have put the individual functions and their derivatives together using the formula above, the calculus is done. The rest of the work involves using algebra to simplify the derivative.

EXAMPLES

Use the product rule to find the derivative.

- $y = (2x+1)(x^2-4)$
  The two functions are $f(x) = 2x + 1$ and $g(x) = x^2 - 4$. Their derivatives are $f'(x) = 2$ and $g'(x) = 2x$. Then $y' = f'(x)g(x) + f(x)g'(x)$ becomes

$$\begin{aligned} y' &= \overbrace{2}^{f'}\,\overbrace{(x^2-4)}^{g} + \overbrace{(2x+1)}^{f}\,\overbrace{(2x)}^{g'} \\ &= 2x^2 - 8 + 4x^2 + 2x = 6x^2 + 2x - 8. \end{aligned}$$

- $y = (x + 4)(2x - 3)$

$$f(x) = x + 4 \qquad g(x) = 2x - 3$$
$$f'(x) = 1 \qquad g'(x) = 2$$

$$y' = f'(x)g(x) + f(x)g'(x)$$
$$y' = 1(2x - 3) + (x + 4)(2)$$
$$= 2x - 3 + 2x + 8 = 4x + 5$$

- $y = (t^3 + t^2 - 4)(2t + 5)$

$$f(t) = t^3 + t^2 - 4 \qquad g(t) = 2t + 5$$
$$f'(t) = 3t^2 + 2t \qquad g'(t) = 2$$

$$y' = f'(t)g(t) + f(t)g'(t)$$
$$y' = (3t^2 + 2t)(2t + 5) + (t^3 + t^2 - 4)(2)$$
$$= 6t^3 + 15t^2 + 4t^2 + 10t + 2t^3 + 2t^2 - 8 = 8t^3 + 21t^2 + 10t - 8$$

- $y = \left(\frac{8}{x^3} + \frac{2}{x}\right)(3x^2 - 6\sqrt{x})$

  Rewrite as $y = (8x^{-3} + 2x^{-1})(3x^2 - 6x^{1/2})$.

$$f(x) = 8x^{-3} + 2x^{-1} \qquad g(x) = 3x^2 - 6x^{1/2}$$
$$f'(x) = -24x^{-4} - 2x^{-2} \qquad g'(x) = 6x - 3x^{-1/2}$$

  When simplifying y', we will use the exponent facts $a^m a^n = a^{m+n}$ and $a^0 = 1$.

$$y' = f'(x)g(x) + f(x)g'(x)$$
$$y' = (-24x^{-4} - 2x^{-2})(3x^2 - 6x^{1/2}) + (8x^{-3} + 2x^{-1})(6x - 3x^{-1/2})$$
$$= -72x^{-4}x^2 + 144x^{-4}x^{1/2} - 6x^{-2}x^2 + 12x^{-2}x^{1/2} + 48x^{-3}x$$
$$- 24x^{-3}x^{-1/2} + 12x^{-1}x - 6x^{-1}x^{-1/2}$$
$$= -72x^{-2} + 144x^{-7/2} - 6x^0 + 12x^{-3/2} + 48x^{-2} - 24x^{-7/2} + 12x^0 - 6x^{-3/2}$$
$$= -24x^{-2} + 120x^{-7/2} + 6 + 6x^{-3/2} \text{ or } -\frac{24}{x^2} + \frac{120}{\sqrt{x^7}} + \frac{6}{\sqrt{x^3}} + 6$$

PRACTICE

For Problems 1–4, use the product rule to find the derivative.

1. $y = (4x^5+2x^3+6)(x^2+1)$
2. $y = (x^3 - 2x + 5)(x^2 + x + 2)$
3. $y = (x^{-3} - x^{-1})(x^{-2} - 2x^{-1})$
4. $y = \left(\frac{1}{\sqrt{x}} + \frac{1}{x}\right)(2x - \sqrt{x})$
5. Find an equation of the tangent line for $y = (x^2 - 4x - 1)(x + 2)$ at $x = -1$.

SOLUTIONS

1.

$$f(x) = 4x^5 + 2x^3 + 6 \qquad g(x) = x^2 + 1$$

$$f'(x) = 20x^4 + 6x^2 \qquad g'(x) = 2x$$

$$\begin{aligned} y' &= (20x^4 + 6x^2)(x^2 + 1) + (4x^5 + 2x^3 + 6)(2x) \\ &= 20x^6 + 20x^4 + 6x^4 + 6x^2 + 8x^6 + 4x^4 + 12x \\ &= 28x^6 + 30x^4 + 6x^2 + 12x \end{aligned}$$

2.

$$f(x) = x^3 - 2x + 5 \qquad g(x) = x^2 + x + 2$$

$$f'(x) = 3x^2 - 2 \qquad g'(x) = 2x + 1$$

$$\begin{aligned} y' &= (3x^2 - 2)(x^2 + x + 2) + (x^3 - 2x + 5)(2x + 1) \\ &= 3x^4 + 3x^3 + 6x^2 - 2x^2 - 2x - 4 + 2x^4 + x^3 - 4x^2 - 2x + 10x + 5 \\ &= 5x^4 + 4x^3 + 6x + 1 \end{aligned}$$

3.

$$f(x) = x^{-3} - x^{-1} \qquad g(x) = x^{-2} - 2x^{-1}$$

$$f'(x) = -3x^{-4} + x^{-2} \qquad g'(x) = -2x^{-3} + 2x^{-2}$$

$$y' = (-3x^{-4} + x^{-2})(x^{-2} - 2x^{-1}) + (x^{-3} - x^{-1})(-2x^{-3} + 2x^{-2})$$

$$= -3x^{-6} + 6x^{-5} + x^{-4} - 2x^{-3} - 2x^{-6} + 2x^{-5} + 2x^{-4} - 2x^{-3}$$

$$= -5x^{-6} + 8x^{-5} + 3x^{-4} - 4x^{-3}$$

4. $y = (x^{-1/2} + x^{-1})(2x - x^{1/2})$

$$f(x) = x^{-1/2} + x^{-1} \qquad g(x) = 2x - x^{1/2}$$

$$f'(x) = -\frac{1}{2}x^{-3/2} - x^{-2} \qquad g'(x) = 2 - \frac{1}{2}x^{-1/2}$$

$$y' = \left(-\frac{1}{2}x^{-3/2} - x^{-2}\right)(2x - x^{1/2}) + (x^{-1/2} + x^{-1})\left(2 - \frac{1}{2}x^{-1/2}\right)$$

$$= -x^{-3/2}x + \frac{1}{2}x^{-3/2}x^{1/2} - 2x^{-2}x + x^{-2}x^{1/2} + 2x^{-1/2}$$

$$- \frac{1}{2}x^{-1/2}x^{-1/2} + 2x^{-1} - \frac{1}{2}x^{-1}x^{-1/2}$$

$$= -x^{-1/2} + \frac{1}{2}x^{-1} - 2x^{-1} + x^{-3/2} + 2x^{-1/2} - \frac{1}{2}x^{-1} + 2x^{-1}$$

$$- \frac{1}{2}x^{-3/2}$$

$$= x^{-1/2} + 0x^{-1} + \frac{1}{2}x^{-3/2} \text{ or } \frac{1}{\sqrt{x}} + \frac{1}{2\sqrt{x^3}}$$

5.

$$f(x) = x^2 - 4x - 1 \qquad g(x) = x + 2$$

$$f'(x) = 2x - 4 \qquad g'(x) = 1$$

$$y' = (2x - 4)(x + 2) + (x^2 - 4x - 1)(1) = 3x^2 - 4x - 9$$

$$m = 3(-1)^2 - 4(-1) - 9 = -2$$

$$y = [(-1)^2 - 4(-1) - 1)](-1 + 2) = 4$$

$$4 = -2(-1) + b$$

$$2 = b$$

$$y = -2x + 2$$

The Quotient Rule

The process for finding the derivative of a quotient of two functions is similar to finding the derivative of a product of two functions. We need to identify the numerator function and denominator function and their derivatives. Next, we will put them into the quotient rule, which is the derivative of the numerator times the denominator minus the numerator times the derivative of the denominator, all divided by the denominator squared. Finally, we will use algebra to simplify the derivative.

> If $f(x)$ and $g(x)$ are differentiable and $y = \dfrac{f(x)}{g(x)}$,
>
> then $y' = \dfrac{f'(x)g(x) - f(x)g'(x)}{[g(x)]^2}$.

EXAMPLES

Use the quotient rule to find the derivative.

- $y = \dfrac{4x - 1}{x^2 + 3}$

  The numerator function is $f(x) = 4x - 1$. The denominator function is $g(x) = x^2 + 3$.

$$f(x) = 4x - 1 \qquad g(x) = x^2 + 3$$

$$f'(x) = 4 \qquad g'(x) = 2x$$

$$y' = \frac{f'(x)g(x) - f(x)g'(x)}{[g(x)]^2}$$

$$y' = \frac{4(x^2 + 3) - (4x - 1)(2x)}{(x^2 + 3)^2}$$

$$= \frac{4x^2 + 12 - 8x^2 + 2x}{(x^2 + 3)^2}$$

$$= \frac{-4x^2 + 2x + 12}{(x^2 + 3)^2}$$

The denominator is usually not expanded.

- $y = \dfrac{5x^3 + \sqrt{x}}{x}$

$$f(x) = 5x^3 + x^{1/2} \qquad g(x) = x$$

$$f'(x) = 15x^2 + \frac{1}{2}x^{-1/2} \qquad g'(x) = 1$$

$$y' = \frac{f'(x)g(x) - f(x)g'(x)}{[g(x)]^2}$$

$$y' = \frac{\left(15x^2 + \frac{1}{2}x^{-1/2}\right)(x) - (5x^3 + x^{1/2})(1)}{x^2}$$

$$= \frac{15x^3 + \frac{1}{2}x^{-1/2}x - 5x^3 - x^{1/2}}{x^2} = \frac{15x^3 + \frac{1}{2}x^{1/2} - 5x^3 - x^{1/2}}{x^2}$$

$$= \frac{10x^3 - \frac{1}{2}x^{1/2}}{x^2} \text{ or } \frac{10x^3 - \frac{1}{2}\sqrt{x}}{x^2} \cdot \frac{2}{2} = \frac{20x^3 - \sqrt{x}}{2x^2}$$

- $y = \dfrac{3}{x^4 + 6}$

$$f(x) = 3 \qquad g(x) = x^4 + 6$$

$$f'(x) = 0 \qquad g'(x) = 4x^3$$

$$y' = \frac{f'(x)g(x) - f(x)g'(x)}{[g(x)]^2}$$

$$y' = \frac{0(x^4 + 6) - 3(4x^3)}{(x^4 + 6)^2} = \frac{-12x^3}{(x^4 + 6)^2}$$

PRACTICE

For Problems 1–5, use the quotient rule to find the derivative.

1. $y = \dfrac{7x^3}{4x^2 + x - 5}$
2. $y = \dfrac{-6x^4 + 2x^3 + 9}{x^2 - x - 2}$
3. $y = \dfrac{8\sqrt{x}}{x + 3}$
4. $y = \dfrac{2}{x^3 + 1}$

5. $y = \dfrac{x - \sqrt{x}}{x + \sqrt{x}}$

6. Find an equation for the tangent line for $y = \dfrac{x^2 - x + 3}{x + 2}$ at $x = -3$

SOLUTIONS

1.

$$f(x) = 7x^3 \qquad g(x) = 4x^2 + x - 5$$

$$f'(x) = 21x^2 \qquad g'(x) = 8x + 1$$

$$y' = \frac{21x^2(4x^2 + x - 5) - 7x^3(8x + 1)}{(4x^2 + x - 5)^2}$$

$$= \frac{84x^4 + 21x^3 - 105x^2 - 56x^4 - 7x^3}{(4x^2 + x - 5)^2} = \frac{28x^4 + 14x^3 - 105x^2}{(4x^2 + x - 5)^2}$$

2.

$$f(x) = -6x^4 + 2x^3 + 9 \qquad g(x) = x^2 - x - 2$$

$$f'(x) = -24x^3 + 6x^2 \qquad g'(x) = 2x - 1$$

$$y' = \frac{(-24x^3 + 6x^2)(x^2 - x - 2) - (-6x^4 + 2x^3 + 9)(2x - 1)}{(x^2 - x - 2)^2}$$

$$= \frac{-24x^5 + 24x^4 + 48x^3 + 6x^4 - 6x^3 - 12x^2}{(x^2 - x - 2)^2}$$

$$- \frac{(-12x^5 + 6x^4 + 4x^4 - 2x^3 + 18x - 9)}{(x^2 - x - 2)^2}$$

$$= \frac{-12x^5 + 20x^4 + 44x^3 - 12x^2 - 18x + 9}{(x^2 - x - 2)^2}$$

3.

$$f(x) = 8\sqrt{x} = 8x^{1/2} \qquad g(x) = x + 3$$

$$f'(x) = 4x^{-1/2} \qquad g'(x) = 1$$

$$y' = \frac{4x^{-1/2}(x+3) - 8x^{1/2}(1)}{(x+3)^2} = \frac{4x^{-1/2}x + 12x^{-1/2} - 8x^{1/2}}{(x+3)^2}$$

$$= \frac{4x^{1/2} + 12x^{-1/2} - 8x^{1/2}}{(x+3)^2} = \frac{-4x^{1/2} + 12x^{-1/2}}{(x+3)^2}$$

$$\text{or } \frac{-4\sqrt{x} + \frac{12}{\sqrt{x}}}{(x+3)^2} = \frac{\frac{-4\sqrt{x}}{1} \cdot \frac{\sqrt{x}}{\sqrt{x}} + \frac{12}{\sqrt{x}}}{(x+3)^2}$$

$$= \frac{\frac{-4\sqrt{x}\cdot\sqrt{x}}{\sqrt{x}} + \frac{12}{\sqrt{x}}}{(x+3)^2} = \frac{\frac{-4x+12}{\sqrt{x}}}{(x+3)^2}$$

$$= \frac{-4x+12}{\sqrt{x}} \div \frac{(x+3)^2}{1} = \frac{-4x+12}{\sqrt{x}} \cdot \frac{1}{(x+3)^2}$$

$$= \frac{-4x+12}{\sqrt{x}(x+3)^2}$$

4.

$$f(x) = 2 \qquad g(x) = x^3 + 1$$

$$f'(x) = 0 \qquad g'(x) = 3x^2$$

$$y' = \frac{0(x^3+1) - 2(3x^2)}{(x^3+1)^2} = \frac{-6x^2}{(x^3+1)^2}$$

5.

$$f(x) = x - \sqrt{x} = x - x^{1/2} \qquad g(x) = x + \sqrt{x} = x + x^{1/2}$$

$$f'(x) = 1 - \frac{1}{2}x^{-1/2} \qquad g'(x) = 1 + \frac{1}{2}x^{-1/2}$$

$$y' = \frac{\left(1 - \frac{1}{2}x^{-1/2}\right)(x + x^{1/2}) - (x - x^{1/2})\left(1 + \frac{1}{2}x^{-1/2}\right)}{(x+x^{1/2})^2}$$

$$= \frac{x + x^{1/2} - \frac{1}{2}x^{-1/2}x - \frac{1}{2}x^{-1/2}x^{1/2}}{(x+x^{1/2})^2}$$

$$- \frac{\left(x + \frac{1}{2}xx^{-1/2} - x^{1/2} - \frac{1}{2}x^{1/2}x^{-1/2}\right)}{(x+x^{1/2})^2}$$

$$= \frac{x + x^{1/2} - \frac{1}{2}x^{1/2} - \frac{1}{2}x^0 - (x + \frac{1}{2}x^{1/2} - x^{1/2} - \frac{1}{2}x^0)}{(x + x^{1/2})^2}$$

$$= \frac{x + \frac{1}{2}x^{1/2} - \frac{1}{2} - x + \frac{1}{2}x^{1/2} + \frac{1}{2}}{(x + x^{1/2})^2}$$

$$= \frac{x^{1/2}}{(x + x^{1/2})^2} \text{ or } \frac{\sqrt{x}}{(x + \sqrt{x})^2}$$

6.

$$f(x) = x^2 - x + 3 \qquad g(x) = x + 2$$

$$f'(x) = 2x - 1 \qquad g'(x) = 1$$

$$y' = \frac{(2x-1)(x+2) - (x^2 - x + 3)(1)}{(x+2)^2} = \frac{x^2 + 4x - 5}{(x+2)^2}$$

$$m = \frac{(-3)^2 + 4(-3) - 5}{(-3+2)^2} = -8$$

$$y = \frac{(-3)^2 - (-3) + 3}{-3 + 2} = -15$$

$$-15 = -8(-3) + b$$

$$-39 = b$$

$$y = -8x - 39$$

CHAPTER 4 REVIEW

1. $f(x) = 5x^2 - x + 3$
   (a) $f'(x) = 10x - 1$
   (b) $f'(x) = 10x + 2$
   (c) $f'(x) = 10x$
   (d) $f'(x) - 4x + 2$

2. $y = \sqrt{x^3}$
   (a) $y' = \sqrt{3x^2}$
   (b) $y' = \frac{2}{3}x^{-1/3}$
   (c) $y' = 3\sqrt{x^2}$
   (d) $y' = \frac{3}{2}x^{1/2}$

3. $f(x) = 8x + 9$
   (a) $f'(x) = 8x$
   (b) $f'(x) = 17$
   (c) $f'(x) = 8$
   (d) $f'(x)$ does not exist.

4.
$$y = \frac{16}{x^2}$$
   (a) $y' = \frac{8}{x}$
   (b) $y' = \frac{0}{2x}$
   (c) $y' = -32x^{-3}$
   (d) $y' = -32x^{-1}$

5. $f(x) = 12x^3 + 4x^2 - 9$
   (a) $f'(x) = 36x^2 + 8x - 9$
   (b) $f'(x) = 36x^2 + 8x$
   (c) $f'(x) = 36x^2 + 4x - 9x^{-1}$
   (d) $f'(x) = 36x^2 + 8x - 9x^{-1}$

6. $f(x) = (4x - 1)(3x^2 + x + 1)$
   (a) $f'(x) = 4(3x^2 + x + 1) + (4x - 1)(6x + 1)$
   (b) $f'(x) = 4(6x + 1)$
   (c) $f'(x) = 4(3x^2 + x + 1) - (4x - 1)(6x + 1)$
   (d) $f'(x) = 4(3x^2 + x + 1) - (4x - 1)(6x + 2)$

7. $y = \dfrac{8x^2 + 2x}{x - 1}$

   (a) $y' = \dfrac{(16x + 2)(x - 1) + (8x^2 + 2x)(1)}{x - 1}$

   (b) $y' = \dfrac{(16x + 2)(x - 1) - (8x^2 + 2x)(1)}{x - 1}$

(c) $y' = \dfrac{(16x+2)(x-1)+(8x^2+2x)(1)}{(x-1)^2}$

(d) $y' = \dfrac{(16x+2)(x-1)-(8x^2+2x)(1)}{(x-1)^2}$

8. Find the tangent line for $f(x) = -3x^4 + 4x^2 + x$ at $(1, 2)$.
   (a) $y = 2x - 3$
   (b) $y = 2x$
   (c) $y = -3x + 7$
   (d) $y = -3x + 5$

9. Find the tangent line for $y = \frac{x-2}{x+1}$ at $x = 0$.
   (a) $y = 3x - 2$
   (b) $y = 3x + 6$
   (c) $y = -x - 2$
   (d) $y = -x + 6$

SOLUTIONS

1. a	2. d	3. c	4. c	5. b
6. a	7. d	8. d	9. a	

Instantaneous Rates of Change

A common problem involving the rate of change and the derivative concerns velocity. We can find the average velocity by dividing the distance traveled by the time it took to travel this distance. For example, if a car covered 90 miles in two hours, then its average velocity was 90 miles/2 hour, which reduces to 45 miles/1 hour or 45 mph. If it covers a total of 120 miles in three hours, then its average velocity is 120 miles/3 hours = 40 miles/1 hour or 40 mph.

In the following examples, we will be given a formula that gives us the distance an object has traveled in terms of time. We will be asked to find the average velocity over a period of time. We will first compute the distance covered. And then we will divide this distance by the time traveled to get the average velocity.

EXAMPLES

- If an object is dropped, the distance it has fallen can be approximated by $d = 16t^2$ (ignoring air resistance), where d is in feet and t is in seconds. Suppose an object is dropped from a height of 300 feet.

    1. What was the object's average velocity between 3 and 4 seconds?
    2. What was the object's average velocity between 3 and 3.5 seconds?
    3. What was the object's average velocity between 3 and 3.1 seconds?

    1. After 3 seconds, the object had fallen $d = 16(3^2) = 144$ feet. After 4 seconds, it had fallen $d = 16(4^2) = 256$ feet. Between the third and fourth second, the object had fallen $256 - 144 = 112$ feet. It took $4-3 = 1$ second to fall this distance, so its average velocity was 112 feet/1 second = 112 feet per second.
    2. After 3.5 seconds, the object had fallen $d = 16(3.5^2) = 196$ feet. Between 3 and $3\frac{1}{2}$ seconds, the object had fallen $196 - 144 = 52$ feet. Its average velocity was 52 feet/$\frac{1}{2}$ second = 52(2) feet/1 second = 104 feet per second.
    3. After 3.1 seconds, the object had fallen $d = 16(3.1^2) = 153.76$ feet. Between 3 and 3.1 seconds, the object had fallen $153.76 - 144 = 9.76$ feet. Its average velocity was 9.76 feet/0.1 second = 9.76(10) feet/1 second = 97.6 feet per second.

- A particle travels in a straight line. Its distance (in meters) after t seconds is $d(t) = t^2 - t$.

    1. What was the object's average velocity between 4 and 5 seconds?
    2. What was the object's average velocity between 4 and 4.5 seconds?
    3. What was the object's average velocity between 4 and 4.1 seconds?
    4. What was the object's average velocity between 4 and 4.01 seconds?

We will answer these questions by filling in Table 5.1; t is 4 and h is the length of the time interval. The first time interval is $h = 5 - 4 = 1$ second; the second is $h = 4.5 - 4 = 0.5$ seconds; the third, $h = 4.1 - 4 = 0.1$; and the fourth, $h = 4.01 - 4 = 0.01$. The distance traveled over this interval is computed as $d(t+h) - d(t)$. This is the distance traveled in $t+h$ seconds minus the distance traveled in t seconds. In the last column, $d(t+h) - d(t)$ is divided by h, the length of the time interval.

The instantaneous velocity is the velocity at an exact moment in time. It is the limit of the average velocity as h (the length of the time interval) goes to 0. In other

Table 5.1

t	$t+h$	h	$d(t+h)$ $= (t+h)^2 - (t+h)$	$d(t)$	$d(t+h) - d(t)$	$\frac{d(t+h) - d(t)}{h}$
			Distance traveled in $t+h$ seconds	Distance traveled in t seconds	Distance traveled during the interval	Average velocity
4	5	1	$5^2 - 5 = 20$	$4^2 - 4 = 12$	$20 - 12 = 8$	$\frac{8}{1} = 8$
4	4.5	0.5	$4.5^2 - 4.5 = 15.75$	$4^2 - 4 = 12$	$15.75 - 12 = 3.75$	$\frac{3.75}{0.5} = 7.5$
4	4.1	0.1	$4.1^2 - 4.1 = 12.71$	$4^2 - 4 = 12$	$12.71 - 12 = 0.71$	$\frac{0.71}{0.1} = 7.1$
4	4.01	0.01	$4.01^2 - 4.01 = 12.0701$	$4^2 - 4 = 12$	$12.0701 - 12 = 0.0701$	$\frac{0.0701}{0.01} = 7.01$

words, the instantaneous velocity is the derivative of the distance function. In the above example, the particle appears to be moving at the rate of 7 meters per second at the instant the particle has traveled 4 seconds.

$$
\begin{aligned}
d'(t) &= \lim_{h \to 0} \frac{d(t+h) - d(t)}{h} = \lim_{h \to 0} \frac{(t+h)^2 - (t+h) - (t^2 - t)}{h} \\
&= \lim_{h \to 0} \frac{t^2 + 2ht + h^2 - t - h - t^2 + t}{h} \\
&= \lim_{h \to 0} \frac{2ht + h^2 - h}{h} = \lim_{h \to 0} \frac{h(2t + h - 1)}{h} \\
&= \lim_{h \to 0} (2t + h - 1) = 2t - 1
\end{aligned}
$$

When we evaluate the derivative at $t = 4$ seconds, we have the velocity of $2(4) - 1 = 7$ meters per second.

EXAMPLE

- Find the velocity of a falling object at 5 seconds.
  The derivative of the function $d(t) = 16t^2$ is $d'(t) = 32t$. The velocity of the falling object at 5 seconds is $d'(5) = 32(5) = 160$ feet per second.

PRACTICE

1. A particle is moving in a straight line. The distance it has traveled from its initial point is given by $d(t) = 2t^2 - 10t + 13$, where d is in feet, and t is in seconds.

   (a) Find the particle's average velocity between 4 and 5 seconds.

   (b) Find the particle's average velocity between 4 and 4.5 seconds.

   (c) Find the particle's instantaneous velocity at 4 seconds.

SOLUTIONS

1. (a) The average velocity is $\frac{\text{distance traveled}}{\text{length of time}} = \frac{d(5) - d(4) \text{ feet}}{5 - 4 \text{ sec}}$.

$$d(5) = 2(5^2) - 10(5) + 13 = 13$$

$$d(4) = 2(4^2) - 10(4) + 13 = 5$$

$$\frac{d(5) - d(4)}{5 - 4} = \frac{13 - 5}{1} = 8$$

The average velocity between 4 and 5 seconds is 8 feet per second.

(b) The average velocity is $\frac{d(4.5) - d(4)}{4.5 - 4}$.

$$d(4.5) = 2(4.5^2) - 10(4.5) + 13 = 8.5$$

$$\frac{d(4.5) - d(4)}{4.5 - 4} = \frac{8.5 - 5}{0.5} = 7$$

The average velocity between 4 and 4.5 seconds is 7 feet per second.

(c) The velocity at 4 seconds can be found by evaluating the derivative of $d(t) = 2t^2 - 10t + 13$ at $t = 4$.

$$d'(t) = 4t - 10$$

$$d'(4) = 4(4) - 10 = 6$$

The instantaneous velocity at 4 seconds is 6 feet per second.

In business, the derivative is used to find the *marginal revenue*, the *marginal cost*, and the *marginal profit*. The marginal revenue is the amount of revenue gained by selling the "next" unit. For example, suppose the price for a movie ticket is \$8. The marginal revenue for the first ticket is \$8, the marginal revenue for the 50th ticket is \$8, and so on. Suppose the marginal cost function for a product is given by $MC = 3x - 10$. The cost to produce the 101st unit can be found by evaluating the marginal cost function at $x = 100$: $3(100) - 10 = 290$. After having produced the first 100 units, the cost to produce the 101st unit is \$290. Later, we will use the marginal cost, marginal revenue, and marginal profit functions to find the level of production that makes the most of the revenue while minimizing the cost.

In the following problems, we will be given the cost and revenue functions. These functions tell us the cost to produce x units of a product and the revenue from selling x units. We can find the profit function by subtracting the cost function from the revenue function. The marginal cost function is the derivative of the cost function; the marginal revenue function is the derivative of the revenue function; and so on. Evaluating these marginal functions at $x = a$ units tells us the cost, revenue, and profit from selling/producing the $(a + 1)$th unit.

EXAMPLES

- The revenue for selling x units of a product is given by $R(x) = -0.01x^2 + 2x + 2000$, and the cost is given by $C(x) = 0.08x + 1000$. Find the marginal revenue, marginal cost, and marginal profit functions. Find marginal revenue, marginal cost, and marginal profit for a production level of 55 units.

  The marginal revenue function is the derivative of $R(x) = -0.01x^2 + 2x + 2000$: $R'(x) = -0.02x + 2$. When we evaluate $R'(x)$ at $x = 55$ units, we will have the revenue for selling the 56th unit: $R'(55) = -0.02(55) + 2 = 0.90$. The revenue for selling the 56th unit is \$0.90. The marginal cost function is the derivative of $C(x) = 0.08x + 1000$: $C'(x) = 0.08$. This is a constant function, which means that each unit costs \$0.08 to produce, regardless of the production level. The profit function is found by subtracting cost from revenue: $P(x) = R(x) - C(x) = -0.01x^2 + 2x + 2000 - (0.08x + 1000) = -0.01x^2 + 1.92x - 1000$. The marginal profit function is the derivative of this function: $P'(x) = -0.02x + 1.92$. The marginal profit for the production level of 55 units is $P'(55) = -0.02(55) + 1.92 = 0.82$. The profit for producing/selling the 56th unit is \$0.82. Notice that we can find the marginal profit by subtracting the marginal cost from the marginal revenue.

  When the revenue and cost are given in terms of price, the marginal revenue and marginal cost functions describe what happens to the revenue and cost when there is a small increase in the price at different price levels.

- The revenue function for an office complex is $R(r) = -0.005r^2 + 55.5r$, where r is the monthly rent. The monthly cost function is $C(r) = 42{,}400 - 12r$. The cost depends on the rent because when the rent increases, the number of vacancies also increases. Find the marginal revenue and marginal cost for a monthly rent of \$5000 and \$7000.

  The marginal revenue is $R'(r) = -0.01r + 55.5$, and the marginal cost is $C'(r) = -12$. When the monthly rent is \$5000, the marginal revenue is $R'(5000) = -0.01(5000) + 55.5 = 5.5$, and the marginal cost is $C'(5000) = -12$. These numbers mean that, theoretically, when the rent is increased by \$1, the revenue is increasing at the rate of \$5.50 per month and the cost is decreasing at the rate of \$12 per month. When the monthly rent is \$7000, the marginal revenue is $R'(7000) = -0.01(7000) + 55.5 = -14.5$; a \$1 increase in the rent causes the revenue to decrease at the rate \$14.50 per month and the cost to decrease at the rate of \$12 per month.

  Keep in mind that these numbers are rates of change and not an indication of the change in reality. For example, increasing the rents from \$7000 to \$7001 will not really cause a loss in revenue because it is unlikely that the dollar increase will cause a tenant to leave. However, raising the rent by \$500 could very well affect both revenue and cost as a tenant might leave. The size and sign of the marginal numbers tell us how an increase in the price can affect revenue and cost. A large positive marginal revenue tells us that a small increase in the price results in a large increase in revenue, which means that we are probably under-priced. A large negative marginal revenue tells us that a small increase in the price results in a large decrease in revenue, which means that we are probably over-priced. There will be more on this topic later when we cover price elasticity.
- The value of an investment over a 20-year period can be approximated by the function $V(t) = 0.8x^4 - 13x^3 + 75x^2 + 2150x + 6800$, where t is the number of years after 1980.

  1. How fast was the value of the investment increasing in the year 1984?
  2. How fast was the value of the investment increasing in the year 1995?

  We will answer these questions by finding the derivative of the value function and evaluating the derivative at $t = 4$, for the first question, and $t = 15$, for the second.

$$V'(t) = 3.2x^3 - 39x^2 + 150x + 2150$$

$$V'(4) = 3.2(4^3) - 39(4^2) + 150(4) + 2150 = 2330.8$$

$$V'(15) = 3.2(15)^3 - 39(15^2) + 150(15) + 2150 = 6425$$

In the year 1984, the investment was increasing at the rate of \$2330.80 per year. In the year 1995, the investment was increasing at the rate of \$6425 per year.

PRACTICE

1. The cost for producing x units of a product is given by $C(x) = 50x + 20{,}000$, and its revenue is given by $R(x) = -0.25x^2 + 247.5x + 2500$.
   (a) Find the marginal cost for producing 300 units.
   (b) Find the marginal revenue for producing 300 units.
   (c) Find the marginal profit for producing 300 units.
   (d) Interpret these numbers.
2. The number of units sold after spending x thousand dollars on advertising a product can be approximated by the function $s(x) = 0.11x^4 - 2.9x^3 - 18.82x^2 + 929x - 590$. Find the marginal sales function. Find the marginal sales when \$2000 ($x = 2$) and \$10,000 ($x = 10$) are spent on advertising. Interpret these numbers.
3. The demand function for a product is $D(p) = 0.007p^4 - 0.16p^3 + 1.3p^2 - 4.9p + 10$, where D is the number (in thousands) when the price per unit is p (valid up to \$10 per unit). How fast is the demand decreasing when the price is \$2 per unit? \$6 per unit?

SOLUTIONS

1. (a) The marginal cost function is $C'(x) = 50$, and the marginal cost at 300 units is $C'(300) = 50$.
   (b) The marginal revenue function is $R'(x) = -0.5x + 247.5$, and the marginal revenue at 300 units is $R'(300) = -0.5(300) + 247.5 = 97.50$.
   (c) The marginal profit can be found in one of two ways, by finding the profit function and finding its derivative or by subtracting the marginal cost from the marginal revenue: $P'(x) = R'(x) - C'(x) = -0.5x + 247.5 - 50 = -0.5x + 197.5$. The marginal profit at 300 units is $P'(300) = -0.5(300) + 197.5 = 47.50$.
   (d) As production/sales increase from 300 to 301 units, the cost increases by \$50. As production/sales increase from 300 to 301 units, the revenue increases by \$97.50 and the profit increases by \$47.50.

2.

$$s'(x) = 0.44x^3 - 8.7x^2 - 37.64x + 929$$

$$s'(2) = 0.44(2)^3 - 8.7(2^2) - 37.64(2) + 929 = 822.44$$

$$s'(10) = 0.44(10^3) - 8.7(10^2) - 37.64(10) + 929 = 122.6$$

$s'(2) = 822.44$ means that as spending on advertising increases from \$2000 to \$3000, the number of units sold will increase by about 822. $s'(10) = 122.6$ means that as spending on advertising increases from \$10,000 to \$11,000, the number of units sold will increase by about 123.

3. The derivative of the demand function tells us how fast the demand is decreasing at each price level.

$$D'(p) = 0.028p^3 - 0.48p^2 + 2.6p - 4.9$$

$$D'(2) = 0.028(2^3) - 0.48(2^2) + 2.6(2) - 4.9 = -1.396$$

$$D'(6) = 0.028(6^3) - 0.48(6^2) + 2.6(6) - 4.9 = -0.532$$

If the price per unit increases from \$2 to \$3, sales will decrease by 1396 units. If the price per unit increases from \$6 to \$7, sales will decrease by 532 units.

CHAPTER 5 REVIEW

1. Recall that a falling object falls d feet after t seconds, where $d = 16t^2$. What is a falling object's average velocity between 5 and 6 seconds?
   (a) 160 feet per second
   (b) 400 feet per second
   (c) 576 feet per second
   (d) 176 feet per second

2. What is a falling object's instantaneous velocity at 5 seconds?
   (a) 160 feet per second
   (b) 400 feet per second
   (c) 576 feet per second
   (d) 176 feet per second

3. The cost to produce x units of a product is given by $C(x) = x^2 + 5.7x + 104$. Find the marginal cost for 20 units.

(a) \$618
(b) \$45.70
(c) \$96.10
(d) \$214

4. The annual revenue for a product during its first ten years can be approximated by the function $R(t) = 0.93t^4 - 39t^3 + 562t^2 - 3332t + 8233$, where the revenue R is in dollars and t is the number of years after the product is introduced. What happened to the revenue at two years?
   (a) The revenue increased at the rate of about \$3519 per year.
   (b) The revenue decreased at the rate of about \$3519 per year.
   (c) The revenue increased at the rate of about \$1522 per year.
   (d) The revenue decreased at the rate of about \$1522 per year.

5. The revenue for a product is $R(x) = -0.005x^2 + 11x - 5400$ and the cost is $C(x) = 0.015x + 15$, for x units produced and sold. Find the marginal profit for 800 units.
   (a) \$0.015
   (b) \$18.985
   (c) \$3
   (d) \$2.985

6. Suppose the marginal revenue for 200 units is \$10. What does this mean?
   (a) 200 units cost \$10 to produce.
   (b) The cost to produce the 201st unit is \$10.
   (c) It costs \$2000 to produce 200 units.
   (d) It does not mean anything.

SOLUTIONS

1. d 2. a 3. b 4. d 5. d 6. b

6

CHAPTER

Chain Rule

The derivative for some functions can be difficult to find using only the rules we have so far. Imagine how much work it would be to find the derivative for $y = (x^2 - 5)^4$. We could multiply $(x^2 - 5)(x^2 - 5)(x^2 - 5)(x^2 - 5)$. This would give us $y = x^8 - 20x^6 + 150x^4 - 500x^2 + 625$, which we would differentiate term by term. Or, we could use the product rule a few times. Instead, we will use the generalized power rule, which comes from the *chain rule*, to find the derivative in one quick step. We will discuss the chain rule later in the chapter. For now, we will concentrate on the generalized power rule. Suppose $f(x)$ is a differentiable function, and n is any real number. The derivative of a function to a power is the power times the function raised to the old power minus one, times the derivative of the function.

> If $y = [f(x)]^n$, then $y' = n[f(x)]^{n-1} f'(x)$.

This rule allows us to easily differentiate the following functions.

- $y = (x^2 - 5)^4$
- $y = (x^3 - x^2 + 3x + 10)^5$
- $y = \dfrac{1}{(x+4)^6} = (x+4)^{-6}$
- $y = \sqrt{6x - 2} = (6x - 2)^{1/2}$

EXAMPLES

Find y'.

- $y = (x^2 - 5)^4$

  We will begin by identifying n and $f(x)$: $n = 4$ and $f(x) = x^2 - 5$. According to the formula, we need $f'(x)$ and $n - 1$: $f'(x) = 2x$ and $n - 1 = 3$. Now we will put these into the generalized power rule.

$$\begin{aligned} y' &= n[f(x)]^{n-1} f'(x) \\ y' &= 4(x^2 - 5)^3(2x) \\ &= 8x(x^2 - 5)^3 \end{aligned}$$

- $y = (x^3 - x^2 + 3x + 10)^5$

  According to the formula, we need $f(x)$, $f'(x)$, n, and $n - 1$.

  $f(x) = x^3 - x^2 + 3x + 10$, $f'(x) = 3x^2 - 2x + 3$, $n = 5$, and $n - 1 = 4$

$$\begin{aligned} y' &= n[f(x)]^{n-1} f'(x) \\ y' &= 5(x^3 - x^2 + 3x + 10)^4(3x^2 - 2x + 3) \\ &= 5(3x^2 - 2x + 3)(x^3 - x^2 + 3x + 10)^4 \\ &= (15x^2 - 10x + 15)(x^3 - x^2 + 3x + 10)^4 \end{aligned}$$

- $y = \frac{1}{(x+4)^6}$

  In order to use the generalized power rule, we need to rewrite this function using the fact that $\frac{1}{a^n} = a^{-n}$. The function becomes $y = (x+4)^{-6}$. Now we can see that $f(x) = x+4$, $f'(x) = 1$, $n = -6$, and $n - 1 = -6 - 1 = -7$.

$$y' = -6(x+4)^{-7}(1)$$
$$= -6(x+4)^{-7} \text{ or } \frac{-6}{(x+4)^7}$$

- $y = \sqrt{6x-2}$

  Again, we need to rewrite this function. We will use the fact that $\sqrt[n]{a} = a^{1/n}$. It becomes $y = (6x-2)^{1/2}$. Now we can see that $f(x) = 6x-2$, $f'(x) = 6$, $n = \frac{1}{2}$, and $n - 1 = \frac{1}{2} - 1 = -\frac{1}{2}$.

$$y' = \frac{1}{2}(6x-2)^{-1/2}(6)$$
$$= \frac{1}{2}(6)(6x-2)^{-1/2} = 3(6x-2)^{-1/2}$$
$$\text{or } \frac{3}{(6x-2)^{1/2}} = \frac{3}{\sqrt{6x-2}}$$

- $f(x) = \sqrt[3]{(x^2+1)^4}$

  We will begin by rewriting the function as $f(x) = (x^2+1)^{4/3}$.

$$F(x) = x^2 + 1 \qquad F'(x) = 2x \qquad n = \frac{4}{3} \qquad n - 1 = \frac{4}{3} - 1 = \frac{1}{3}$$

$$f'(x) = \frac{4}{3}(x^2+1)^{1/3}(2x) = \frac{4}{3}(2x)(x^2+1)^{1/3}$$
$$= \frac{8x}{3}(x^2+1)^{1/3} \text{ or } \frac{8x\sqrt[3]{x^2+1}}{3}$$

Forgetting to include "$f'(x)$" in "$nf(x)^{n-1}f'(x)$" is very easy to do, so take extra care to write it down, even if it is only 1.

PRACTICE

Find the derivative.

1. $y = (3x^2 - 4)^7$
2. $y = (5x^3 - x^2 + 1)^4$

3. $f(x) = \frac{1}{15x+3}$
4. $f(x) = \sqrt{x^2 + x + 1}$
5. $y = \frac{1}{\sqrt[3]{x+7}}$

SOLUTIONS

1. $f(x) = 3x^2 - 4,\ f'(x) = 6x,\ n = 7,\ n - 1 = 6$

$$y' = 7(3x^2 - 4)^6(6x) = 7(6x)(3x^2 - 4)^6 = 42x(3x^2 - 4)^6$$

2. $f(x) = 5x^3 - x^2 + 1,\ f'(x) = 15x^2 - 2x,\ n = 4,\ n - 1 = 3$

$$y' = 4(5x^3 - x^2 + 1)^3(15x^2 - 2x) = 4(15x^2 - 2x)(5x^3 - x^2 + 1)^3$$
$$= (60x^2 - 8x)(5x^3 - x^2 + 1)^3$$

3. $f(x) = (15x + 3)^{-1}$,

$F(x) = 15x + 3,\ F'(x) = 15,\ n = -1,\ n - 1 = -1 - 1 = -2$

$$f'(x) = -1(15x + 3)^{-2}(15) = -1(15)(15x + 3)^{-2}$$
$$= -15(15x + 3)^{-2} \text{ or } \frac{-15}{(15x + 3)^2}$$

4. $f(x) = (x^2 + x + 1)^{1/2},\ F(x) = x^2 + x + 1,\ F'(x) = 2x + 1,\ n = \frac{1}{2}$, $n - 1 = \frac{1}{2} - 1 = -\frac{1}{2}$

$$f'(x) = \frac{1}{2}(x^2 + x + 1)^{-1/2}(2x + 1) = \frac{1}{2}(2x + 1)(x^2 + x + 1)^{-1/2}$$
$$\text{or} \quad \frac{2x + 1}{2(x^2 + x + 1)^{1/2}} = \frac{2x + 1}{2\sqrt{x^2 + x + 1}}$$

5. $y = (x+7)^{-1/3},\ f(x) = x+7,\ f'(x) = 1,\ n = -\frac{1}{3},\ n-1 = -\frac{1}{3} - 1 = -\frac{4}{3}$

$$y' = -\frac{1}{3}(x + 7)^{-4/3}(1) = -\frac{1}{3}\frac{1}{(x + 7)^{4/3}} = -\frac{1}{3\sqrt[3]{(x + 7)^4}}$$

We are ready to find the derivative of functions using a combination of the power rule and the product or quotient rule. We begin by deciding which rule to use first. We have to decide if we have a function to a power, where the function is a product or quotient. Or if we have a product or quotient where one or more parts is itself a power.

A	$y = \left[\dfrac{\text{numerator function}}{\text{denominator function}}\right]^n$	Begin with the power rule.
B	$y = \dfrac{(\text{numerator function})^n}{\text{denomintor function}}$	
	or $\dfrac{\text{numerator function}}{(\text{denominator function})^n}$	Begin with the quotient rule.
C	$y = [(\text{first function})(\text{second function})]^n$	Begin with the power rule.
D	$y = [(\text{first function})^n(\text{second function})]$	
	or $[(\text{first function})(\text{second function})^n]$	Begin with the product rule.

EXAMPLES

Determine which rule should be used first.

- $y = \frac{2x+1}{(x-3)^2}$

  This function has the same form as B, so we would begin with the quotient rule.
- $y = \left(\frac{5x-14}{2x+1}\right)^3$

  This function has the same form as A, so we would begin with the power rule.
- $y = \sqrt{(2x+5)(x-6)}$

Rewriting this function, we have $y = ((2x+5)(x-6))^{1/2}$. This function has the same form as C, so we would begin with the power rule.

PRACTICE

Determine which rule should be used first.

1. $y = \frac{(x-1)^3}{x+1}$
2. $y = (16x+5)(4x-3)^2$
3. $y = \sqrt[4]{(x+1)(x-3)}$

4. $y = \left(\frac{7x+4}{3x-1}\right)^5$

5. $y = \frac{\sqrt{x+2}}{x+8}$

SOLUTIONS

1. Quotient rule
2. Product rule
3. Power rule
4. Power rule
5. Quotient rule

Now that we know where to begin, we are ready to find the derivatives. Because the algebra can be tedious, we will leave our answers unsimplified.

EXAMPLES

Find y'.

- $y = \frac{2x+1}{(x-3)^2}$

We will begin with the quotient rule.

$$f(x) = 2x + 1 \qquad g(x) = (x-3)^2$$

$$f'(x) = 2 \qquad g'(x) = 2(x-3)^1(1) \quad \text{Power rule}$$

$$y' = \frac{f'(x)g(x) - f(x)g'(x)}{[g(x)]^2}$$

$$= \frac{2(x-3)^2 - (2x+1)(2)(x-3)}{[(x-3)^2]^2}$$

- $y = \left(\frac{5x-14}{2x+1}\right)^3$

$$y' = n[f(x)]^{n-1} f'(x)$$

$$= 3\left(\frac{5x-14}{2x+1}\right)^2 \cdot f'(x)$$

Next, we will find $f'(x)$ using the quotient rule on $f(x) = \frac{5x-14}{2x+1}$.

$$f'(x) = \frac{F'(x)G(x) - F(x)G'(x)}{[G(x)]^2}$$

(where $F(x) = 5x - 14,\ F'(x) = 5,\ \ G(x) = 2x + 1,\ G'(x) = 2$)

$$= \frac{5(2x+1) - (5x-14)(2)}{(2x+1)^2}$$

$y' = n[f(x)]^{n-1} f'(x)$ becomes

$$y' = 3\left(\frac{5x-14}{2x+1}\right)^2 \cdot \frac{5(2x+1) - (5x-14)(2)}{(2x+1)^2}$$

- $y = 6x(x+4)^5$

  We will begin with the product rule.

$$f(x) = 6x \qquad g(x) = (x+4)^5$$

$$f'(x) = 6 \qquad g'(x) = 5(x+4)^4(1) \quad \text{Power rule}$$

$$y' = f'(x)g(x) + f(x)g'(x)$$

$$y = 6(x+4)^5 + 6x(5)(x+4)^4$$

- $y = \sqrt{(2x+5)(x-6)}$

  Rewriting the function, we have $y = [(2x+5)(x-6)]^{1/2}$. We will begin with the power rule, where $f(x) = (2x+5)(x-6)$ and $n = \frac{1}{2}$ (so $n - 1 = -\frac{1}{2}$).

$$y' = \frac{1}{2}[(2x+5)(x-6)]^{-1/2} f'(x)$$

  Now we will find $f'(x)$ using the product rule on $(2x+5)(x-6)$.

$$F(x) = 2x + 5 \qquad G(x) = x - 6$$

$$F'(x) = 2 \qquad G'(x) = 1$$

$$f'(x) = 2(x-6) + (2x+5)(1) = 2x - 12 + 2x + 5 = 4x - 7$$

$$y' = \frac{1}{2}[(2x+5)(x-6)]^{-1/2} f'(x)$$

$$= \frac{1}{2}[(2x+5)(x-6)]^{-1/2}(4x-7)$$

PRACTICE

Find y'.

1. $y = \frac{(x-1)^3}{x+1}$

2. $y = (16x + 5)(4x - 3)^2$

3. $y = \sqrt[4]{(x + 1)(x - 3)}$

4. $y = \left(\frac{7x+4}{3x-1}\right)^5$

5. $y = \frac{\sqrt{x+2}}{x+8}$

SOLUTIONS

1. For $y = \frac{f(x)}{g(x)}$,

$$f(x) = (x - 1)^3 \qquad g(x) = x + 1$$

$$f'(x) = 3(x - 1)^2(1) \qquad g'(x) = 1$$

$$y' = \frac{3(x - 1)^2(x + 1) - (x - 1)^3}{(x + 1)^2}.$$

2. For $y = f(x)g(x)$,

$$f(x) = 16x + 5 \qquad g(x) = (4x - 3)^2$$

$$f'(x) = 16 \qquad g'(x) = 2(4x - 3)^1(4)$$

$$y' = 16(4x - 3)^2 + (16x + 5)(2)(4x - 3)(4).$$

3. Rewrite the function as $y = [(x + 1)(x - 3)]^{1/4}$. We have $y = [f(x)]^n$, where $f(x) = (x + 1)(x - 3)$ and $n = \frac{1}{4}$ (so $n - 1 = -\frac{3}{4}$).

$$f'(x) = \overbrace{1}^{F'}\ \overbrace{(x - 3)}^{G} + \overbrace{(x + 1)}^{F}\ \overbrace{(1)}^{G'} = 2x - 2$$

$$y' = n[f(x)]^{n-1} f'(x)$$

$$= \frac{1}{4}[(x + 1)(x - 3)]^{-3/4}(2x - 2)$$

4. We have $y = [f(x)]^n$, where $f(x) = \frac{7x+4}{3x-1}$, and $n = 5$,

$$y' = n[f(x)]^{n-1} f'(x)$$
$$= 5\left(\frac{7x+4}{3x-1}\right)^4 f'(x).$$

We need $f'(x)$ for $f(x) = \frac{7x+4}{3x-1}$.

$$f'(x) = \frac{\overbrace{7}^{F'}\overbrace{(3x-1)}^{G} - \overbrace{(7x+4)}^{F}\overbrace{(3)}^{G'}}{\underbrace{(3x-1)^2}_{G^2}}$$

$$y' = 5\left(\frac{7x+4}{3x-1}\right)^4 f'(x)$$
$$= 5\left(\frac{7x+4}{3x-1}\right)^4 \cdot \frac{7(3x-1)-(7x+4)(3)}{(3x-1)^2}$$

5. $y = \frac{(x+2)^{1/2}}{x+8}$

Begin with the quotient rule where $f(x) = (x+2)^{1/2}$ and $g(x) = x + 8$.

$$y' = \frac{\overbrace{\frac{1}{2}(x+2)^{-1/2}}^{f'}\overbrace{(x+8)}^{g} - \overbrace{(x+2)^{1/2}}^{f}\overbrace{(1)}^{g'}}{\underbrace{(x+8)^2}_{g^2}}$$

The Chain Rule

We use the chain rule to find the rate of change of one variable with respect to a second variable, which is itself a function of a third variable. For example, suppose the sales level of a product depends on the amount of money spent on advertising, and the amount of money spent on advertising depends on the previous year's profit. Then the sales level of a product ultimately depends on the previous year's profit.

According to the chain rule, we can find the rate of change of the first variable with respect to the third variable by multiplying the rate of change of the first and second variable with the rate of change of the second and third variable. Suppose y is a function of u and u is a function of x.

$$\overbrace{\frac{dy}{dx}}^{\text{Rate of change of } y \text{ with respect to } x} = \overbrace{\frac{dy}{du}}^{\text{Rate of change of } y \text{ with respect to } u} \times \overbrace{\frac{du}{dx}}^{\text{Rate of change of } u \text{ with respect to } x}$$

These expressions are not exactly fractions, instead they involve limits, but they do "cancel" in a way.

In the following problems, we will be given two separate functions, one where y is a function of u, and the other where u is a function of x. We can find $\frac{dy}{dx}$ by finding the individual derivatives and multiplying them together. We will then make a substitution for u.

EXAMPLES

Find $\frac{dy}{dx}$.

- $y = 4u^2 + 6u + 3$ and $u = 5x - 2$.

  The individual derivatives are $\frac{dy}{du} = 8u + 6$ (from $4u^2 + 6u + 3$) and $\frac{du}{dx} = 5$ (from $5x - 2$).

$$\begin{aligned}\frac{dy}{dx} &= \frac{dy}{du} \cdot \frac{du}{dx} \\ &= (8u + 6)5 \\ &= 40u + 30 \\ &= 40\overbrace{(5x - 2)}^{u=5x-2} + 30 \qquad \text{Replace } u \text{ with } 5x - 2. \\ &= 200x - 80 + 30 = 200x - 50\end{aligned}$$

- $y = \sqrt{2u + 9}$ and $u = 6x^3 - 5x + 2$

$$y = (2u + 9)^{1/2}$$

$$\begin{aligned}\frac{dy}{du} &= \frac{1}{2}(2u + 9)^{-1/2}(2) \\ &= \frac{1}{2}(2)\frac{1}{(2u + 9)^{1/2}} = \frac{1}{\sqrt{2u + 9}}\end{aligned}$$

$$\frac{du}{dx} = 18x^2 - 5$$

$$\frac{dy}{dx} = \frac{dy}{du} \cdot \frac{du}{dx}$$

$$= \frac{1}{\sqrt{2u+9}} \cdot (18x^2 - 5) = \frac{18x^2 - 5}{\sqrt{2u+9}}$$

$$= \frac{18x^2 - 5}{\sqrt{2(6x^3 - 5x + 2) + 9}}$$ Replace u with $6x^3 - 5x + 2$.

$$= \frac{18x^2 - 5}{\sqrt{12x^3 - 10x + 13}}$$

PRACTICE

Find $\frac{dy}{dx}$.

1. $y = u^2 - 6$ and $u = 5x^2 + 1$
2. $y = u^3 + u^2 - 5$ and $u = 4x + 6$
3. $y = \frac{4}{u}$ and $u = 14x + 9$

SOLUTIONS

1. $\frac{dy}{du} = 2u$ and $\frac{du}{dx} = 10x$

$$\frac{dy}{dx} = \frac{dy}{du} \cdot \frac{du}{dx} = (2u)(10x)$$

$$= [2(5x^2 + 1)](10x)$$ Replace u with $5x^2 + 1$.

$$= (10x^2 + 2)(10x) = 100x^3 + 20x$$

2. $\frac{dy}{du} = 3u^2 + 2u$ and $\frac{du}{dx} = 4$

$$\frac{dy}{dx} = \frac{dy}{du} \cdot \frac{du}{dx} = (3u^2 + 2u)(4)$$

$$= [3(4x + 6)^2 + 2(4x + 6)](4)$$

$$= [3(4x + 6)(4x + 6) + 2(4x + 6)](4)$$

$$= [3(16x^2 + 48x + 36) + 2(4x + 6)](4)$$

$$= (48x^2 + 144x + 108 + 8x + 12)(4)$$

$$= 192x^2 + 608x + 480$$

3. $y = 4u^{-1}$, so $\frac{dy}{du} = 4(-1)u^{-2} = -4u^{-2}$ and $\frac{du}{dx} = 14$

$$\frac{dy}{dx} = \frac{dy}{du} \cdot \frac{du}{dx} = -4u^{-2} \cdot 14 = -\frac{4}{u^2} \cdot 14 = -\frac{56}{u^2}$$

$$= -\frac{56}{(14x + 9)^2}$$

Chain Rule Notation

In the previous problems, we could have avoided some of the steps by substituting for u *before* finding any derivatives. In problem 1 above, we found $\frac{dy}{dx} = 100x^3 + 20x$ for $y = u^2 - 6$ and $u = 5x^2 + 1$. We will work this problem differently by substituting $5x^2 + 1$ for u in $y = u^2 - 6$ as the first step. Then we will use the generalized power rule to find the derivative.

$$y = (5x^2 + 1)^2 - 6$$

$$y' = 2(5x^2 + 1)^1(10x)$$

$$= 2(10x)(5x^2 + 1) = 20x(5x^2 + 1) = 100x^3 + 20x$$

Think of $y = u^2 - 6$ and $u = 5x^2 + 1$ as one function composed from two separate functions, $y = f(u) = u^2 - 6$ and $u = g(x) = 5x^2 + 1$. Then $y = f(u) = f(g(x))$ (with $g(x)$ substituted for u). Then $\frac{dy}{dx}$ becomes

$$\frac{dy}{dx} = \frac{dy}{du} \cdot \frac{du}{dx}$$

$= f'(u) \cdot g'(x)$ Replace $\frac{dy}{du}$ with $f'(u)$ and $\frac{du}{dx}$ with $g'(x)$.

$= f'(g(x)) \cdot g'(x)$ Replace u with $g(x)$.

The notation $\frac{dy}{dx} = f'(g(x))g'(x)$ allows us to find the derivative for more complicated functions. The generalized power rule is one example. If $y = [g(x)]^n$, then $\frac{dy}{dx} = n[g(x)]^{n-1}g'(x)$. Here, $f(u) = u^n$ and $u = g(x)$, giving us $y = f(u) = f(g(x)) = (g(x))^n$.

The notation $\frac{dy}{dx} = \frac{dy}{du} \cdot \frac{du}{dx}$ allows us to find the rate of change between two variables (y and x) that are each related to a third variable (u).

For example, suppose the profit for selling x thousand units of a product can be found by using the formula $P = -500x^2 + 20{,}000x$, and the demand function is $x = \frac{1}{p}$ (x thousand units can be sold when the selling price is p dollars). The profit depends on the number of units sold, and the number of units sold depends on the price, which makes the profit depend on the price. $\frac{dP}{dx}$ is the rate of change of the profit with respect to the number of units sold, and $\frac{dx}{dp}$ is the rate of change of the number of units sold with respect to the price. The quantity $\frac{dP}{dp}$ is the rate of change of the profit with respect to the price.

$$\frac{dP}{dx} = -1000x + 20{,}000$$

For example, when 10,000 units are sold ($x = 10$), $\frac{dP}{dx} = -1000(10) + 20{,}000 = 10{,}000$. This means that the profit is increasing at the rate of \$10,000 per 1000 units sold.

$$\frac{dx}{dp} = \frac{-1}{p^2}$$

For example, when the price is \$2, $\frac{dx}{dp} = \frac{-1}{2^2} = \frac{-1}{4}$. This means that at a price of \$2, demand is dropping at the rate of 250 (one-fourth of 1000 units) per \$1 increase in the price.

$$\frac{dP}{dp} = \frac{dP}{dx} \cdot \frac{dx}{dp} = (-1000x + 20{,}000)\left(\frac{-1}{p^2}\right)$$

$$= \left(-1000\left(\frac{1}{p}\right) + 20{,}000\right)\left(\frac{-1}{p^2}\right) \qquad \text{Substitute } \frac{1}{p} \text{ for } x.$$

This derivative tells us how the profit changes at different prices. When the price is \$2, $\frac{dP}{dp} = (-1000(\frac{1}{2}) + 20{,}000)(\frac{-1}{2^2}) = -4875$. This means that when the price is \$2, the profit is decreasing at the rate of \$4875 per dollar increase in the price.

EXAMPLES

- A company sells all it can produce of a product. The revenue function is $R = 16x$, for x units sold. The weekly production function is $x = 500n$, where n is the number of employees (up to 100). Find and interpret $\frac{dR}{dn}$.

  We have two functions, $R = 16x$ and $x = 500n$, and three variables, R, x, and n. Although R and n are not directly related, they are related through x. We can find $\frac{dR}{dn}$ with the chain rule: $\frac{dR}{dn} = \frac{dR}{dx} \cdot \frac{dx}{dn}$. This is the amount of revenue generated by each employee.

$$\frac{dR}{dx} = 16 \qquad \text{Each unit generates \$16 of revenue.}$$

$$\frac{dx}{dn} = 500 \qquad \text{Each employee produces 500 units.}$$

$$\frac{dR}{dn} = \frac{dR}{dx} \cdot \frac{dx}{dn} = 16(500) = 8000$$

Each employee produces $8000 per week in revenue.

- The profit function for selling x units of a product is $P = -0.01x^2 + 100x + 600$. The company produces 900 units per day, making $x = 900t$ the production function. Find $\frac{dP}{dt}$. Evaluate $\frac{dP}{dt}$ at $t = 1$ and at $t = 10$ days. Interpret these numbers.

  The profit depends on the quantity produced. The quantity produced depends on how long the product is in production. This makes the profit depend on how long the product is in production.

$$\frac{dP}{dx} = -0.02x + 100 \qquad \text{The marginal profit for } x \text{ units.}$$

$$\frac{dx}{dt} = 900 \qquad \text{Production increases 900 per day.}$$

$$\frac{dP}{dt} = \frac{dP}{dx} \cdot \frac{dx}{dt} \qquad \text{The marginal profit after } t \text{ days.}$$

$$= (-0.02x + 100)(900) = -18x + 90{,}000$$

$$= -18(900t) + 90{,}000 \qquad \text{Replace } x \text{ with } 900t.$$

$$= -16{,}200t + 90{,}000$$

First, we will let $t = 1$: $-16{,}200(1) + 90{,}000 = 73{,}800$. On the first day, the profit is increasing at the rate of $73,800 per day. Now we will let $t = 10$: $-16{,}200(10) + 90{,}000 = -72{,}000$. On the tenth day, the profit is decreasing at the rate of $72,000 per day.

PRACTICE

1. An author receives 10% royalty on the price of a book. The price of the book is $15 but will increase $0.75 per year for the next ten years. The

royalty function is $r = 0.10p$, where p is the price of the book, and the price function is $p = 15 + 0.75t$, where t is the number of years after the book is released. Find and interpret $\frac{dr}{dt}$.

2. A landscaping company charges \$0.25 per square foot to maintain landscaping for a summer. The cost is $C = 0.25A$. A city manager is considering hiring the landscaping company to maintain part of a park. The width of the area under consideration is 200 feet, and the length can vary. The area is $A = 200l$, where l is the length, in feet, of the maintained area. Find and interpret $\frac{dC}{dl}$.
3. When a company spends a dollars on advertising per month, x units of a product are sold, where $x = -(a-40)^2 + 50$. The monthly advertising budget is 1% of the previous year's profit, P, on the product, making $a = 0.01P$. Find $\frac{dx}{dP}$. Interpret $\frac{dx}{dP}$ for $P = 2000$ and $P = 3000$.

SOLUTIONS

1. $\frac{dr}{dp} = 0.10$ and $\frac{dp}{dt} = 0.75$.

$$\frac{dr}{dt} = \frac{dr}{dp} \cdot \frac{dp}{dt} = (0.10)(0.75) = 0.075$$

The author's royalty will increase \$0.075 per year (per book).

2. $\frac{dC}{dA} = 0.25$ and $\frac{dA}{dl} = 200$

$$\frac{dC}{dl} = \frac{dC}{dA} \cdot \frac{dA}{dl} = (0.25)(200) = 50$$

Each foot in the length of the area to be maintained costs \$50.

3. $\frac{dx}{da} = -2(a-40) = -2a + 80$ and $\frac{da}{dP} = 0.01$

$$\begin{aligned}\frac{dx}{dP} &= \frac{dx}{da} \cdot \frac{da}{dP} \\ &= (-2a+80)(0.01) = [-2(0.01P) + 80](0.01) \\ &= -0.0002P + 0.8\end{aligned}$$

Let $P = 2000$: $-0.0002(2000) + 0.8 = 0.4$. When the previous year's profit was \$2000, sales will increase at the rate of 0.4 units per dollar of profit. Let $P = 3000$: $-0.0002(3000) + 0.8 = 0.2$. When the previous year's profit was \$3000, sales will increase at the rate of 0.2 units per dollar of profit.

CHAPTER 6 REVIEW

1. Find $f'(x)$ for $f(x) = (4x^2 + 2x + 5)^3$.
   (a) $f'(x) = 3(4x^2 + 2x + 5)^2(8x + 2)$
   (b) $f'(x) = 3(4x^2 + 2x + 5)^2$
   (c) $f'(x) = 3(8x + 2)$
   (d) $f'(x) = 3(4x^2 + 2x + 5)^2(8x)$

2. Find $\frac{dy}{dx}$ for $y = 8u^3 - 2u$ and $u = 2x^3$.
   (a) $\frac{dy}{dx} = (24x^2 - 2)(6x^2)$
   (b) $\frac{dy}{dx} = [24(2x^3)^2 - 2](6x^2)$
   (c) $\frac{dy}{dx} = 24x^2 - 2$
   (d) $\frac{dy}{dx} = 24(2x^3)^2 - 2(2x^3)$

3. Find $\frac{dy}{dx}$ for $y = \frac{10}{(x^2-1)^3}$ using the power rule.
   (a) $\frac{dy}{dx} = -30(x^2 - 1)^{-4}$
   (b) $\frac{dy}{dx} = -60x(x^2 - 1)^{-4}$
   (c) $\frac{dy}{dx} = -30(x^2 - 1)^{-2}$
   (d) $\frac{dy}{dx} = -60x(x^2 - 1)^{-2}$

4. The revenue function for a product is $R = 8x$, where R is in dollars and x is the number of units sold. The demand function is $x = -\frac{1}{4}p + 10{,}000$, where x units can be sold when the selling price is p. What is $\frac{dR}{dp}$?
   (a) 8
   (b) $\frac{-1}{4}$
   (c) -2
   (d) 4

5. Find y' for $y = \frac{\sqrt{2x+9}}{x-4}$
   (a) $y' = \frac{(2x+9)^{-1/2}(x-4)-(2x+9)^{1/2}}{(x-4)^2}$
   (b) $y' = \frac{\frac{1}{2}(2x+9)^{-1/2}(x-4)-(2x+9)^{1/2}}{(x-4)^2}$
   (c) $y' = \frac{\frac{1}{2}(2x+9)^{-1/2}(x-4)+(2x+9)^{1/2}}{(x-4)^2}$
   (d) $y' = \frac{(2x+9)^{-1/2}(x-4)-\frac{1}{2}(2x+9)^{1/2}(x-4)}{(x-4)^2}$

6. Find $\frac{dy}{dx}$ for $y = \frac{1}{u^2}$ and $u = 12x + 9$
   (a) $\frac{dy}{dx} = \frac{1}{(12x+9)^2}$
   (b) $\frac{dy}{dx} = \frac{12}{(12x+9)^2}$
   (c) $\frac{dy}{dx} = \frac{-24}{(12x+9)^3}$
   (d) $\frac{dy}{dx} = \frac{-12}{(12x+9)^3}$

7. Find y' for $y = (x^2 + 1)^3(4x - 8)^2$.
   (a) $y' = 6x(x^2 + 1)^2(4x - 8)^2 + 8(x^2 + 1)^3(4x - 8)$
   (b) $y' = 3(x^2 + 1)^2(4x - 8)^2 + 2(x^2 + 1)^3(4x - 8)$
   (c) $y' = 3(x^2 + 1)^2(4x - 8)^2 \cdot 2(4x - 8)$
   (d) $y' = 3(x^2 + 1)^2 + 2(4x - 8)$

8. Find $f'(x)$ for $f(x) = \sqrt{(x + 2)(x^3 + 1)}$.
   (a) $f'(x) = \sqrt{3x^2}$
   (b) $f'(x) = \sqrt{(x^3 + 1) + (x + 2)(3x^2)}$
   (c) $f'(x) = \frac{\frac{1}{2}}{\sqrt{(x^3+1)+(x+2)(3x^2)}}$
   (d) $f'(x) = \frac{\frac{1}{2}}{\sqrt{(x+2)(x^3+1)}}[(x^3 + 1) + 3x^2(x + 2)]$

9. The revenue function for a product is $R = 25x$, where R is in dollars and x is the number of units sold. During the first year, the number of units sold after t months is $x = \frac{1000}{t^2}$. How fast is the revenue decreasing after two months?
   (a) At the rate of \$20 per month
   (b) At the rate of \$6250 per month
   (c) At the rate of \$250 per month
   (d) At the rate of \$3125 per month

SOLUTIONS

1. a 2. b 3. b 4. c 5. a 6. c 7. a 8 d 9. b

Implicit Differentiation and Related Rates

In most of this book, we use formulas to find the derivative of y with respect to x, when y is a function of x. For many of the equations in this chapter, y will not be a function of x. For example, y is not a function of x in the equation of $x^2 + y^2 = 4$. The graph of this equation is a circle, which you might remember from algebra fails the vertical line test. We can still find equations of tangent lines and even $\frac{dy}{dx}$ for such equations. The slope of a tangent line to $x^2 + y^2 = 4$ can be found by computing $\frac{dy}{dx} = -\frac{x}{y}$, where (x, y) is a point on the circle. For example, the slope of the tangent line for the point $(\sqrt{2}, \sqrt{2})$ is $-\frac{\sqrt{2}}{\sqrt{2}} = -1$ (see Figure 7.1).

Finding $\frac{dy}{dx}$ for these kinds of equations is called *implicit differentiation*.

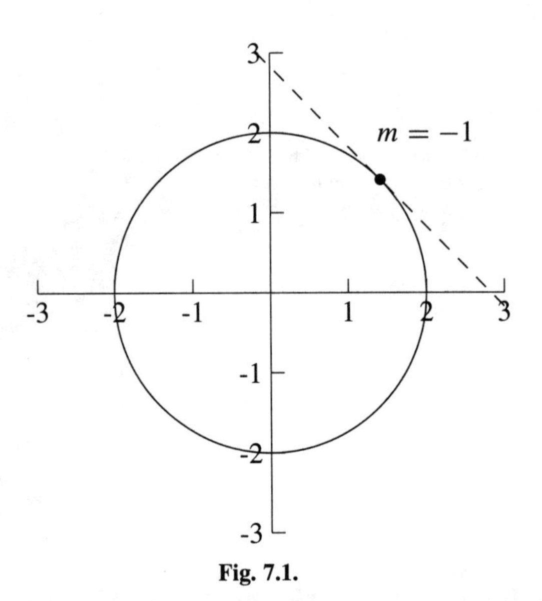

Fig. 7.1.

In order to make some of the work a little easier to follow, we will use the notation $\frac{d}{dx}(\qquad)$, which means the derivative, with respect to x, of the quantity in the parentheses. For example, $\frac{d}{dx}(2x^3 + 6x - 4)$ is simply $6x^2 + 6$. Using this notation, we can rewrite the derivative formulas without using y.

$\frac{d}{dx}(f(x) \pm g(x)) = \frac{d}{dx}(f(x)) \pm \frac{d}{dx}(g(x))$	Sum/difference rule
$\frac{d}{dx}(f(x) \cdot g(x)) = f'(x)g(x) + f(x)g'(x)$	Product rule
$\frac{d}{dx}\left(\frac{f(x)}{g(x)}\right) = \frac{f'(x)g(x) - f(x)g'(x)}{(g(x))^2}$	Quotient rule
$\frac{d}{dx}\left([f(x)]^n\right) = n[f(x)]^{n-1}f'(x)$	Power rule
$\frac{d}{dx}(f(g(x))) = f'(g(x))g'(x)$	Chain rule

We will begin with the simplest derivatives, y to a power. For $\frac{d}{dx}(y^n)$, y replaces $f(x)$ in the power rule and $\frac{dy}{dx}$ replaces $f'(x)$. Then $\frac{d}{dx}([f(x)]^n) = n[f(x)]^{n-1}f'(x)$ becomes $\frac{d}{dx}(y^n) = ny^{n-1}\frac{dy}{dx}$.

EXAMPLES

Evaluate the expression.

- $\frac{d}{dx}(y^4) = 4y^3\frac{dy}{dx}$
- $\frac{d}{dx}(y^{5/3}) = \frac{5}{3}y^{2/3}\frac{dy}{dx}$
- $\frac{d}{dx}\left(\frac{1}{y^2}\right) = \frac{d}{dx}(y^{-2}) = -2y^{-3}\frac{dy}{dx}$ or $-2\frac{1}{y^3}\frac{dy}{dx} = \frac{-2}{y^3}\frac{dy}{dx}$
- $\frac{d}{dx}(\sqrt{y}) = \frac{d}{dx}(y^{1/2}) = \frac{1}{2}y^{-1/2}\frac{dy}{dx}$ or $\frac{1}{2}\frac{1}{y^{1/2}}\frac{dy}{dx} = \frac{1}{2\sqrt{y}}\frac{dy}{dx}$

PRACTICE

Evaluate the expression.

1. $\frac{d}{dx}(y^6)$
2. $\frac{d}{dx}(y^{-3})$
3. $\frac{d}{dx}(y^{2/3})$
4. $\frac{d}{dx}(\frac{1}{y^5})$
5. $\frac{d}{dx}(\sqrt[3]{y})$

SOLUTIONS

1. $\frac{d}{dx}(y^6) = 6y^5\frac{dy}{dx}$
2. $\frac{d}{dx}(y^{-3}) = -3y^{-4}\frac{dy}{dx}$ or $-3\frac{1}{y^4}\frac{dy}{dx} = \frac{-3}{y^4}\frac{dy}{dx}$
3. $\frac{d}{dx}(y^{2/3}) = \frac{2}{3}y^{-1/3}\frac{dy}{dx}$ or $\frac{2}{3}\cdot\frac{1}{y^{1/3}}\frac{dy}{dx} = \frac{2}{3\sqrt[3]{y}}\frac{dy}{dx}$
4. $\frac{d}{dx}\left(\frac{1}{y^5}\right) = \frac{d}{dx}(y^{-5}) = -5y^{-6}\frac{dy}{dx}$ or $\frac{-5}{y^6}\frac{dy}{dx}$
5. $\frac{d}{dx}(\sqrt[3]{y}) = \frac{d}{dx}(y^{1/3}) = \frac{1}{3}y^{-2/3}\frac{dy}{dx}$ or $\frac{1}{3}\cdot\frac{1}{y^{2/3}}\frac{dy}{dx} = \frac{1}{3\sqrt[3]{y^2}}\frac{dy}{dx}$

For some equations, we need to use the product rule, quotient rule, power rule, or some combination of these rules. We will use the product rule when we have the product of two quantities, one with an x and the other with a y. We will let the x-expression be $f(x)$ in the formula, and the y-expression be $g(x)$. $f'(x)$ will be computed in the usual way, and $g'(x)$ will computed as above.

EXAMPLES

Evaluate the expression.

- $\frac{d}{dx}(2xy)$

  The expression $2xy$ is the product of two functions, one of them is $2x$ and the other is y. In the product rule, we will let $f(x)$ be $2x$ and $g(x)$, be y. This makes $f'(x) = 2$ and $g'(x) = \frac{dy}{dx}$.

$$\frac{d}{dx}(f(x) \cdot g(x)) = \frac{d}{dx}(2x \cdot y) = \overbrace{2 \cdot y}^{f'g} + \overbrace{2x \cdot \frac{dy}{dx}}^{fg'}$$

- $\frac{d}{dx}(3x^2y)$

$$f(x) = 3x^2 \qquad g(x) = y$$

$$f'(x) = 6x \qquad g'(x) = \frac{dy}{dx}$$

$$\frac{d}{dx}(3x^2y) = 6xy + 3x^2\frac{dy}{dx}$$

- $\frac{d}{dx}(x^4y^3)$

$$f(x) = x^4 \qquad g(x) = y^3$$

$$f'(x) = 4x^3 \qquad g'(x) = 3y^2\frac{dy}{dx}$$

$$\frac{d}{dx}(x^4y^3) = 4x^3y^3 + x^4 \cdot 3y^2\frac{dy}{dx} = 4x^3y^3 + 3x^4y^2\frac{dy}{dx}$$

- $\frac{d}{dx}(x^2 + y^2)$

  The derivative of x^2 is $2x$ and the derivative of y^2 is $2y\frac{dy}{dx}$.

$$\frac{d}{dx}(x^2 + y^2) = 2x + 2y\frac{dy}{dx}$$

- $\frac{d}{dx}(4x^3y^2 + x^2y^5)$

  We will differentiate each term individually, and then we will add the derivatives in the last step.

$$\frac{d}{dx}(4x^3y^2) = 12x^2y^2 + 4x^3 \cdot 2y\frac{dy}{dx} = 12x^2y^2 + 8x^3y\frac{dy}{dx}$$

$$\frac{d}{dx}(x^2y^5) = 2xy^5 + x^2 \cdot 5y^4\frac{dy}{dx} = 2xy^5 + 5x^2y^4\frac{dy}{dx}$$

$$\frac{d}{dx}(4x^3y^2 + x^2y^5) = \frac{d}{dx}(4x^3y^2) + \frac{d}{dx}(x^2y^5)$$

$$= 12x^2y^2 + 8x^3y\frac{dy}{dx} + 2xy^5 + 5x^2y^4\frac{dy}{dx}$$

PRACTICE

Evaluate the expression.

1. $\frac{d}{dx}(xy^2)$
2. $\frac{d}{dx}(2x^6y^3)$
3. $\frac{d}{dx}(x^3\sqrt{y})$
4. $\frac{d}{dx}(4x^7y^{-2})$
5. $\frac{d}{dx}(x^3 + xy^2)$

SOLUTIONS

1. $\frac{d}{dx}(xy^2) = 1 \cdot y^2 + x \cdot 2y\frac{dy}{dx} = y^2 + 2xy\frac{dy}{dx}$

2. $\frac{d}{dx}(2x^6y^3) = 12x^5y^3 + 2x^6 \cdot 3y^2\frac{dy}{dx} = 12x^5y^3 + 6x^6y^2\frac{dy}{dx}$

3. $\frac{d}{dx}(x^3\sqrt{y}) = \frac{d}{dx}(x^3y^{1/2}) = 3x^2y^{1/2} + x^3 \cdot \frac{1}{2}y^{-1/2}\frac{dy}{dx}$

$$= 3x^2\sqrt{y} + \frac{1}{2}x^3\frac{1}{y^{1/2}}\frac{dy}{dx} = 3x^2\sqrt{y} + \frac{x^3}{2\sqrt{y}}\frac{dy}{dx}$$

4. $\frac{d}{dx}(4x^7y^{-2}) = 28x^6y^{-2} + 4x^7(-2)y^{-3}\frac{dy}{dx}$

$= 28x^6y^{-2} - 8x^7y^{-3}\frac{dy}{dx}$ or $\frac{28x^6}{y^2} - \frac{8x^7}{y^3}\frac{dy}{dx}$

5. $\frac{d}{dx}(x^3 + xy^2) = 3x^2 + 1 \cdot y^2 + x \cdot 2y\frac{dy}{dx} = 3x^2 + y^2 + 2xy\frac{dy}{dx}$

We will use one or more of the other two formulas, the quotient rule and the power rule, on the next set of problems.

EXAMPLES

Evaluate the expression.

- $\frac{d}{dx}\left(\frac{x^2}{y^3}\right)$

  We will use the quotient rule where $f(x)$ is x^2 and $g(x)$ is y^3, making $f'(x) = 2x$ and $g'(x) = 3y^2\frac{dy}{dx}$.

  $$\frac{d}{dx}\left(\frac{x^2}{y^3}\right) = \frac{2xy^3 - x^2(3y^2)\frac{dy}{dx}}{(y^3)^2} = \frac{2xy^3 - 3x^2y^2\frac{dy}{dx}}{(y^3)^2}$$

- $\frac{d}{dx}\left(\frac{x^2+x}{y^2-1}\right)$

  We will use the quotient rule where $f(x) = x^2 + x$ and $g(x) = y^2 - 1$, making $f'(x) = 2x + 1$ and $g'(x) = 2y\frac{dy}{dx}$.

  $$\frac{d}{dx}\left(\frac{x^2+x}{y^2-1}\right) = \frac{(2x+1)(y^2-1) - (x^2+x)(2y)\frac{dy}{dx}}{(y^2-1)^2}$$

- $\frac{d}{dx}((x+y)^4)$

  We will begin with the power rule where $f(x) = x + y$, making $f'(x) = 1 + \frac{dy}{dx}$. Then $nf(x)^{n-1}f'(x)$ becomes $4(x+y)^3(1+\frac{dy}{dx})$.

- $\frac{d}{dx}((xy)^3)$

  We will begin with the power rule, where $f(x) = xy$. By the product rule, $f'(x) = 1 \cdot y + x \cdot \frac{dy}{dx} = y + x\frac{dy}{dx}$. Then $nf(x)^{n-1}f'(x)$ becomes $3(xy)^2(y + x\frac{dy}{dx})$.

PRACTICE

Evaluate the expression.

1. $\frac{d}{dx}\left(\frac{x+1}{y+1}\right)$

2. $\frac{d}{dx}\left(\frac{x^2-x}{4y^3+2y}\right)$

3. $\frac{d}{dx}((2x+3y)^2)$

4. $\frac{d}{dx}(\sqrt{x-y})$

5. $\frac{d}{dx}((3x^4y)^5)$

SOLUTIONS

1. $\frac{d}{dx}\left(\frac{x+1}{y+1}\right) = \frac{y+1-(x+1)\frac{dy}{dx}}{(y+1)^2}$

2. $\frac{d}{dx}\left(\frac{x^2-x}{4y^3+2y}\right) = \frac{(2x-1)(4y^3+2y)-(x^2-x)\left(12y^2\frac{dy}{dx}+2\frac{dy}{dx}\right)}{(4y^3+2y)^2}$

3. $\frac{d}{dx}((2x+3y)^2) = 2(2x+3y)\left(2+3\frac{dy}{dx}\right)$

4. $\frac{d}{dx}(\sqrt{x-y}) = \frac{d}{dx}((x-y)^{1/2}) = \frac{1}{2}(x-y)^{-1/2}\left(1-\frac{dy}{dx}\right)$

   $= \frac{1}{2}\frac{1}{(x-y)^{1/2}}\left(1-\frac{dy}{dx}\right) = \frac{1}{2\sqrt{x-y}}\left(1-\frac{dy}{dx}\right)$

5. $\frac{d}{dx}((3x^4y)^5) = 5(3x^4y)^4\frac{d}{dx}(3x^4y)$ Power rule

   $\frac{d}{dx}(3x^4y) = 12x^3y + 3x^4\frac{dy}{dx}$ Product rule

$$\frac{d}{dx}((3x^4y)^5) = 5(3x^4y)^4\left(12x^3y + 3x^4\frac{dy}{dx}\right) \qquad \text{Replace } \tfrac{d}{dx}(3x^4y) \text{ with } 12x^3y + 3x^4\tfrac{dy}{dx}.$$

We are ready to use implicit differentiation to find $\frac{dy}{dx}$ for an equation. We will differentiate both sides of the equation with respect to x. After differentiating, we will use algebra to solve the equation for $\frac{dy}{dx}$. This will be pretty straightforward for the next set of problems. However, we will polish our algebra skills to be able to find $\frac{dy}{dx}$ for the equations that come up later. To solve an equation for a variable means to isolate the variable on one side of the equation, and it can only appear on that side. For example, $x = 6y - 1$ is solved for x, but $x = 6y - x$ is not because x appears on both sides of the equation.

EXAMPLES

Find $\frac{dy}{dx}$.

- $3x^2 + 6y^5 = 2x$

  We will differentiate both sides of the equation with respect to x.

$$\frac{d}{dx}(3x^2 + 6y^5) = \frac{d}{dx}(2x)$$

$$6x + 30y^4\frac{dy}{dx} = 2$$

  Now we need to use algebra to isolate $\frac{dy}{dx}$ on one side of the equation.

$$6x + 30y^4\frac{dy}{dx} = 2$$

$$30y^4\frac{dy}{dx} = 2 - 6x$$

$$\frac{30y^4\frac{dy}{dx}}{30y^4} = \frac{2-6x}{30y^4}$$

$$\frac{dy}{dx} = \frac{2-6x}{30y^4} = \frac{2(1-3x)}{30y^4} = \frac{1-3x}{15y^4}$$

- $x^2 - y^2 = 9$

$$\frac{d}{dx}(x^2 - y^2) = \frac{d}{dx}(9) \qquad \text{Differentiate both sides.}$$

$$2x - 2y\frac{dy}{dx} = 0$$

$$-2y\frac{dy}{dx} = -2x \qquad \text{Solve for } \frac{dy}{dx}.$$

$$\frac{-2y\frac{dy}{dx}}{-2y} = \frac{-2x}{-2y}$$

$$\frac{dy}{dx} = \frac{x}{y}$$

PRACTICE

Find $\frac{dy}{dx}$.

1. $7x - 2y^3 = x^2 - 3$
2. $x^3 + xy^2 = 5$
3. $\sqrt[3]{x} - \sqrt[3]{y} = 6x$

SOLUTIONS

1.

$$\frac{d}{dx}(7x - 2y^3) = \frac{d}{dx}(x^2 - 3)$$

$$7 - 6y^2\frac{dy}{dx} = 2x - 0$$

$$-6y^2\frac{dy}{dx} = 2x - 7$$

$$\frac{dy}{dx} = \frac{2x - 7}{-6y^2} \text{ or } -\frac{2x - 7}{6y^2}$$

2.

$$\frac{d}{dx}(x^3 + xy^2) = \frac{d}{dx}(5)$$

$$3x^2 + 1 \cdot y^2 + x \cdot 2y\frac{dy}{dx} = 0 \qquad \text{Use the product rule on } xy^2.$$

$$2xy\frac{dy}{dx} = -3x^2 - y^2$$

$$\frac{dy}{dx} = \frac{-3x^2 - y^2}{2xy}$$

3.
$$\frac{d}{dx}(x^{1/3} - y^{1/3}) = \frac{d}{dx}(6x)$$

$$\frac{1}{3}x^{-2/3} - \frac{1}{3}y^{-2/3}\frac{dy}{dx} = 6$$

$$\frac{1}{3}\frac{1}{x^{2/3}} - \frac{1}{3}\frac{1}{y^{2/3}}\frac{dy}{dx} = 6$$

$$\frac{1}{3\sqrt[3]{x^2}} - \frac{1}{3\sqrt[3]{y^2}}\frac{dy}{dx} = 6$$

$$-\frac{1}{3\sqrt[3]{y^2}}\frac{dy}{dx} = 6 - \frac{1}{3\sqrt[3]{x^2}}$$

$$\frac{dy}{dx} = -3\sqrt[3]{y^2}\left(6 - \frac{1}{3\sqrt[3]{x^2}}\right) \text{ or } -18\sqrt[3]{y^2} + \frac{\sqrt[3]{y^2}}{\sqrt[3]{x^2}}$$

Before finding $\frac{dy}{dx}$ for more complicated problems, let us review how to solve equations having more than one variable. We will solve for t in the following problems. First, we will move each term with a t in it to one side of the equation and the terms without a t in them to the other. Second, we will factor t. Finally, we will divide both sides of the equation by the coefficient of t. This is the quantity in the parentheses next to t.

EXAMPLES

- $x^2 + y^2t = 18y + 18xt$

We will put the t-terms, y^2t and $18xt$, on the left side of the equation and the terms without t, x^2 and $18y$, on the right side.

$$y^2t - 18xt = 18y - x^2$$

Now we will factor t from y^2t and $-18xt$, leaving y^2 and $-18x$.

$$t(y^2 - 18x) = 18y - x^2$$

We will divide both sides of the equation by the coefficient of t, $y^2 - 18x$.

$$t = \frac{18y - x^2}{y^2 - 18x}$$

- $2x + 2y + 2xt = 6yt$

  If we move $2xt$ to the right side of the equation, we will have t-terms on one side of the equation and terms without t on the other.

$$2x + 2y = 6yt - 2xt \qquad \text{Factor } t.$$

$$2x + 2y = t(6y - 2x) \qquad \text{Divide by } 6y - 2x.$$

$$\frac{2x + 2y}{6y - 2x} = t$$

$$\text{or } t = \frac{2(x + y)}{2(3y - x)} = \frac{x + y}{3y - x}$$

PRACTICE

Solve for t.

1. $2yt - 2 - 4t = 0$
2. $4y^3t - 8yt = 4x^3 - 18x$
3. $2x - \frac{1}{2}\sqrt{x}y - \sqrt{x}t = 4yt$
4. $2xy + x^2t + y^2 + 2xyt = 0$

SOLUTIONS

1.

$$2yt - 2 - 4t = 0$$

$$2yt - 4t = 2$$

$$t(2y - 4) = 2$$

$$t = \frac{2}{2y - 4} = \frac{2}{2(y - 2)} = \frac{1}{y - 2}$$

2.

$$4y^3t - 8yt = 4x^3 - 18x$$

$$t(4y^3 - 8y) = 4x^3 - 18x$$

$$t = \frac{4x^3 - 18x}{4y^3 - 8y} \text{ or } \frac{2(2x^3 - 9x)}{2(2y^3 - 4y)} = \frac{2x^3 - 9x}{2y^3 - 4y}$$

3.

$$2x - \frac{1}{2}\sqrt{x}y - \sqrt{x}t = 4yt$$

$$2x - \frac{1}{2}\sqrt{x}y = 4yt + \sqrt{x}t$$

$$2x - \frac{1}{2}\sqrt{x}y = t(4y + \sqrt{x})$$

$$\frac{2x - \frac{1}{2}\sqrt{x}y}{4y + \sqrt{x}} = t$$

4.

$$2xy + x^2t + y^2 + 2xyt = 0$$

$$x^2t + 2xyt = -2xy - y^2$$

$$t(x^2 + 2xy) = -2xy - y^2$$

$$t = \frac{-2xy - y^2}{x^2 + 2xy}$$

We will put together our ability to differentiate implicitly and our ability to solve an equation to find $\frac{dy}{dx}$ for more complicated equations. As before, we will differentiate both sides of the equation implicitly with respect to x. Then we will solve for $\frac{dy}{dx}$.

EXAMPLES

Find $\frac{dy}{dx}$.

- $3x^2y + y^4 = 6x$

$$\frac{d}{dx}(3x^2y + y^4) = \frac{d}{dx}(6x)$$

$$6xy + 3x^2\frac{dy}{dx} + 4y^3\frac{dy}{dx} = 6$$

We will solve for $\frac{dy}{dx}$ by moving $6xy$ to the right side of the equation, and then by factoring $\frac{dy}{dx}$ on the left side.

$$3x^2\frac{dy}{dx} + 4y^3\frac{dy}{dx} = 6 - 6xy$$

$$\frac{dy}{dx}(3x^2 + 4y^3) = 6 - 6xy \qquad \text{Divide by } 3x^2 + 4y^3.$$

$$\frac{dy}{dx} = \frac{6 - 6xy}{3x^2 + 4y^3}$$

- $x^2 - y^2 = x^3 + y^3$

$$\frac{d}{dx}(x^2 - y^2) = \frac{d}{dx}(x^3 + y^3)$$

$$2x - 2y\frac{dy}{dx} = 3x^2 + 3y^2\frac{dy}{dx}$$

$$2x - 3x^2 = 2y\frac{dy}{dx} + 3y^2\frac{dy}{dx}$$

$$2x - 3x^2 = \frac{dy}{dx}(2y + 3y^2)$$

$$\frac{2x - 3x^2}{2y + 3y^2} = \frac{dy}{dx}$$

PRACTICE

Find $\frac{dy}{dx}$.

1. $x^2y^2 - y = 7$
2. $4x^3y^2 - x^2 + 2y = 9x$
3. $(x + y)^2 = 7x$
4. $x\sqrt{y} + y^3 = 5x - 1$

SOLUTIONS

1.

$$2xy^2 + x^2 \cdot 2y\frac{dy}{dx} - \frac{dy}{dx} = 0$$

$$2x^2y\frac{dy}{dx} - \frac{dy}{dx} = -2xy^2$$

$$\frac{dy}{dx}(2x^2y - 1) = -2xy^2$$

$$\frac{dy}{dx} = \frac{-2xy^2}{2x^2y - 1}$$

2.

$$12x^2y^2 + 4x^3 \cdot 2y\frac{dy}{dx} - 2x + 2\frac{dy}{dx} = 9$$

$$8x^3y\frac{dy}{dx} + 2\frac{dy}{dx} = 9 - 12x^2y^2 + 2x$$

$$\frac{dy}{dx}(8x^3y + 2) = 9 - 12x^2y^2 + 2x$$

$$\frac{dy}{dx} = \frac{9 - 12x^2y^2 + 2x}{8x^3y + 2}$$

3.

$$2(x+y)^1\left(1 + \frac{dy}{dx}\right) = 7$$ The derivative of $x + y$ is $1 + \frac{dy}{dx}$.

$$(2x+2y)\left(1 + \frac{dy}{dx}\right) = 7$$ Distribute $2x + 2y$ in the parentheses.

$$2x + 2y + (2x + 2y)\frac{dy}{dx} = 7$$

$$(2x + 2y)\frac{dy}{dx} = 7 - 2x - 2y$$

$$\frac{dy}{dx} = \frac{7 - 2x - 2y}{2x + 2y}$$

4.

$$\frac{d}{dx}(xy^{1/2}+y^3)=\frac{d}{dx}(5x-1)$$

$$1\cdot y^{1/2}+x\cdot\frac{1}{2}y^{-1/2}\frac{dy}{dx}+3y^2\frac{dy}{dx}=5-0$$

$$\sqrt{y}+\frac{1}{2}x\frac{1}{\sqrt{y}}\frac{dy}{dx}+3y^2\frac{dy}{dx}=5$$

$$\frac{x}{2\sqrt{y}}\frac{dy}{dx}+3y^2\frac{dy}{dx}=5-\sqrt{y}$$

$$\frac{dy}{dx}\left(\frac{x}{2\sqrt{y}}+3y^2\right)=5-\sqrt{y}$$

$$\frac{dy}{dx}=\frac{5-\sqrt{y}}{\dfrac{x}{2\sqrt{y}}+3y^2}$$

Once we know how to find $\frac{dy}{dx}$, we can find an equation of the tangent line to the graph of an equation at a given point. Remember that once we find $\frac{dy}{dx}$, we will use the coordinates of the point in the derivative to find the slope. Once we have the slope, we will use the point and m in $y = mx + b$ to find b.

EXAMPLE

- Find an equation of the tangent line to the graph of $2y^2 - xy^2 = x^3$ at the point $(1, 1)$.
  We will begin by differentiating both sides of the equation with respect to x.

$$4y\frac{dy}{dx}-\left(1\cdot y^2+x\cdot 2y\frac{dy}{dx}\right)=3x^2$$

$$4y\frac{dy}{dx}-y^2-2xy\frac{dy}{dx}=3x^2$$

$$4y\frac{dy}{dx}-2xy\frac{dy}{dx}=3x^2+y^2$$

$$\frac{dy}{dx}(4y-2xy)=3x^2+y^2$$

$$\frac{dy}{dx} = \frac{3x^2 + y^2}{4y - 2xy}$$

$$m = \frac{3(1)^2 + 1^2}{4(1) - 2(1)(1)} \quad \text{Let } x = 1,\ y = 1 \text{ in } \frac{dy}{dx}.$$

$$= \frac{4}{2} = 2$$

Now we will put $x = 1,\ y = 1,\ m = 2$ in $y = mx + b$ to find b.

$$1 = 2(1) + b$$

$$-1 = b$$

The tangent line is $y = 2x - 1$. The curve and tangent line are shown in Figure 7.2.

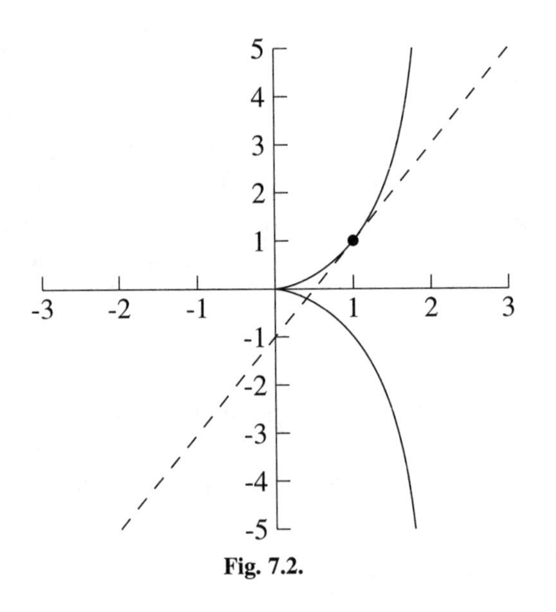

Fig. 7.2.

PRACTICE

Find an equation of the tangent line.

1. $x^3 + 2y^2 + y = 23$ at $(2, -3)$

2. $x^3 + y^3 + 9xy = 27$ at $(1, 2)$
3. $16x^3 + 16y^2 = 3$ at $(\frac{1}{2}, \frac{1}{4})$

SOLUTIONS

1.

$$3x^2 + 4y\frac{dy}{dx} + \frac{dy}{dx} = 0$$

$$4y\frac{dy}{dx} + \frac{dy}{dx} = -3x^2$$

$$\frac{dy}{dx}(4y + 1) = -3x^2$$

$$\frac{dy}{dx} = \frac{-3x^2}{4y + 1}$$

$$m = \frac{-3(2)^2}{4(-3) + 1} = \frac{-12}{-11} = \frac{12}{11}$$

$$-3 = \frac{12}{11}(2) + b \qquad \text{Let } x = 2,\ y = -3, \text{ and } m = \frac{12}{11} \text{ in } y = mx + b.$$

$$-3 = \frac{24}{11} + b$$

$$-3 - \frac{24}{11} = -\frac{57}{11} = b$$

The tangent line is $y = \frac{12}{11}x - \frac{57}{11}$.

2.

$$3x^2 + 3y^2\frac{dy}{dx} + 9y + 9x\frac{dy}{dx} = 0$$

$$3y^2\frac{dy}{dx} + 9x\frac{dy}{dx} = -3x^2 - 9y$$

$$\frac{dy}{dx}(3y^2 + 9x) = -3x^2 - 9y$$

$$\frac{dy}{dx} = \frac{-3x^2 - 9y}{3y^2 + 9x}$$

$$= \frac{3(-x^2 - 3y)}{3(y^2 + 3x)} = \frac{-x^2 - 3y}{y^2 + 3x}$$

$$m = \frac{-1^2 - 3(2)}{2^2 + 3(1)} = \frac{-7}{7} = -1$$

$$2 = -1(1) + b \qquad \text{Put } x = 1,\ y = 2,$$

$$m = -1 \text{ in } y = mx + b.$$

$$3 = b$$

The tangent line is $y = -x + 3$.

3.

$$48x^2 + 32y\frac{dy}{dx} = 0$$

$$32y\frac{dy}{dx} = -48x^2$$

$$\frac{dy}{dx} = \frac{-48x^2}{32y} = -\frac{3x^2}{2y}$$

$$m = \frac{-3(\frac{1}{2})^2}{2(\frac{1}{4})} = \frac{-\frac{3}{4}}{\frac{1}{2}}$$

$$= -\frac{3}{4} \div \frac{1}{2} = -\frac{3}{4} \cdot \frac{2}{1} = -\frac{3}{2}$$

$$\frac{1}{4} = -\frac{3}{2}\left(\frac{1}{2}\right) + b$$

$$1 = b$$

The tangent line is $y = -\frac{3}{2}x + 1$.

Using implicit differentiation, we can differentiate an expression with respect to a variable that does not appear in the formula. Usually that variable is t, representing time. In the problems below, we will differentiate both sides of the equation with respect to t. We will be given some values to use in the equation to find the rate of change (with respect to t). Later, we will use this technique to solve applied problems.

EXAMPLES

- Find $\frac{dy}{dt}$ for $y = 6x^2 + 5x - 1$ when $x = -4$ and $\frac{dx}{dt} = 2$.
  We will begin by implicitly differentiating both sides of the equation with respect to t.

$$\frac{d}{dt}(y) = \frac{d}{dt}(6x^2 + 5x - 1)$$

$$\frac{dy}{dt} = 12x\frac{dx}{dt} + 5\frac{dx}{dt}$$

Now we will substitute $x = -4$ and $\frac{dx}{dt} = 2$ in this equation.

$$\frac{dy}{dt} = 12(-4)(2) + 5(2) = -86$$

- Find $\frac{dy}{dt}$ for $x^2 + y^2 = 169$ when $x = 12$ and $\frac{dx}{dt} = 10$.

$$2x\frac{dx}{dt} + 2y\frac{dy}{dt} = 0$$

$$2(12)(10) + 2y\frac{dy}{dt} = 0 \qquad \text{Let } x = 12 \text{ and } \frac{dx}{dt} = 10.$$

$$240 + 2y\frac{dy}{dt} = 0$$

We cannot find $\frac{dy}{dt}$ until we know what value y has. We can use the equation $x^2 + y^2 = 169$ to find y.

$$12^2 + y^2 = 169$$

$$y^2 = 169 - 144 = 25$$

$$y = \pm 5$$

In the applications covered later in this chapter, we will only be concerned with positive values for x and y because they will represent real-world

numbers—quantities, dollars, distances, etc. If we use $y = 5$ only, $240 + 2y\frac{dy}{dx} = 0$ becomes $240 + 2(5)\frac{dy}{dt} = 0$. Now we can solve for $\frac{dy}{dt}$.

$$240 + 2(5)\frac{dy}{dt} = 0$$

$$10\frac{dy}{dt} = -240$$

$$\frac{dy}{dt} = -24$$

- Find $\frac{dz}{dt}$ for $z = xy^3 - 7x^2$, when $x = 2$, $y = 1$, $\frac{dx}{dt} = -3$, and $\frac{dy}{dt} = 2$.

$$\frac{dz}{dt} = y^3\frac{dx}{dt} + x \cdot 3y^2\frac{dy}{dt} - 14x\frac{dx}{dt}$$

$$\frac{dz}{dt} = 1^3(-3) + 2(3 \cdot 1^2)(2) - 14(2)(-3) = 93$$

PRACTICE

1. Find $\frac{dy}{dt}$ for $y = 9 - x^2$ when $x = 4$ and $\frac{dx}{dt} = 3$.
2. Find $\frac{dy}{dt}$ for $x^3 - y^2 = 109$ when $x = 5$, $y = 4$, and $\frac{dx}{dt} = -2$.
3. Find $\frac{dy}{dt}$ for $x^2 - y^2 = 16$ when $x = 5$, $\frac{dx}{dt} = 10$ and y is positive.

SOLUTIONS

1.

$$\frac{dy}{dt} = -2x\frac{dx}{dt}$$

$$\frac{dy}{dt} = -2(4)(3) = -24$$

2.

$$3x^2\frac{dx}{dt} - 2y\frac{dy}{dt} = 0$$

$$3(5^2)(-2) - 2(4)\frac{dy}{dt} = 0$$

$$-150 - 8\frac{dy}{dt} = 0$$

$$-8\frac{dy}{dt} = 150$$

$$\frac{dy}{dt} = \frac{150}{-8} = -\frac{75}{4}$$

3.

$$2x\frac{dx}{dt} - 2y\frac{dy}{dt} = 0$$

$$2(5)(10) - 2y\frac{dy}{dt} = 0$$

$$-2y\frac{dy}{dt} = -100$$

$$\frac{dy}{dt} = \frac{-100}{-2y} = \frac{50}{y}$$

We can find y using the fact that $x^2 - y^2 = 16$ and $x = 5$.

$$5^2 - y^2 = 16$$

$$-y^2 = 16 - 25 = -9$$

$$y^2 = 9$$

$y = \pm 3$ Use positive 3 since y must be positive.

Now we have $\frac{dy}{dt} = \frac{50}{3}$.

Related Rates

The demand for most products decreases when the price increases. How fast will demand decrease if there is a monthly price increase? A container is being filled, how fast is the level rising? A person is walking away from a lamp post, how fast is his shadow lengthening? All of these quantities are based on *time*. We can use implicit differentiation with respect to time to find how fast a quantity is changing. We will begin with business problems.

In the following applications, we will be given an equation, usually with two variables, and told how fast one of the variables is changing. We will be asked how fast the other variable is changing. To answer the question, we will implicitly differentiate both sides of the equation with respect to t. And as we did above, we will substitute known values into the differentiated equation to find the unknown rate of change. Information on the rate of change will be given in phrases such as, "the price will increase each month by \$0.25" and, "the container is being drained at the rate of 5 cubic feet per minute." As with any applied problem, units of measure must be consistent. If the information on the rate of change is given in months, t should represent months. If it is given in terms of minutes, then t should represent minutes.

EXAMPLES

- A buyer for a department store determines that demand for a certain fabric is given by $q = \frac{1000}{\sqrt{p}}$, where q yards are demanded when the price per yard is p. How fast will demand decrease if, at the current price of \$4 per yard, the price increases by \$0.05 per month?

$$\frac{d}{dt}(q) = \frac{d}{dt}\left(\frac{1000}{\sqrt{p}}\right) = \frac{d}{dt}(1000p^{-1/2})$$

$$\frac{dq}{dt} = \frac{-1}{2} \cdot 1000p^{-3/2}\frac{dp}{dt} = -\frac{500}{p^{3/2}}\frac{dp}{dt} = -\frac{500}{\sqrt{p^3}}\frac{dp}{dt}$$

The quantity $\frac{dq}{dt}$ is the rate of change in demand, the number we are looking for, and the quantity $\frac{dp}{dt}$ is the rate of change in price, which is \$0.05 per month. We will use $p = 4$ and $\frac{dp}{dt} = 0.05$ in the derivative to find $\frac{dq}{dt}$.

$$\frac{dq}{dt} = -\frac{500}{\sqrt{4^3}}(0.05) = -\frac{25}{\sqrt{64}} = -\frac{25}{8} = -3.125$$

Demand will decrease at the rate of 3.125 yards per month.

- The profit (in \$ thousand) for selling x units of a product is given by $P = -0.003x^2 + 4.8x + 18{,}080$. How much will the profit increase if currently 400 units have been produced and sold and 25 units will be produced per week?

$$\frac{dP}{dt} = -0.006x\frac{dx}{dt} + 4.8\frac{dx}{dt}$$

Production is 400 units, so x is 400. The production is increasing at the rate of 25 units per week, so $\frac{dx}{dt}$ is 25.

$$\frac{dP}{dt} = -0.006(400)(25) + 4.8(25) = 60$$

Profit will increase at the rate of \$60,000 thousand per week.

- The number of sales for a medical device that a manufacturer sells depends on the number of sales representatives. When the number of sales representatives is between 2 and 20, the number of units sold when there are x representatives can be approximated by $y = -1.7x^4 + 67x^3 - 895x^2 + 5527x - 4651$. The company has five representatives selling this device and plans to increase the number of representatives by two per month. How will this affect the number sold?

$$\frac{dy}{dt} = -6.8x^3\frac{dx}{dt} + 201x^2\frac{dx}{dt} - 1790x\frac{dx}{dt} + 5527\frac{dx}{dt}$$

Let $x = 5$ and $\frac{dx}{dt} = 2$.

$$\frac{dy}{dt} = -6.8(5^3)(2) + 201(5^2)(2) - 1790(5)(2) + 5527(2) = 1504$$

The number of devices sold will increase at the rate of 1504 per month.

PRACTICE

1. The revenue for selling x units of a product is given by $R = 100{,}000 - \frac{40{,}000}{\sqrt{x}}$. When 15 units are sold, daily production is 5 units. How fast is revenue increasing?
2. When the price is p dollars for a product, q units are demanded, where $q = \frac{500}{p}$. When the price is \$5, a distributor decides to increase the price by \$0.10 per month. How does this price increase affect the demand?
3. The profit for selling x units of a product is given by $P = -(x - 150)^2 + 5000$ (when at least 80 units are sold). When 100 units are sold, 15 per day are produced. How much is the profit increasing per day?
4. A distributor for a cleaning product believes that when \$$a$ thousand is spent on advertising, $y = 100 - \frac{500}{a+1}$ thousand units are sold (when at least \$2000 is spent). The company has spent \$4000 on advertising and plans to increase its advertising budget by \$500 per month. How will the increase in advertising affect sales?

SOLUTIONS

1. Daily production is 5 units, so $\frac{dx}{dt} = 5$.

$$R = 100{,}000 - 40{,}000x^{-1/2}$$

$$\frac{dR}{dt} = -\frac{1}{2}(-40{,}000)x^{-3/2}\frac{dx}{dt} = \frac{20{,}000}{x^{3/2}}\frac{dx}{dt}$$

$$= \frac{20{,}000}{\sqrt{x^3}}\frac{dx}{dt}$$

$$= \frac{20{,}000}{\sqrt{15^3}}(5) \qquad \text{Let } x = 15, \text{ and } \frac{dx}{dt} = 5.$$

$$\approx 1721$$

Revenue is increasing at the rate of \$1721 per day.

2. The price increase is \$0.10/month, so $\frac{dp}{dt} = 0.10$.

$$q = 500p^{-1}$$

$$\frac{dq}{dt} = (-1)(500)p^{-2}\frac{dp}{dt} = -\frac{500}{p^2}\frac{dp}{dt}$$

$$= -\frac{500}{5^2}(0.10) \qquad \text{Let } p = 5 \text{ and } \frac{dp}{dt} = 0.10.$$

$$= -2$$

Demand will decrease at the rate of two units per month.

3. 15 per day are produced, so $\frac{dx}{dt} = 15$.

$$P = -(x - 150)^2 + 5000$$

$$\frac{dP}{dt} = -2(x - 150)\frac{dx}{dt} \qquad \text{Power rule}$$

$$= -2(100 - 150)(15) \qquad \text{Let } x = 100 \text{ and } \frac{dx}{dt} = 15.$$

$$= 1500$$

Profit is increasing at the rate of \$1500 per day.

4. Advertising is increasing at \$500/month, so $\frac{da}{dt} = 0.5$

$$y = 100 - \frac{500}{a+1} = 100 - 500(a+1)^{-1}$$

$$\frac{dy}{dt} = (-1)(-500)(a+1)^{-2}\frac{da}{dt}$$

$$\text{or } = \frac{500}{(a+1)^2}\frac{da}{dt}$$

$$= \frac{500}{(4+1)^2}(0.5) \quad \text{Let } a = 4 \text{ and } \frac{da}{dt} = 0.5.$$

$$= 10$$

Sales will increase at the rate of 10 thousand units per month.

We will differentiate formulas from geometry for the rest of the problems in this chapter. The first step in solving these problems is to identify the shape involved and to determine which formula to use. Usually, we will differentiate this formula, but there will be times when we will need to make a substitution, based on information given in the problems, before differentiating.

It is important to know *when* to use the numbers given in the problem, before or after differentiating. Rates of change cannot be used until after differentiating. In fact, all numbers could wait until after differentiating, but waiting to use numbers sometimes can cause the differentiation to be more complicated than it needs to be. If the value of a variable changes with time, we must wait until after differentiating before making the substitution. If the value remains constant through time, then we might be able to substitute the number for the variable before differentiating to make the calculations easier. For example, if we are pouring water into a cup with straight sides (the top and bottom have the same radius), then the radius of the water's shape does not change. We would be safe in using the cup's radius in the formula before differentiating. If we are pouring water into a cup shaped like a cone, the radius of the shape of the water does change. We would not be able to substitute for the radius until after differentiating.

EXAMPLES

- A pebble is dropped into a still pond. The radius of the ripple is expanding at the rate of 4 inches per second. How fast is the area increasing 3 seconds after the pebble is dropped, when the radius is 12 inches?

Because the rate of change is given in seconds, t represents seconds. The question asks how fast a circular area is increasing, telling us that we should begin with the formula $A = \pi r^2$. It would *not* be safe to substitute 12 for the radius in this formula until after we have differentiated.

$$\frac{d}{dt}(A) = \frac{d}{dt}(\pi r^2)$$

$$\frac{dA}{dt} = 2\pi r\frac{dr}{dt}$$

$$= 2\pi(12)(4) = 96\pi$$

Three seconds after the pebble is dropped, the area is increasing at the rate of 96π square inches per second.

- Two men leave a park at the same time. One man rode his bicycle southward at 15 mph. The other ran eastward at $6\frac{1}{4}$ mph. How fast were they moving from each other 48 minutes after leaving the park?

  Let us call the distance traveled by the man on the bicycle y and the distance traveled by the runner x. Let s represent the distance traveled between them (see Figure 7.3). Because the shape is a right triangle, we can use the Pythagorean theorem as our equation: $x^2 + y^2 = s^2$. The rates of change are given in miles per hour, so t represents hours.

$$\frac{d}{dt}(x^2 + y^2) = \frac{d}{dt}(s^2)$$

$$2x\frac{dx}{dt} + 2y\frac{dy}{dt} = 2s\frac{ds}{dt} \qquad \text{Divide by 2.}$$

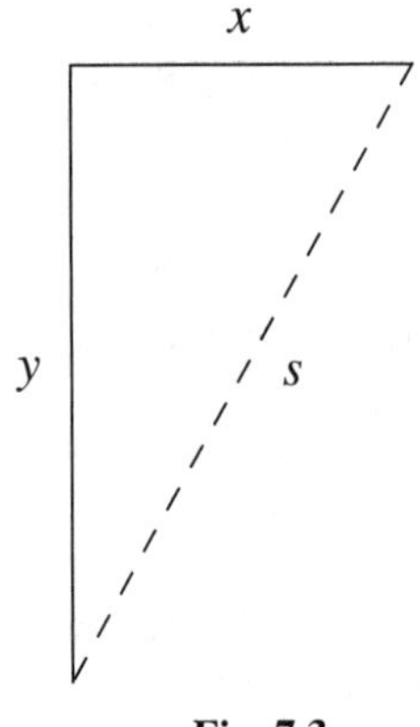

Fig. 7.3.

$$x\frac{dx}{dt} + y\frac{dy}{dt} = s\frac{ds}{dt}$$

The runner's speed is $6\frac{1}{4}$ mph, so $\frac{dx}{dt} = 6\frac{1}{4}$. The cyclist's speed is 15 mph, so $\frac{dy}{dt} = 15$.

$$6\frac{1}{4}x + 15y = 6.25x + 15y = s\frac{ds}{dt}$$

We can find x and y by multiplying the men's speed by $\frac{48}{60} = 0.8$ hours.

$$x = 6.25t = (6.25)(0.8) = 5 \text{ miles}$$

$$y = 15t = 15(0.8) = 12 \text{ miles}$$

Using the original equation, $x^2 + y^2 = s^2$, and the fact that $x = 5$ and $y = 12$, we can find s: $5^2 + 12^2 = s^2$, making $s = 13$ miles.

$$6.25x + 15y = s\frac{ds}{dt}$$

$$6.25(5) + 15(12) = 13\frac{ds}{dt}$$

$$211.25 = 13\frac{ds}{dt}$$

$$16.25 = \frac{ds}{dt}$$

48 minutes after leaving the park, the distance between the runner and the cyclist is increasing at the rate of 16.25 miles per hour.

When a ladder is leaning against a wall, the ladder, the wall, and ground form a right triangle. When the ladder slides down the wall, we can find the rate at which the top of the ladder is sliding downward if we know how fast the bottom is sliding away (or vice versa). As before, we will begin with the Pythogorean theorem. We will let x represent the distance between the base of the ladder and the wall, and y the distance between the top of the ladder and the ground. Usually, we are told how fast the base is moving away from the wall. This tells us $\frac{dx}{dt}$. Then we are asked how fast the top of the ladder is moving. This is $\frac{dy}{dt}$, which is negative because the ladder is moving downward.

EXAMPLES

- A 20-foot ladder is leaning against a wall. The base of the ladder slips and the top of the ladder starts to slide against the wall. When the base of the ladder is 12 feet from the wall, it is moving at the rate of 5 feet per second. At this instant, how fast is the top of the ladder moving? See Figure 7.4.

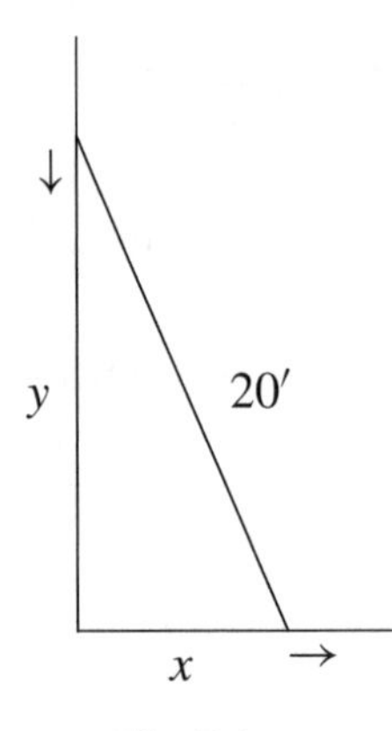

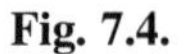
Fig. 7.4.

From the Pythagorean theorem, we have $x^2 + y^2 = 20^2$.

$$\frac{d}{dt}(x^2 + y^2) = \frac{d}{dt}(20^2)$$

$$2x\frac{dx}{dt} + 2y\frac{dy}{dt} = 0 \qquad \text{Divide by 2.}$$

$$x\frac{dx}{dt} + y\frac{dy}{dt} = 0$$

$$12(5) + y\frac{dy}{dt} = 0 \qquad \text{We know } x \text{ is 12 and } \frac{dx}{dt} \text{ is 5.}$$

We can find y using the fact that $x = 12$ and $x^2 + y^2 = 20^2$: $12^2 + y^2 = 20^2$ gives us $y = 16$.

$$12(5) + 16\frac{dy}{dt} = 0$$

$$16\frac{dy}{dt} = -60$$

$$\frac{dy}{dt} = \frac{-60}{16} = -3.75$$

When the base of the ladder is 12 feet from the wall, the top of the ladder is sliding down the wall at the rate of 3.75 feet per second.

- A large cannister in the shape of a right circular cylinder is being drained. The radius of the cannister is 3 feet. At the instant that it is draining at the rate of 5 cubic feet per minute, how fast is the level dropping? (See Figure 7.5.)

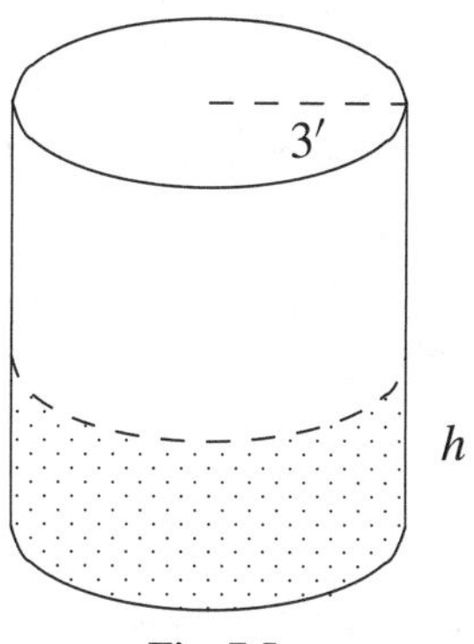

Fig. 7.5.

The volume in the cannister is changing, which tells us to use the formula for the volume of a right circular cylinder, $V = \pi r^2 h$. When the cannister is not empty, the radius of the cannister's contents is always 3 feet, so we can use $r = 3$ in the formula: $V = \pi(3^2)h = 9\pi h$. This substitution saves us from having to use the product rule on $r^2 h$. (Following this example is an explanation why it is safe to use this substitution *before* differentiating.)

$$\frac{d}{dt}(V) = \frac{d}{dt}(9\pi h)$$

$$\frac{dV}{dt} = 9\pi \frac{dh}{dt}$$

The cannister is begin drained, so the volume is decreasing. This makes $\frac{dV}{dt}$ negative. It is decreasing at the rate of 5 cubic feet per minute, so $\frac{dV}{dt}$ is -5.

$$-5 = 9\pi \frac{dh}{dt}$$

$$\frac{-5}{9\pi} = \frac{dh}{dt}$$

The level of the cannister is decreasing at the rate of $\frac{5}{9\pi}$ feet per minute.

Let us see what happens in the above example if we do not substitute $r = 3$ in $V = \pi r^2 h$ before differentiating.

$$\frac{d}{dt}(V) = \frac{d}{dt}(\pi r^2 h)$$

$$\frac{dV}{dt} = \pi\left(2rh\frac{dr}{dt} + r^2\frac{dh}{dt}\right) \qquad \text{Using the product rule on } r^2h.$$

Because the radius is not changing, $\frac{dr}{dt}$ is 0. This gives us

$$\frac{dV}{dt} = \pi\left(2rh(0) + r^2\frac{dh}{dt}\right) = \pi r^2\frac{dh}{dt}$$

Now we can use the fact that $\frac{dV}{dt} = -5$ and $r^2 = 3^2 = 9$: $-5 = \pi(9\frac{dh}{dt})$, which is what we have above.

PRACTICE

1. A small circular fire is spreading, its radius increasing at the rate of 2 feet per minute. When the radius of the fire is 6 feet, how fast is the burned area growing?
2. Two cars pass through an intersection at about the same time. The northbound car is traveling at 45 mph, and the eastbound car is traveling at 60 mph. After 40 minutes, how fast were the cars moving away from each other?
3. A drum, in the shape of a right circular cylinder, is being filled with liquid cleanser at the rate of 2 cubic feet per second. The radius of the drum is 1 foot. How fast is the level of the cleanser rising?
4. A 25-foot ladder is leaning against a wall when someone starts to pull the base away from the wall, at the rate of 2 feet per second. How fast is the top of the ladder moving down when it is 24 feet above the ground?

SOLUTIONS

1. We begin with the area of a circle: $A = \pi r^2$.

$$\frac{d}{dt}(A) = \frac{d}{dt}(\pi r^2)$$

$$\frac{dA}{dt} = 2\pi r\frac{dr}{dt}$$

$$\frac{dA}{dt} = 2\pi(6)(2) = 24\pi$$

The area is increasing at the rate of 24π square feet per minute at the instant the radius of the fire is 6 feet.

2. Let x represent the distance traveled by the eastbound car; y, the distance traveled by the northbound car; and s, the distance between the cars. We begin with $x^2 + y^2 = s^2$. After differentiating both sides of the equation with respect to t, we have $2x\frac{dx}{dt} + 2y\frac{dy}{dt} = 2s\frac{ds}{dt}$. Dividing through by 2 gives us $x\frac{dx}{dt} + y\frac{dy}{dt} = s\frac{ds}{dt}$. We know that the eastbound car's speed is 60 mph, and the northbound car's speed is 45 mph. This gives us $\frac{dx}{dt} = 60$ and $\frac{dy}{dt} = 45$. Now we have $x(60) + y(45) = s\frac{ds}{dt}$. At $t = \frac{40}{60} = \frac{2}{3}$ hours, the eastbound car has traveled $60(\frac{2}{3}) = 40$ miles, and the northbound car has traveled $45(\frac{2}{3}) = 30$ miles. This gives us $x = 40$ and $y = 30$. Using these numbers in $x^2 + y^2 = s^2$, we have $40^2 + 30^2 = s^2$. From this, we have $s = 50$.

$$x(60) + y(45) = s\frac{ds}{dt}$$

$$40(60) + 30(45) = 50\frac{ds}{dt}$$

$$3750 = 50\frac{ds}{dt}$$

$$75 = \frac{ds}{dt}$$

At 40 minutes, the cars are moving away from each other at the rate of 75 mph.

3. The volume of a right circular cylinder is $V = \pi r^2 h$. Because the radius is always 1, the formula becomes $V = \pi 1^2 h = \pi h$. Differentiating both sides of this equation with respect to t gives us $\frac{dV}{dt} = \pi\frac{dh}{dt}$. The volume is increasing at the rate of 2 cubic feet per second, so $\frac{dV}{dt}$ is postive 2.

$$\frac{dV}{dt} = \pi\frac{dh}{dt}$$

$$2 = \pi\frac{dh}{dt}$$

$$\frac{2}{\pi} = \frac{dh}{dt}$$

The level of cleanser is rising at the rate of $\frac{2}{\pi}$ feet per second.

4. We begin with $x^2 + y^2 = 25^2$, where x is the distance between the base of the ladder and the wall, and y is the distance from the top of the ladder to the ground. Then $\frac{dx}{dt}$ is 2, and y is 24. After differentiating both sides of the equation with respect to t, we have $2x\frac{dx}{dt} + 2y\frac{dy}{dt} = 0$. When we divide through by 2, we have $x\frac{dx}{dt} + y\frac{dy}{dt} = 0$.

$$x\frac{dx}{dt} + y\frac{dy}{dt} = 0$$

$$x(2) + 24\frac{dy}{dt} = 0 \qquad \text{We know } y = 24 \text{ and } \frac{dx}{dt} = 2.$$

$$7(2) + 24\frac{dy}{dt} = 0 \qquad x^2 + 24^2 = 25^2 \text{ gives us } x = 7.$$

$$24\frac{dy}{dt} = -14$$

$$\frac{dy}{dt} = \frac{-14}{24} = -\frac{7}{12}$$

The top of the ladder is moving downward at the rate $\frac{7}{12}$ feet per second at the instant it is 24 feet above the ground.

Two triangles are *similar* if they have the same angles (see Figure 7.6).The ratio of any two sides of one triangle is equal to the ratio of the corresponding sides of a similar triangle. For the triangles in Figure 7.6, we have $a/b = A/B$, $b/c = B/C$, and $a/c = A/C$. We will use this fact in the last two problem types.

The cone problem involves a cup or other vessel in the shape of a cone either being filled or drained. Not only do the volume and height of the cone-shaped contents change, but the radius changes, too. We are told how fast the volume is changing and are asked how fast the level is changing. In the volume formula,

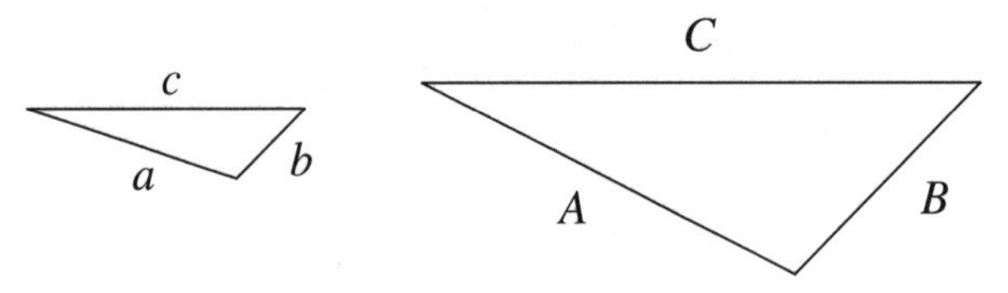

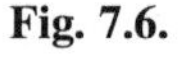
Fig. 7.6.

$V = \frac{1}{3}\pi r^2 h$, we have three variables that are changing with time. We will use similar triangles to replace r with an expression involving h. This reduces the variables from three to two, V and h.

EXAMPLE

- A tank in the shape of a cone is full of water. The top of the tank has a radius of 4 feet, and the tank is 6 feet tall (see Figure 7.7). A pump is draining the tank at the rate of 5 cubic feet per minute. How fast is the water level falling when the water is 3 feet deep?

By similar triangles, for any water level h, we have $\frac{\text{Height when full}}{\text{Radius when full}} = \frac{\text{Any height}}{\text{Any radius}}$, which is $\frac{6}{4} = \frac{h}{r}$. Solving the equation $\frac{6}{4} = \frac{h}{r}$ for r gives us $r = \frac{2}{3}h$. With this substitution, the volume formula $V = \frac{1}{3}\pi r^2 h$ becomes

$$V = \frac{1}{3}\pi\left(\frac{2}{3}h\right)^2 h = \frac{1}{3}\pi\left(\frac{4}{9}h^2\right)h = \frac{4}{27}\pi h^3.$$

$$\frac{d}{dt}(V) = \frac{d}{dt}\left(\frac{4}{27}\pi h^3\right)$$

$$\frac{dV}{dt} = \frac{4}{27}\pi(3)h^2\frac{dh}{dt} = \frac{4}{9}\pi h^2\frac{dh}{dt}$$

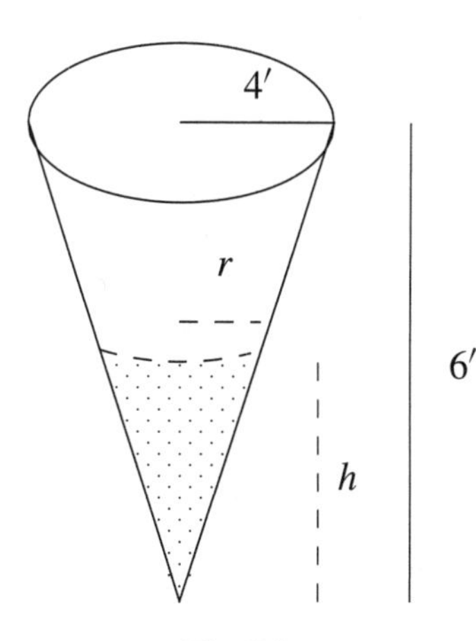

Fig. 7.7.

The volume is decreasing at the rate of 5 feet cubic feet per minute, so $\frac{dV}{dt} = -5$. We want $\frac{dh}{dt}$ when h is 3, so we will let $h = 3$.

$$-5 = \frac{4}{9}\pi(3)^2\frac{dh}{dt} = 4\pi\frac{dh}{dt}$$

$$\frac{-5}{4\pi} = \frac{dh}{dt}$$

The water level is falling at the rate of $\frac{5}{4\pi}$ feet per minute.

When a person walks toward or away from a light, the person's shadow becomes shorter or longer. Two similar triangles are formed. The base of one triangle is formed by the length of the person's shadow, and the height of this triangle is formed by the person's height. The base of the other triangle is formed by the distance from the base of the lamp post to the tip of the shadow, and the height of this triangle is formed by the height of the lamp post. As the person is walking, the heights of the triangles do not change, but their bases do. Let x represent the distance between the person and the lamp post, and let s represent the length of the shadow (see Figure 7.8). By similar triangles, we have

$$\frac{\text{Lamp post's height}}{x + s} = \frac{\text{Person's height}}{s}$$

This gives us a formula to differentiate with respect to t.

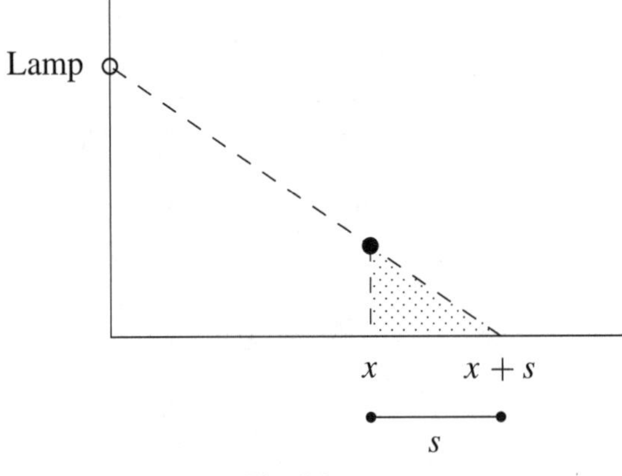

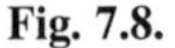

Fig. 7.8.

EXAMPLE

- A woman 5 feet tall walks away from a 30-foot lamp post. How fast is her shadow lengthening if she is walking at the rate of 100 feet per minute?

  In the above equation, the lamp's height can be replaced by 30 and the person's height can be replaced by 5.

$$\frac{30}{x+s} = \frac{5}{s}$$

We will simplify this equation before differentiating.

$$30s = 5(x+s) \qquad \text{Cross-multiply}$$

$$30s = 5x + 5s$$

$$25s = 5x \qquad \text{Subtract } 5s \text{ from each side.}$$

$$\frac{d}{dt}(25s) = \frac{d}{dt}(5x)$$

$$25\frac{ds}{dt} = 5\frac{dx}{dt}$$

Because she is walking away from the lamp post, her distance is increasing, so $\frac{dx}{dt}$ is positive. We know her speed is 100 feet per minute, so $\frac{dx}{dt} = 100$.

$$25\frac{ds}{dt} = 5(100) = 500$$

$$\frac{ds}{dt} = \frac{500}{25} = 20$$

Her shadow is lengthening at the rate of 20 feet per minute.

PRACTICE

1. Someone is pouring coffee into a cone-shaped cup. The radius of the cup is 1.5 inches, and the cup is 4 inches high. The cup is filling up at the rate of 2 cubic inches per second. How fast is the level of coffee rising when it is 3 inches high?
2. A 6-foot man is walking toward a 26-foot-tall lamp post at the rate of 150 feet per minute. How fast is the length of his shadow decreasing?

SOLUTIONS

1. We begin with $V = \frac{1}{3}\pi r^2 h$. By similar triangles, $\frac{r}{h} = \frac{1.5}{4}$. We want $\frac{dh}{dt}$ so we want to keep h and replace r with an expression involving h, so solve this equation for r. When we cross-multiply, we have $4r = 1.5h$, and dividing by 4 gives us $r = \frac{1.5}{4}h = 0.375h$. Because the volume is increasing at the rate of 2 cubic inches per second, $\frac{dV}{dt}$ is 2.

$$V = \frac{1}{3}\pi(0.375h)^2 h = \frac{1}{3}\pi(0.140625h^2)h$$

$$V = 0.046875\pi h^3$$

$$\frac{d}{dt}(V) = \frac{d}{dt}(0.046875\pi h^3)$$

$$\frac{dV}{dt} = 0.140625\pi h^2 \frac{dh}{dt} \qquad \frac{dV}{dt} = 2 \text{ and } h = 3$$

$$2 = 0.140625\pi(3)^2 \frac{dh}{dt}$$

$$\frac{2}{1.265625\pi} = \frac{dh}{dt}$$

$$0.5 \approx \frac{dh}{dt}$$

The coffee is rising at the rate of about 0.5 inches per second at the instant the coffee is 3 inches high.

2. Because the man is walking toward the light, his distance is decreasing, so $\frac{dx}{dt}$ is negative. Because he is walking at the rate of 150 feet per minute, $\frac{dx}{dt} = -150$.

$$\frac{26}{x+s} = \frac{6}{s}$$

$$26s = 6(x+s)$$

$$26s = 6x + 6s$$

$$20s = 6s$$

$$\frac{d}{dt}(20s) = \frac{d}{dt}(6x)$$

$$20\frac{ds}{dt} = 6\frac{dx}{dt}$$

$$20\frac{ds}{dt} = 6(-150) \qquad \frac{dx}{dt} = -150$$

$$\frac{ds}{dt} = -45$$

His shadow is getting shorter at the rate of 45 feet per minute.

CHAPTER 7 REVIEW

For Problems 1–3, find $\frac{dy}{dx}$.

1. $2x - y = x^2 + 3y$

(a)

$$\frac{dy}{dx} = \frac{2x - 1 - x^2}{3}$$

(b)

$$\frac{dy}{dx} = \frac{1 - 2x}{3}$$

(c)

$$\frac{dy}{dx} = \frac{1 - x}{2}$$

(d) $\frac{dy}{dx}$ does not exist.

2. $x^3y^2 + y = 8x$

(a) $\frac{dy}{dx} = 8 - 3x^2y^2$

(b)

$$\frac{dy}{dx} = \frac{8 - x^3}{2y}$$

(c) $\frac{dy}{dx} = 8 - 3x^2y^3$

(d)

$$\frac{dy}{dx} = \frac{8 - 3x^2y^2}{2x^3y + 1}$$

3. $(x + y)^4 = y^3$

(a)

$$\frac{dy}{dx} = \frac{4(x + y)^3}{3y^2}$$

(b)

$$\frac{dy}{dx} = \frac{4(x + y)^3}{3y^2 - 4(x + y)^3}$$

(c)

$$\frac{dy}{dx} = \frac{1}{3y^2 - 1}$$

(d)

$$\frac{dy}{dx} = \frac{4(x + y)^3}{3y^2 - (x + y)^3}$$

4. The profit for selling x units of a product is given by $P = -0.005x^2 + 27x - 28450$. How fast is the profit changing when 2000 units have been sold and weekly production is 90 units?

   (a) Increasing at the rate of \$545 per week
   (b) Increasing at the rate of \$615 per week
   (c) Increasing at the rate of \$630 per week
   (d) Increasing at the rate of \$710 per week

5. A circular puddle is evaporating. When the puddle's radius is 9 inches, it is shrinking at the rate of half an inch per hour. At that instant, how fast is the area decreasing?

   (a) 18π square inches per hour
   (b) 9π square inches per hour
   (c) 6π square inches per hour
   (d) 4π square inches per hour

6. Find an equation of the tangent line to $xy - x^2 = 2y - 8$ at (3, 1).

   (a) $y = 5x - 14$
   (b) $y = -x + 4$

(c) $y = -\frac{3}{2}x + \frac{11}{2}$
(d) $y = -6x + 19$

7. A woman $5\frac{1}{2}$ feet tall walks away from a lamp post that is 35 feet high. If she is walking at the rate of 120 feet per minute, how fast is the length of her shadow growing?

   (a) About 15.6 feet per minute
   (b) About 17.1 feet per minute
   (c) About 18.2 feet per minute
   (d) About 22.4 feet per minute

8. The demand function for a product is given by $q = \frac{12{,}000}{\sqrt{p}}$, where q units are demanded when the price is p dollars. The price is currently \$4 and is expected to increase \$0.06 per month. How will the price increase affect demand?

   (a) Demand will drop at the rate of 40 per month
   (b) Demand will drop at the rate of 55 per month
   (c) Demand will drop at the rate of 50 per month
   (d) Demand will drop at the rate of 45 per month

9. Find an equation of the tangent line to $xy - y^4 = -2$ at $(1, -1)$.

   (a) $y = \frac{1}{5}x - \frac{6}{5}$
   (b) $y = \frac{1}{3}x - \frac{4}{3}$
   (c) $y = -1$
   (d) $y = -\frac{1}{2}x - \frac{1}{2}$

10. A drum in the shape of a right circular cylinder is being drained of its contents. The radius of the drum is 1.5 feet. At the moment that the drum is being drained at the rate of 2 cubic feet per minute, how fast is the level of the contents dropping?

   (a) About 0.21 feet per minute
   (b) About 0.28 feet per minute
   (c) About 0.64 feet per minute
   (d) About 0.42 feet per minute

SOLUTIONS

1. c	2. d	3. b	4. c	5. b
6. a	7. d	8. d	9. a	10. b

CHAPTER 8

Graphing and the First Derivative Test

A function is increasing when an increase in the x-value causes the y-value to increase, too. A function is decreasing when an increase in the x-value causes the y-value to decrease. Many functions are increasing for some values of x and decreasing for others. For example, the function $f(x) = x^2 - 4x + 4$ is increasing to the right of $x = 2$ but decreasing to the left of $x = 2$. If we begin with $x = 3$ and increase to $x = 4$, the y-values increase from $y = f(3) = 3^2 - 4(3) + 4 = 1$ to $y = f(4) = 4^2 - 4(4) + 4 = 4$. If we begin with $x = 0$ and increase to $x = 1$, the y-values decrease from $y = f(0) = 0^2 - 4(0) + 4 = 4$ to $y = f(1) = 1^2 - 4(1) + 4 = 1$. Calculus can help us find where a function is increasing or decreasing. For now, we will use the graph of a function. The function is increasing where the graph goes up and decreasing where it goes down (as we move from left to right).

EXAMPLES

Determine where the functions are increasing and where they are decreasing.

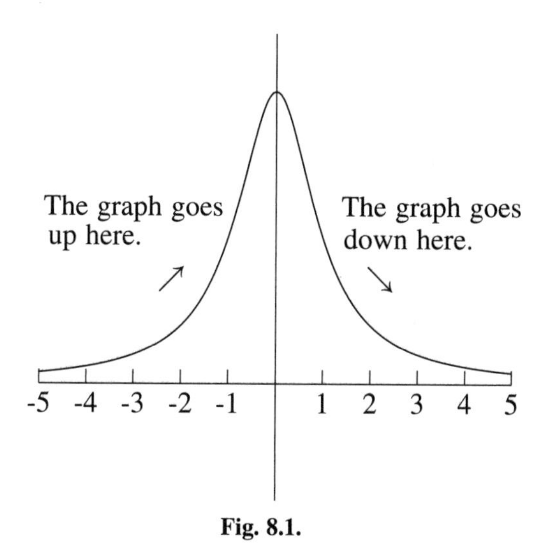

Fig. 8.1.

- The function in Figure 8.1 is increasing to the left of $x = 0$ and decreasing to the right of $x = 0$.

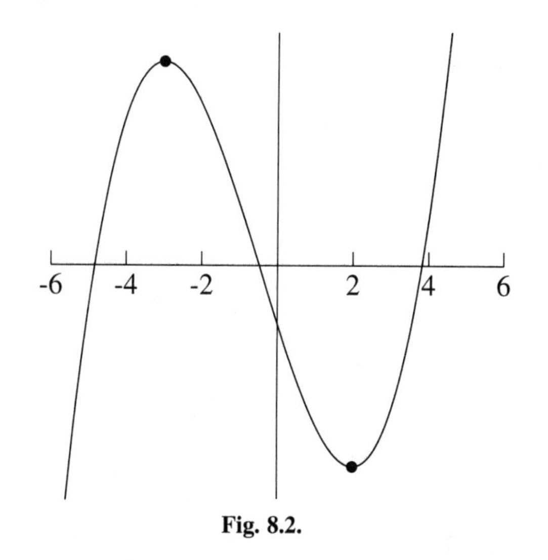

Fig. 8.2.

- The function in Figure 8.2 is increasing to the left of $x = -3$, decreasing between $x = -3$ and $x = 2$, and increasing again to the right of $x = 2$.

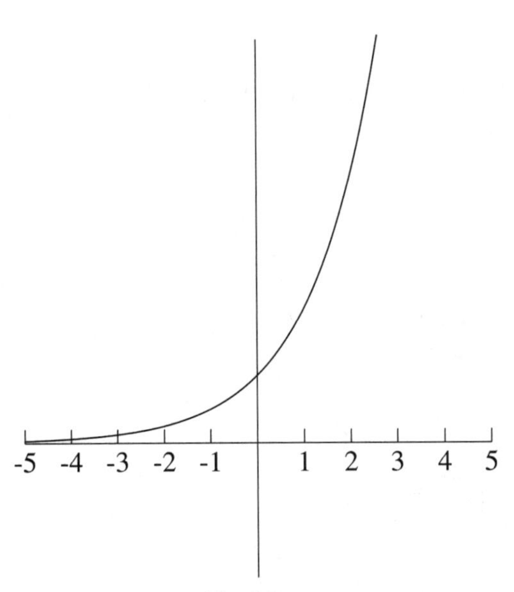

Fig. 8.3.

- The function in Figure 8.3 is increasing for all x-values.

PRACTICE

Determine where the function is increasing and where it is decreasing.

1.

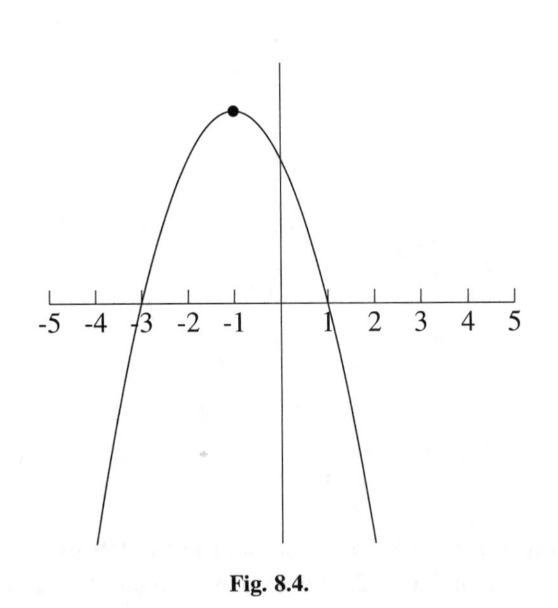

Fig. 8.4.

2.

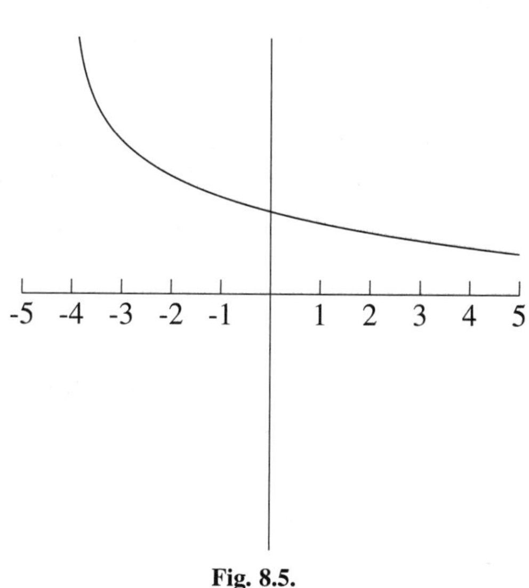

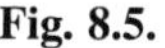
Fig. 8.5.

3.

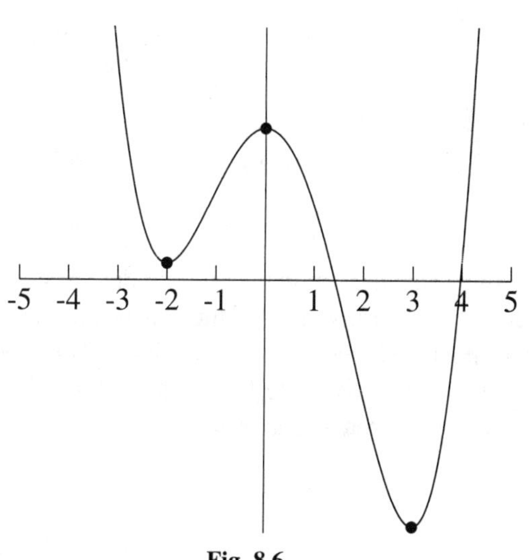

Fig. 8.6.

SOLUTIONS

1. The function is increasing to the left of $x = -1$ and decreasing to the right of $x = -1$.
2. The function is decreasing for all x.

3. The function is decreasing to the left of $x = -2$, increasing between $x = -2$ and $x = 0$, decreasing between $x = 0$ and $x = 3$, and increasing to the right of $x = 3$.

When the slope of a line is positive, the linear function is increasing for all x-values. When the slope is negative, the linear function is decreasing (see Figure 8.7).

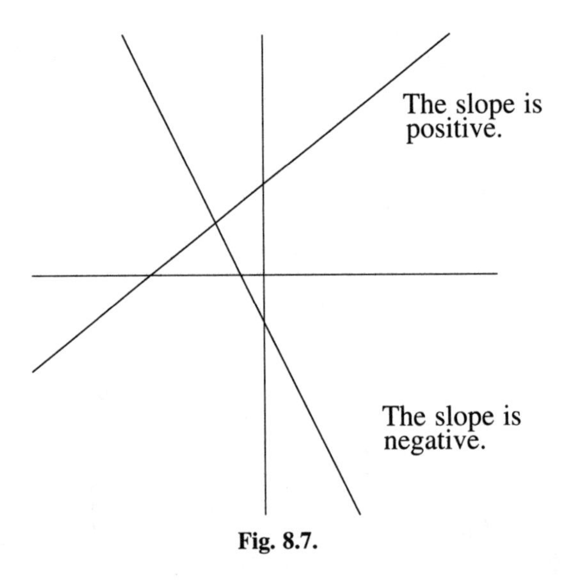

Fig. 8.7.

We can use the slope of the tangent line to describe where a function is increasing and where it is decreasing. A function is increasing where the slope of the tangent line is positive, and it is decreasing where the slope is negative (see Figure 8.8). If $f'(a)$ is positive, the function is increasing at $x = a$. If $f'(a)$ is negative, the function is decreasing at $x = a$.

EXAMPLES

- Determine if $f(x) = x^4 - 2x^2 - 6$ is increasing or decreasing at $x = -2$, $x = -\frac{1}{2}$, and $x = 3$.
  First we will find the derivative: $f'(x) = 4x^3 - 4x$. Now we will evaluate $f'(x)$ at $x = -2$, $-\frac{1}{2}$, and 3.

$$f'(-2) = 4(-2)^3 - 4(-2) = -24$$

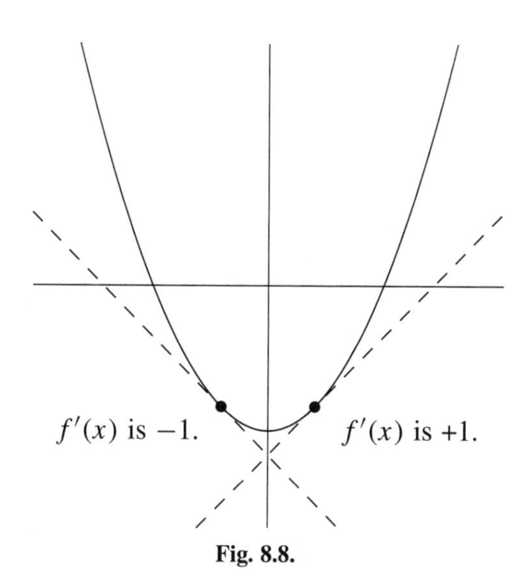

Fig. 8.8.

$$f'\left(-\frac{1}{2}\right)=4\left(-\frac{1}{2}\right)^3-4\left(-\frac{1}{2}\right)=4\left(-\frac{1}{8}\right)-4\left(-\frac{1}{2}\right)=-\frac{4}{8}+2=+\frac{3}{2}$$

$$f'(3)=4(3)^3-4(3)=+96$$

$f'(-2) = -24$ is negative, so $f(x)$ is decreasing at $x = -2$. $f'(-\frac{1}{2}) = \frac{3}{2}$ is positive, so $f(x)$ is increasing at $x = -\frac{1}{2}$. $f'(3) = 96$ is positive, so $f(x)$ is increasing at $x = 3$.

- $f(x) = -10x + 3$
  Because $f'(x) = -10$ is negative for all x-values, $f(x)$ is decreasing for all x.

If we know where a derivative is positive and where it is negative, we can decide where the function is increasing and where it is decreasing. In the following problems, we will be given some graphs and information on where the derivatives are positive and where they are negative. We will match the graphs with the information about the derivatives.

EXAMPLES

Match the derivative information with the graphs in Figures 8.9–8.11.

- $f'(x)$ is positive to the right of $x = -4$.
  The graph is increasing to the right of $x = -4$. This describes Figure 8.11.

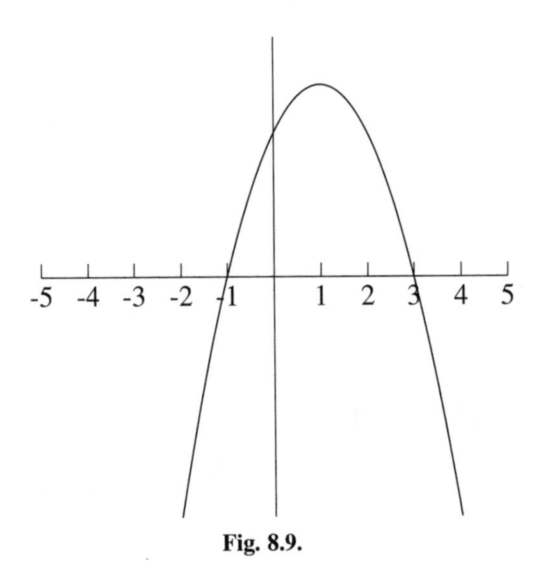

Fig. 8.9.

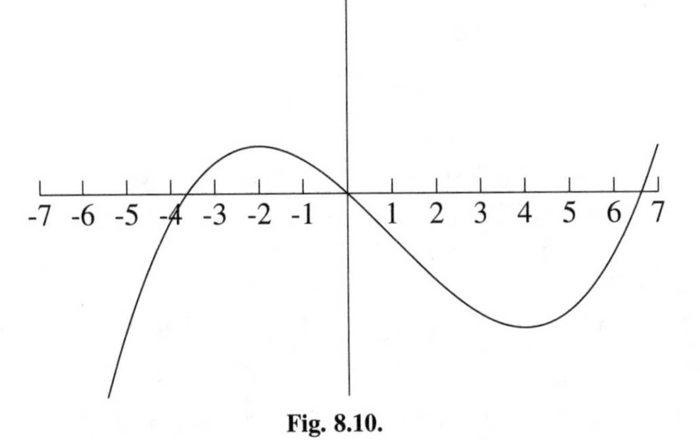

Fig. 8.10.

- $f'(x)$ is positive to the left of $x = 1$ and negative to the right of $x = 1$. The graph is increasing to the left of $x = 1$ and decreasing to the right of $x = 1$. This describes Figure 8.9.
- $f'(x)$ is positive to the left of $x = -2$, negative between $x = -2$ and $x = 4$, and positive to the right of $x = 4$.
  The graph is increasing to the left of $x = -2$, decreasing between $x = -2$ and $x = 4$, and is increasing to the right of $x = 4$. This describes the graph in Figure 8.10.

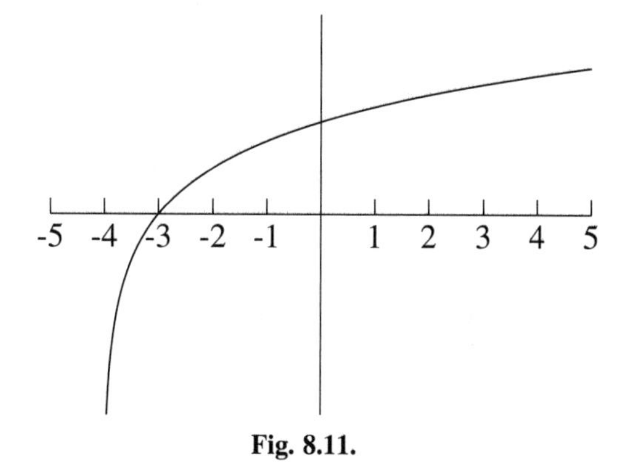

Fig. 8.11.

PRACTICE

For Problems 1–3, determine if the function is increasing or decreasing at the given x-values. For Problems 4–6, match the graphs in Figures 8.12–8.14 with the derivative information.

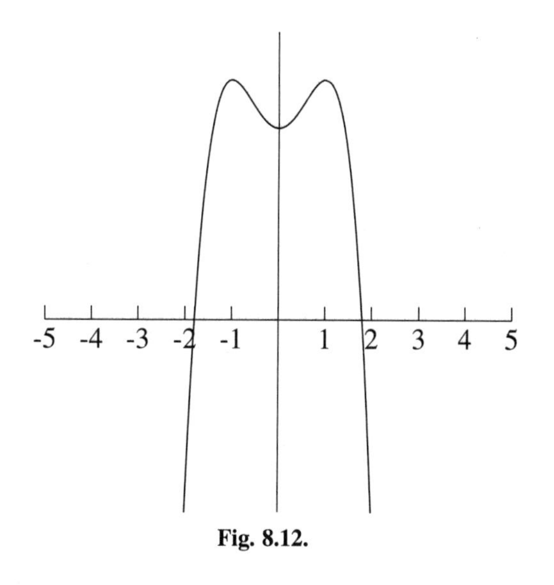

Fig. 8.12.

1. $f(x) = x^4 - 3x^2 - 4$ at $x = 1$ and $x = 3$.
2. $f(x) = \sqrt{x+1}$ at $x = 0$ and $x = 2$.
3. $f(x) = \frac{x}{x^2+1}$ at $x = 0$ and $x = 2$.

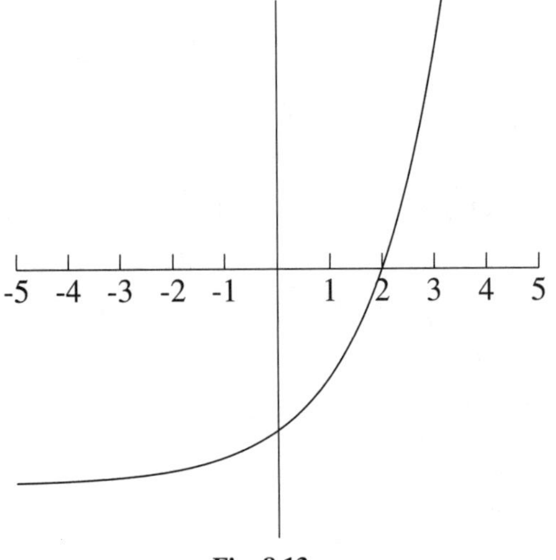

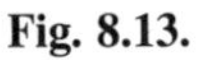
Fig. 8.13.

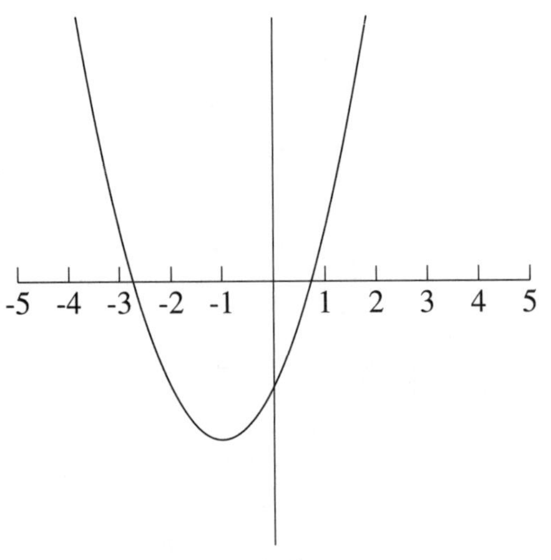

Fig. 8.14.

4. $f'(x)$ is negative to the left of $x = -1$ and positive to the right of $x = -1$.
5. $f'(x)$ is positive for all x.
6. $f'(x)$ is positive to the left of $x = -1$, negative between $x = -1$ and $x = 0$, positive between $x = 0$ and $x = 1$, and negative to the right of $x = 1$.

SOLUTIONS

1. $f'(x) = 4x^3 - 6x$

$$f'(1) = 4(1)^3 - 6(1) = -2 \qquad \text{Decreasing}$$

$$f'(3) = 4(3)^3 - 6(3) = +90 \qquad \text{Increasing}$$

2. $f(x) = (x + 1)^{1/2}$

$$f'(x) = \frac{1}{2}(x + 1)^{-1/2}(1) = \frac{1}{2}\frac{1}{(x + 1)^{1/2}} = \frac{1}{2\sqrt{x + 1}}$$

$$f'(0) = \frac{1}{2\sqrt{0 + 1}} = +\frac{1}{2} \qquad \text{Increasing}$$

$$f'(2) = \frac{1}{2\sqrt{2 + 1}} = +\frac{1}{2\sqrt{3}} \qquad \text{Increasing}$$

3.

$$f'(x) = \frac{1(x^2 + 1) - x(2x)}{(x^2 + 1)^2}$$

$$= \frac{x^2 + 1 - 2x^2}{(x^2 + 1)^2} = \frac{-x^2 + 1}{(x^2 + 1)^2}$$

$$f'(0) = \frac{-0^2 + 1}{(0^2 + 1)^2} = +1 \qquad \text{Increasing}$$

$$f'(2) = \frac{-2^2 + 1}{(2^2 + 1)^2} = \frac{-3}{25} \qquad \text{Decreasing}$$

4. Figure 8.14
5. Figure 8.13
6. Figure 8.12

Rather than say, "$f'(x)$ is positive to the right of $x = 4$," we use mathematical notation. The expression "$f'(x) > 0$" means the derivative is positive. The expression "$f'(x) < 0$" means the derivative is negative. "To the right of $x = a$" is the interval (a, ∞). "To the left of $x = a$" is the interval $(-\infty, a)$. "Between $x = a$ and $x = b$" is the interval (a, b).

Later, we will construct a *sign graph* for the derivative to help us sketch the graph of a function. For the moment, we will see that the sign graph shows us where a function is increasing and where it is decreasing. A sign graph is a number line with plus and/or minus signs on it. The plus signs show the intervals where the derivative is positive. The minus signs show the intervals where the derivative is negative. The sign graph for $f(x) = 4x^3 + 15x^2 - 18x + 6$ is shown in Figure 8.15. This sign graph tells us that the function is increasing to the left of $x = -3$ (the interval $(-\infty, -3)$) because the derivative is positive. The function is decreasing between $x = -3$ and $x = \frac{1}{2}$ (the interval $(-3, 1/2)$) because the derivative is negative. The function is increasing to the right of $x = \frac{1}{2}$ (the interval $(1/2, \infty)$) because the derivative is positive.

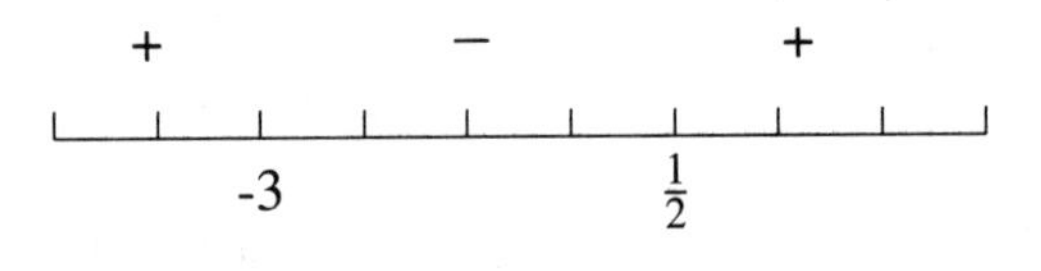

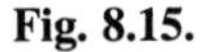

Fig. 8.15.

When a continuous function changes from increasing to decreasing, its graph reaches what we call a *relative maximum.* This is a point that is the highest of the points around it. When a continuous function changes from decreasing to increasing, its graph reaches a *relative miminum.* The graph in Figure 8.16 is the graph of $f(x) = 4x^3 + 15x^2 - 18x + 6$, whose sign graph is given in Figure 8.15. There is a relative maximum at $x = -3$ and a relative minimum at $x = \frac{1}{2}$.

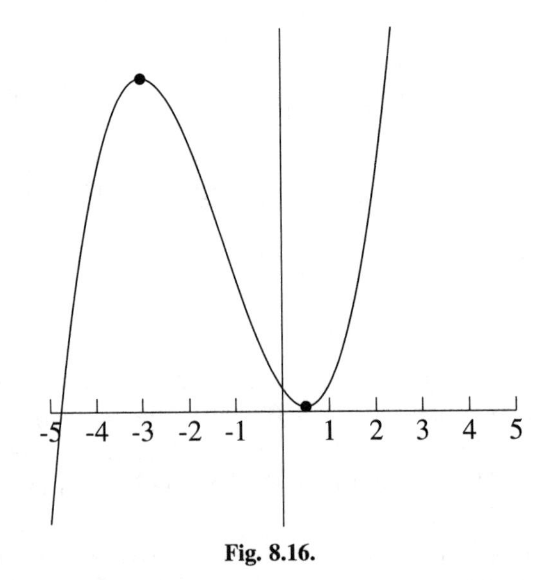

Fig. 8.16.

Another name for the relative minimum is *local* minimum. Similarly, a relative maximum is also called local maximum. Together, they are called *relative extrema,* or local extrema. A graph can have more than one relative maximum or minimum or it might have neither.

EXAMPLES

Determine where the relative extrema are for the graphs in Figures 8.9–8.11.

- The graph in Figure 8.9 has a relative maximum at $x = 1$.
- The graph in Figure 8.10 has a relative maximum at $x = -2$ and a relative minimum at $x = 4$.
- The graph in Figure 8.11 has no relative extrema because it is always increasing.

PRACTICE

Determine where the relative extrema are for the graphs in Figures 8.12–8.14.

1. Refer to Figure 8.12.
2. Refer to Figure 8.13.
3. Refer to Figure 8.14.

SOLUTIONS

1. The graph has a relative maximum at $x = -1$ and another at $x = 1$. It has a relative minimum at $x = 0$.
2. The graph has no relative extrema.
3. The graph has a relative minimum at $x = -1$.

Below are some formal definitions for the ideas in this chapter. Suppose $f(x)$ is a function that is defined on an interval. This interval could be any one of (a, b), $[a, b]$, $(a, b]$, $[a, b)$, $(-\infty, a)$, $(-\infty, a]$, (b, ∞), or $[b, \infty)$.

Definition $f(x)$ is increasing on an interval if for every a and b in the interval, with $a < b$, $f(a) < f(b)$.

This is a formal way of saying that as x gets larger, y gets larger, too.

Definition $f(x)$ is decreasing on an interval if for every a and b in the interval, with $a < b$, $f(a) > f(b)$.

This means that as x gets larger, y gets smaller.

Definition Let c be in the interval. Then $f(c)$ is a relative minimum if $f(c) \leq f(x)$ for every x in the interval.

This means that $f(c)$ is the smallest y-value for all x-values in the interval.

> **Definition** Let c be in the interval. Then $f(c)$ is a relative maximum if $f(c) \geq f(x)$ for every x in the interval.

This means that $f(c)$ is the largest y-value for all x-values in the interval.

If $f(x)$ is also differentiable (the derivative exists on the entire interval) on an open interval (the open intervals are $(-\infty, a)$, (a, ∞), and (a, b)), then we can revise the definitions for increasing and decreasing.

> **Definition** $f(x)$ is increasing on an open interval if $f'(x) > 0$ for every x in the interval. $f(x)$ is decreasing on an open interval if $f'(x) < 0$ for every x in the interval.

What happens when $f'(a) = 0$? It usually means that the function has a relative maximum or a relative minimum at $x = a$. If the tangent line is horizontal, it is likely that the point is a relative extremum. If $f'(a) = 0$, then we call $x = a$ a *critical value* (see Figures 8.17 and 8.18).

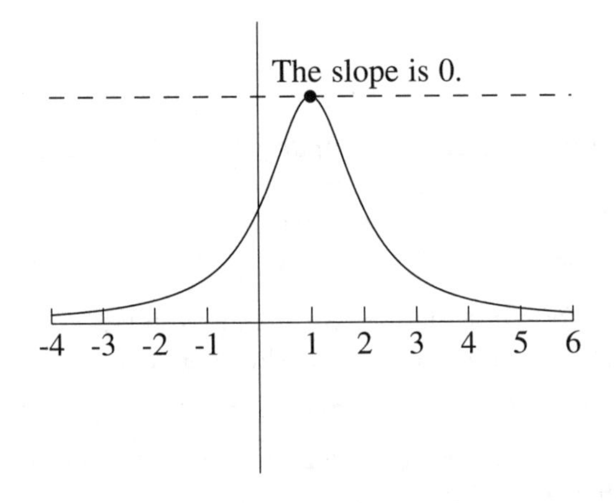

Fig. 8.17.

We can easily find the relative extrema of a function by finding where its derivative is zero. Unfortunately, this will not be enough because the derivative can be zero at points that are not relative extrema. For example, the function $f(x) = x^3 - 3x^2 + 3x$ has no relative extrema but $f'(1) = 0$.

$$f'(x) = 3x^2 - 6x + 3 \text{ and } f'(1) = 3(1)^2 - 6(1) + 3 = 0$$

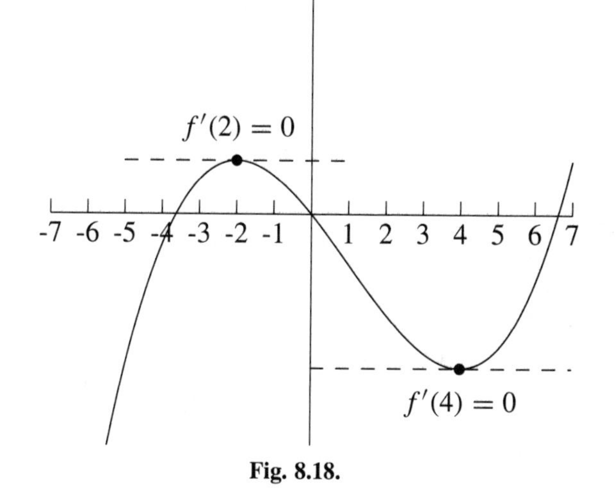

Fig. 8.18.

Figure 8.19 shows the graph of this function. The function is increasing both to the left and right of $x = 1$. Its sign graph is shown in Figure 8.20.

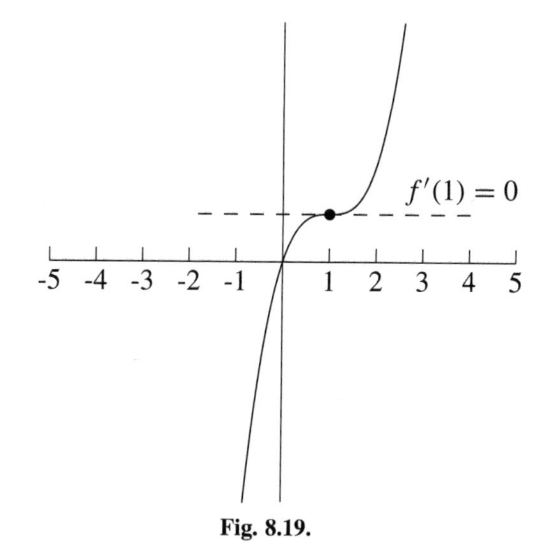

Fig. 8.19.

If the sign graph changes from plus to minus at a critical value, then the graph of the function has a relative maximum there. If the sign graph changes from minus to plus, then the graph of the function has a relative minimum there. Using the sign graph for the derivative to determine where the relative extrema are located is called the *first derivative test*.

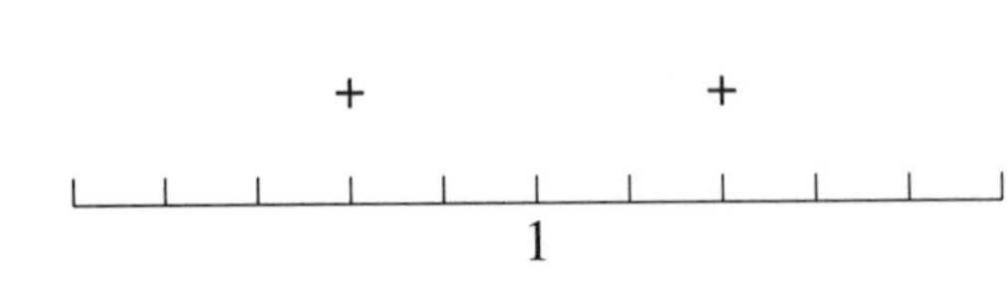

Fig. 8.20.

Step 1 Find the derivative.
Step 2 Find the critical values by setting the derivative equal to zero and solving for x.
Step 3 Begin the sign graph for the derivative by making a number line with the critical value(s) marked on it.
Step 4 Pick any number to the left of the smallest critical value, a number between consecutive critical values (if there are more than one), and number to the right of the largest critical value.
Step 5 Put these numbers in the derivative. If the derivative is positive, put a plus sign over the interval of the test point. If it is negative, put a minus sign over the interval.
Step 6 Identify which value or values, if any, is or are relative extrema. When the sign changes from plus to minus, we have a relative maximum. When the sign changes from minus to plus, we have a relative minimum.
Step 7 Compute the y-value(s) by putting the critical value(s) in the *original* function. The y-values are the extrema.

Later we will see that there are some functions that have no critical values, so there will be no relative extrema.

EXAMPLES

Use the first derivative test to find the local extrema.

- $f(x) = x^3 - 3x - 1$
  Step 1 $f'(x) = 3x^2 - 3$
  Step 2 $0 = 3x^2 - 3$

$$0 = 3(x^2 - 1) = 3(x + 1)(x - 1)$$

$$x + 1 = 0 \qquad x - 1 = 0$$

$$x = -1 \qquad x = 1$$

The critical values are -1 and 1.

Step 3 We mark the critical values on the number line (see Figure 8.21).

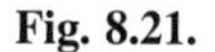

Fig. 8.21.

Step 4 We will use -2 for our number to the left of -1; 0, for our number between -1 and 1; and 2, for our number to the right of 1.
Step 5 Put $x = -2$, 0, and 2 in $f'(x) = 3x^2 - 3$ to see where the derivative is positive or negative: $f'(-2) = 3(-2)^2 - 3 = +9$. Put a plus sign to the left of -1 on the sign graph (Figure 8.22). $f'(0) = 3(0)^2 - 3 = -3$. Put a minus sign between -1 and 1 on the sign graph. $f'(2) = 3(2)^2 - 3 = +9$. Put a plus sign to the right of 1 on the sign graph.

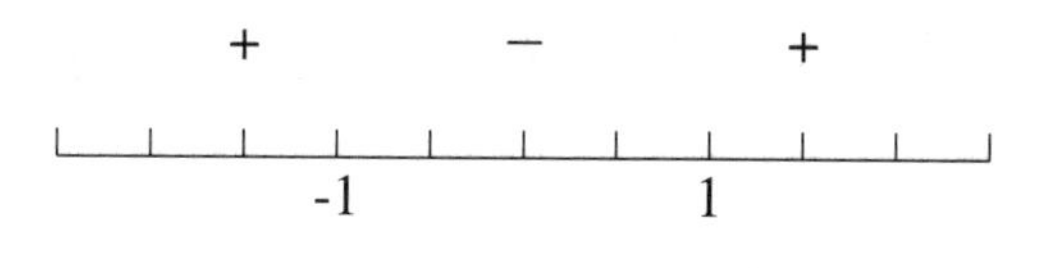

Fig. 8.22.

Step 6 The sign changes from plus to minus at $x = -1$, so there is a relative maximum at $x = -1$. The sign changes from minus to plus at $x = 1$, so there is a relative minimum at $x = 1$.
Step 7 Evaluate $f(x) = x^3 - 3x - 1$ at $x = -1$ and $x = 1$.

$$f(-1) = (-1)^3 - 3(-1) - 1 = 1 \quad \text{Relative maximum}$$

$$f(1) = (1)^3 - 3(1) - 1 = -3 \quad \text{Relative minimum}$$

- $f(x) = 3x^4 + 8x^3 - 4$

Step 1 $f'(x) = 12x^3 + 24x^2$
Step 2

$$0 = 12x^3 + 24x^2$$

$$0 = 12x^2(x + 2)$$

$$12x^2 = 0 \qquad x + 2 = 0$$

$$x = 0 \qquad x = -2$$

Step 3 Mark the critical values on the number line (Figure 8.23).

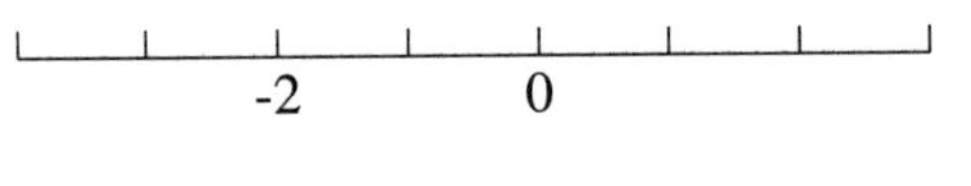

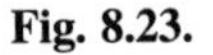
Fig. 8.23.

Step 4 We will use -3 (left of -2), -1 (between -2 and 0), and 1 (right of 0).
Step 5 Evaluate $f'(x) = 12x^3 + 24x^2$ at -3, -1, and 1.

$f'(-3) = 12(-3)^3 + 24(-3)^2 = -108$ Put a minus sign to the left of -2.

$f'(-1) = 12(-1)^3 + 24(-1)^2 = +12$ Put a plus sign between -2 and 0.

$f'(1) = 12(1)^3 + 24(1)^2 = +36$ Put a plus sign to the right of 0.

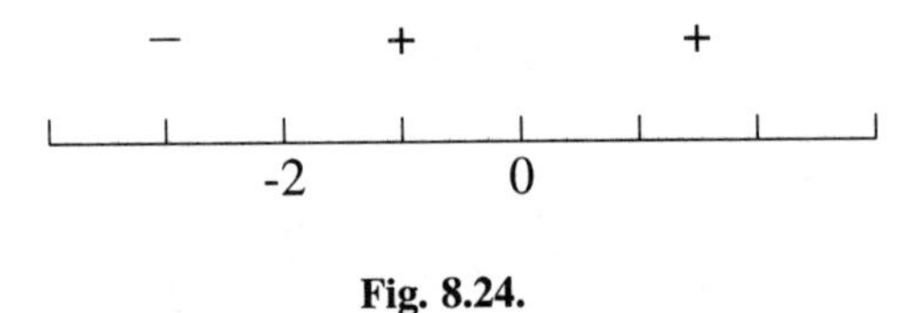

Fig. 8.24.

Step 6 There is only one sign change, so there is only one relative extremum. The sign changes from minus to plus at -2, so there is a relative minimum at $x = -2$.
Step 7 Evaluate the original function at $x = -2$: $f(-2) = 3(-2)^4 + 8(-2)^3 - 4 = -20$. The relative minimum is -20.

PRACTICE

Use the first derivative test to find the local extrema.

1. $f(x) = x^4 - 2x^2 + 3$
2. $f(x) = x^3 + 3x^2 + 3x - 4$
3. $f(x) = \sqrt[3]{(x+1)^4}$

SOLUTIONS

1.

$$f'(x) = 4x^3 - 4x$$

$$0 = 4x^3 - 4x = 4x(x^2 - 1) = 4x(x + 1)(x - 1)$$

$$4x = 0 \qquad x + 1 = 0 \qquad x - 1 = 0$$

$$x = 0 \qquad x = -1 \qquad x = 1$$

The critical values are -1, 0, and 1. We use -2, -0.5, 0.5, and 2 as test points for $f'(x)$.

$f'(-2) = 4(-2)^3 - 4(-2) = -24$	Put "–" to the left of -1.
$f'(-0.5) = 4(-0.5)^3 - 4(-0.5) = +1.5$	Put "+" between -1 and 0.
$f'(0.5) = 4(0.5)^3 - 4(0.5) = -1.5$	Put "–" between 0 and 1.
$f'(2) = 4(2)^3 - 4(2) = +24$	Put "+" to the right of 1.

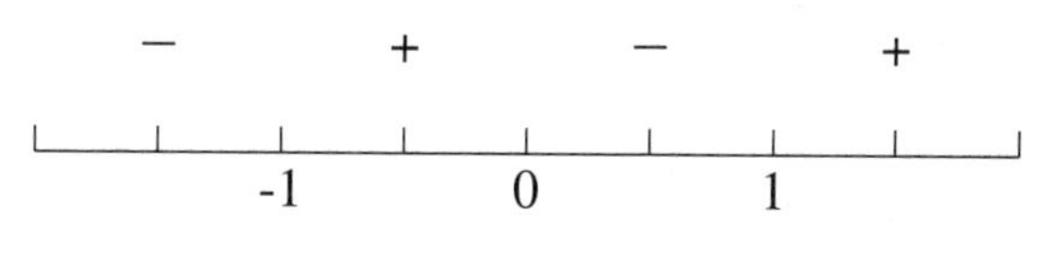

Fig. 8.25.

There is a relative minimum at $x = -1$, a relative maximum at $x = 0$, and a relative minimum at $x = 1$. We will find the relative extrema for this function by putting -1, 0, and 1 in the original function.

$f(-1) = (-1)^4 - 2(-1)^2 + 3 = 2$	The relative minimum is 2.
$f(0) = 0^4 - 2(0)^2 + 3 = 3$	The relative maximum is 3.
$f(1) = (1)^4 - 2(1)^2 + 3 = 2$	The relative minimum is 2.

2.

$$f'(x) = 3x^2 + 6x + 3$$

$$0 = 3x^2 + 6x + 3 = 3(x^2 + 2x + 1) = 3(x + 1)(x + 1)$$

$$x + 1 = 0$$

$$x = -1$$

The only critical value is -1. We will use -2 and 0 as test points in the derivative (Figure 8.26).

$$f'(-2) = 3(-2)^2 + 6(-2) + 3 = +3$$

$$f'(0) = 3(0)^2 + 6(0) + 3 = +3$$

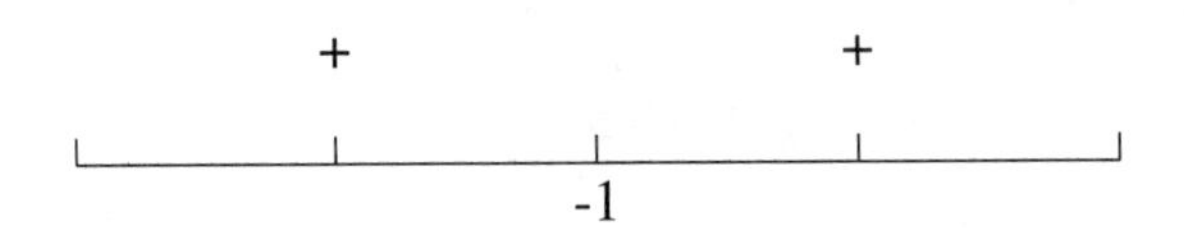

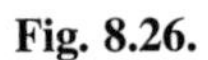

Fig. 8.26.

The signs do not change, so there are no relative extrema.

3. $f(x) = (x + 1)^{4/3}$

$$f'(x) = \frac{4}{3}(x + 1)^{4/3-1} = \frac{4}{3}(x + 1)^{1/3} = \frac{4}{3}\sqrt[3]{x + 1}$$

$$0 = \frac{4}{3}\sqrt[3]{x + 1}$$

$$0 = \sqrt[3]{x + 1}$$

$$0 = x + 1$$

$$-1 = x$$

The critical value is -1. We will test -2 and 0 in the derivative (Figure 8.27).

$$f'(-2) = \frac{4}{3}\sqrt[3]{-2 + 1} = \frac{4}{3}\sqrt[3]{-1} = -\frac{4}{3}$$

$$f'(0) = \frac{4}{3}\sqrt[3]{0 + 1} = \frac{4}{3}\sqrt{1} = \frac{4}{3}$$

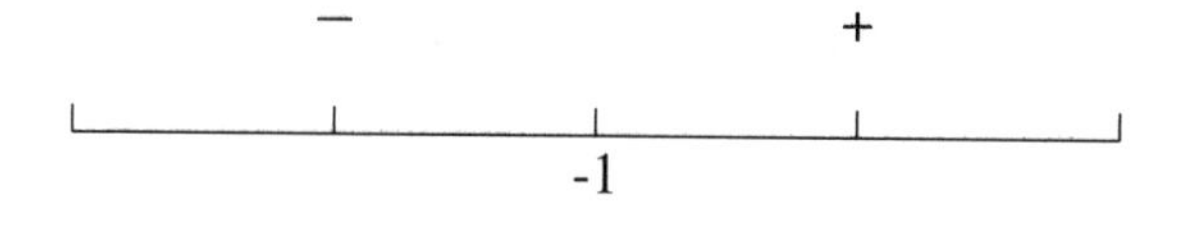

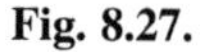

Fig. 8.27.

There is a relative minimum at $x = -1$.
The relative minimum is $f(-1) = \sqrt[3]{(-1+1)^4} = \sqrt[3]{0} = 0$.

While most critical values will come from solving $f'(x) = 0$, a critical value can occur where $f'(x)$ does not exist (but where $f(x)$ does exist). Often this happens when the derivative has a variable in the denominator. To find these critical values, we will set the denominator equal to zero and solve for x.

EXAMPLE

- $f(x) = \sqrt[3]{(x+1)^2}$

$$f(x) = (x+1)^{2/3} \quad f'(x) = \frac{2}{3}(x+1)^{-1/3} = \frac{2}{3}\frac{1}{(x+1)^{1/3}} = \frac{2}{3\sqrt[3]{x+1}}$$

Set the denominator of $f'(x)$ equal to zero to find where $f'(x)$ does not exist.

$$0 = 3\sqrt[3]{x+1}$$

$$0 = \sqrt[3]{x+1}$$

$$0 = x+1$$

$$-1 = x$$

The derivative does not exist at $x = -1$ but the original function is defined for $x = -1$. We will test -2 and 0 in the derivative to see where $f'(x)$ is positive and negative.

$$f'(-2) = \frac{2}{3\sqrt[3]{-2+1}} = \frac{2}{3\sqrt[3]{-1}} = \frac{2}{3(-1)} = -\frac{2}{3}$$

$$f'(0) = \frac{2}{3\sqrt[3]{0+1}} = \frac{2}{3\sqrt[3]{1}} = \frac{2}{3}$$

Because the derivative changes from negative to positive at $x = -1$, we have a relative minimum at -1. The relative minimum of this function is $f(-1) = (-1+1)^{2/3} = \sqrt[3]{0^2} = 0$.

Graphing Functions

Critical values are important points on the graph of a function. Because of this, we want to plot a point for each critical value when sketching the graph of a function. For now, we will sketch the graphs of polynomial functions. We can accurately sketch the graph of a polynomial function by plotting a point for each critical value, a point to the left of the smallest critical value, a point between consecutive critical values (if there are more than one), and a point to the right of the largest critical value.

EXAMPLES

Sketch the graph of the polynomial function.

- $f(x) = x^3 - 3x - 5$

  We will begin by finding the critical values.

$$f'(x) = 3x^2 - 3$$

$$0 = 3x^2 - 3 = 3(x^2 - 1) = 3(x+1)(x-1)$$

$$x + 1 = 0 \qquad x - 1 = 0$$

$$x = -1 \qquad x = 1$$

  We will plot a point for $x = -2$ (for our point to the left of $x = -1$), $x = 0$ (for our point between $x = -1$ and $x = 1$), and $x = 2$ (for our point to the right of $x = 1$) (Table 8.1). The graph can then be sketched as shown in Figure 8.28.

PRACTICE

Sketch the graph of the polynomial functions.

1. $f(x) = \frac{1}{4}x^4 - \frac{1}{3}x^3 - 3x^2$
2. $f(x) = 3x^5 - 5x^3$

Table 8.1

x	$y = x^3 - 3x - 5$	Plot this point
-2	$y = (-2)^3 - 3(-2) - 5 = -7$	$(-2, -7)$
-1	$y = (1)^3 - 3(1) - 5 = -3$	$(-1, -3)$
0	$y = (0)^3 - 3(0) - 5 = -5$	$(0, -5)$
1	$y = (1)^3 - 3(1) - 5 = -7$	$(1, -7)$
2	$y = (2)^3 - 3(2) - 5 = -3$	$(2, -3)$

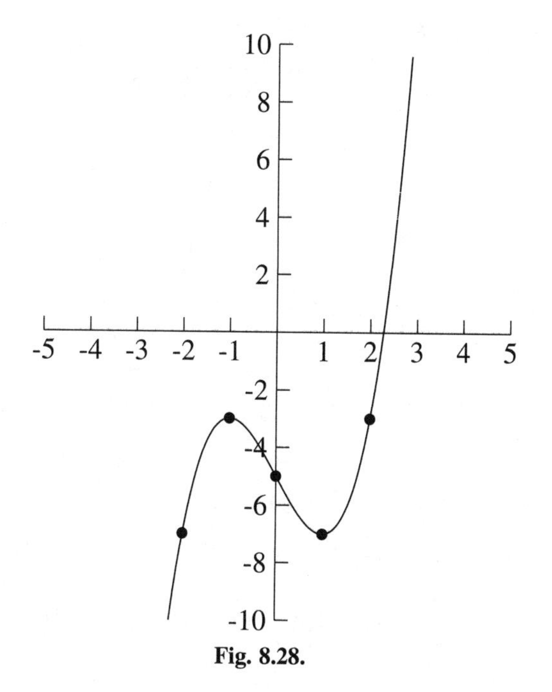

Fig. 8.28.

SOLUTIONS

1.

$$f'(x) = \frac{1}{4}(4)x^3 - \frac{1}{3}(3)x^2 - 6x = x^3 - x^2 - 6x$$

$$0 = x^3 - x^2 - 6x = x(x^2 - x - 6) = x(x+2)(x-3)$$

$$x = 0 \qquad x + 2 = 0 \qquad x - 3 = 0$$

$$x = -2 \qquad x = 3$$

We will plot points for $x = -3, \; -2, \; -1, \; 0, \; 1, \; 3,$ and 4 (Figure 8.29).

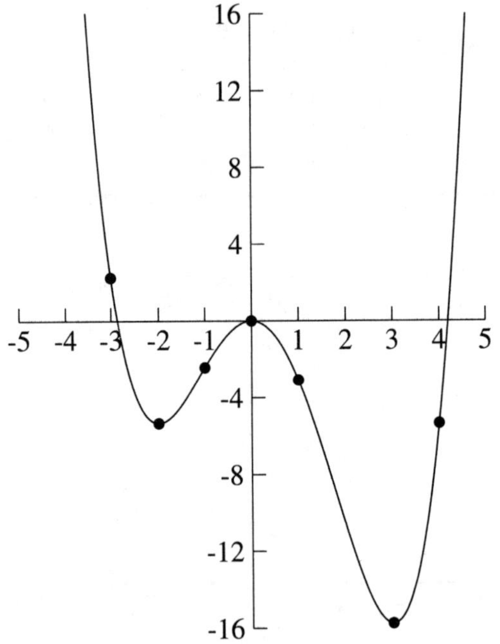

Fig. 8.29.

2.

$$f'(x) = 15x^4 - 15x^2$$

$$0 = 15x^4 - 15x^2$$

$$= 15x^2(x^2 - 1) = 15x^2(x + 1)(x - 1)$$

$$15x^2 = 0 \qquad x + 1 = 0 \qquad x - 1 = 0$$

$$x = 0 \qquad x = -1 \qquad x = 1$$

We will plot points for $x = -1.5, \; -1, \; -0.5, \; 0, \; 0.5, \; 1,$ and 1.5 (Figure 8.30).

Absolute Extrema

Most functions we study in calculus do not have a highest point or a lowest point on their graphs, but some do. In this section we will look at the *absolute*

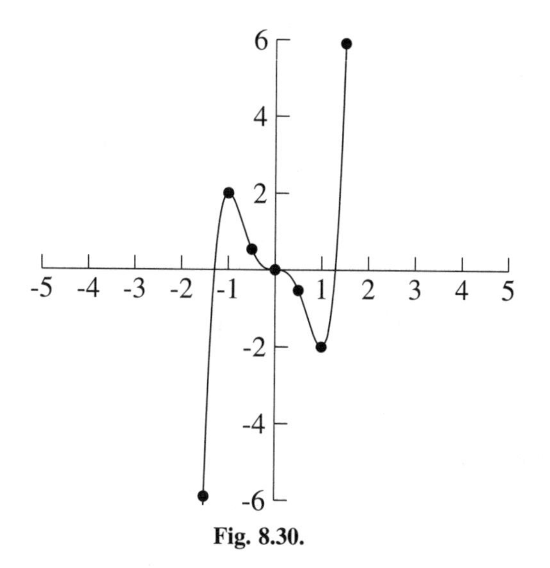

Fig. 8.30.

extrema of a function. A number is an absolute maximum for a function if it is the highest y-value. A number is the absolute minimum if it is the lowest y-value. The function whose graph is shown in Figure 8.31 has an absolute maximum but no absolute minimum. The function whose graph is shown in Figure 8.32 has both an absolute maximum and an absolute minimum.

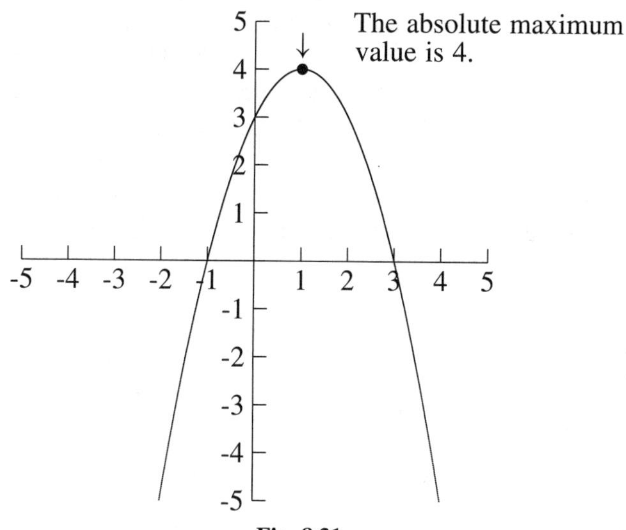

Fig. 8.31.

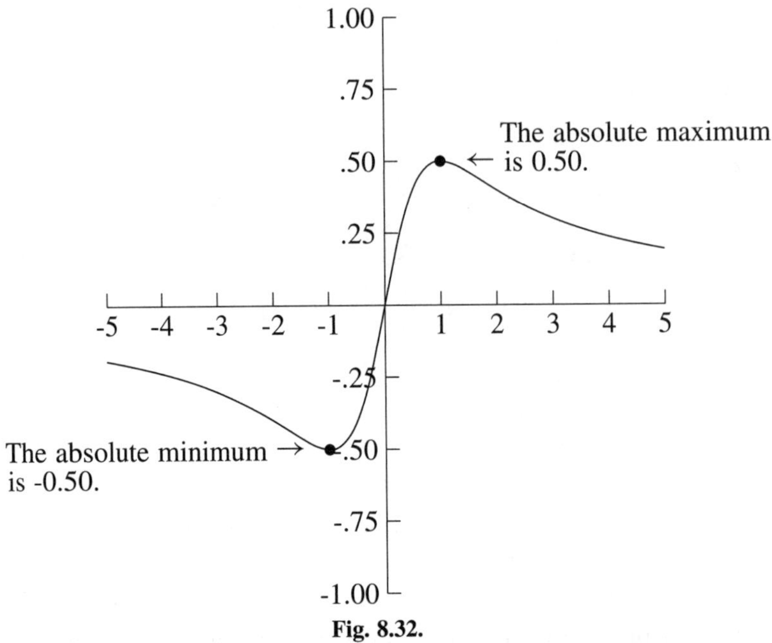

Fig. 8.32.

When a function is continuous on a closed interval, it *will* have an absolute minimum value and an absolute maximum value. In other words, its graph will have a highest point and a lowest point. The graph for a continuous function on a closed interval has a solid dot at each end indicating the left-most and right-most points. The absolute extrema occur at one or both endpoints or somewhere in between. The extrema that occur between the endpoints will be critical values for the derivative. The absolute maximum and absolute minimum occur at the endpoints for the graph in Figure 8.33. The absolute minimum occurs at the endpoints for the graph in Figure 8.34 and the absolute maximum occurs at a relative maximum (where $f'(x) = 0$). The absolute maximum and absolute minimum occur at the relative maximum and relative minimum for the graph in Figure 8.35.

PRACTICE

Identify the absolute extrema from the graph (see Figures 8.36, 8.37, and 8.38).

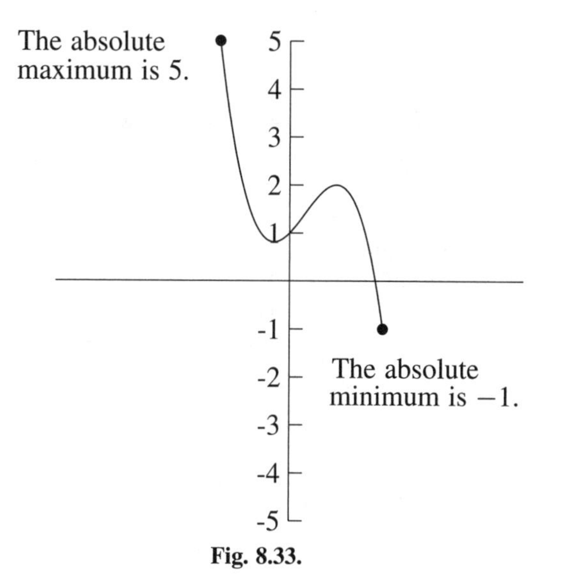

Fig. 8.33.

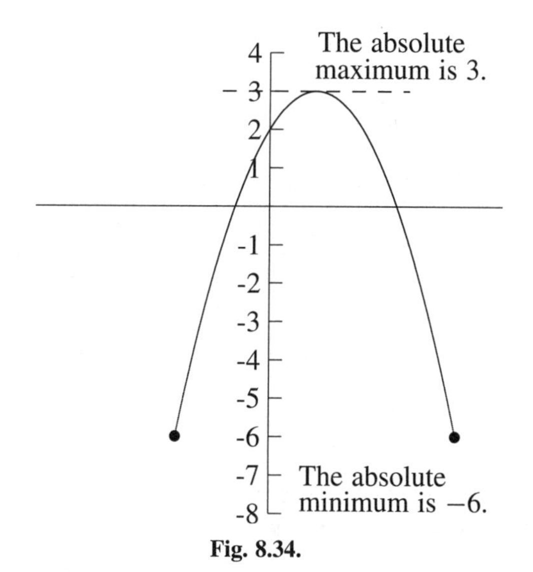

Fig. 8.34.

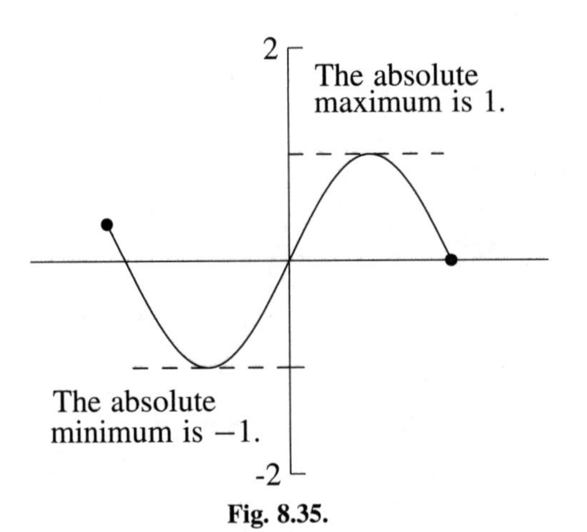

Fig. 8.35.

1.

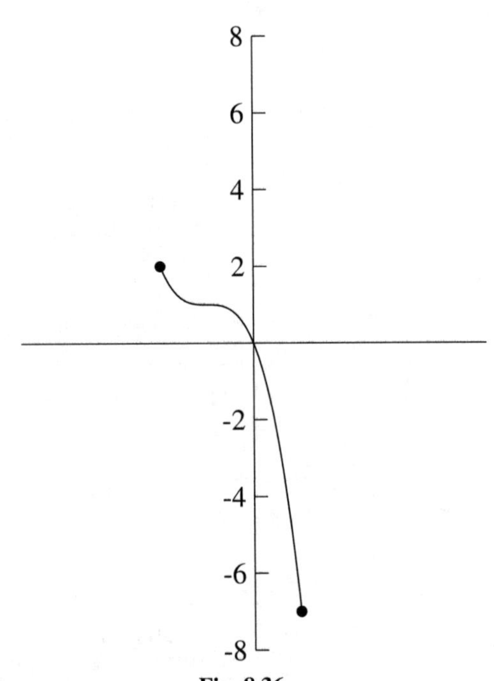

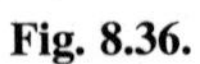
Fig. 8.36.

2.

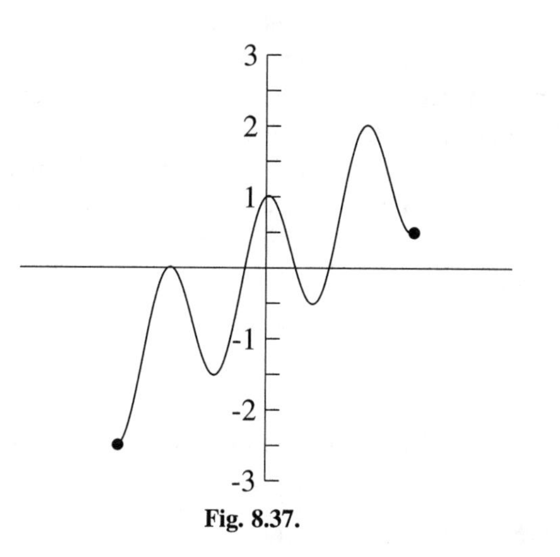

Fig. 8.37.

3.

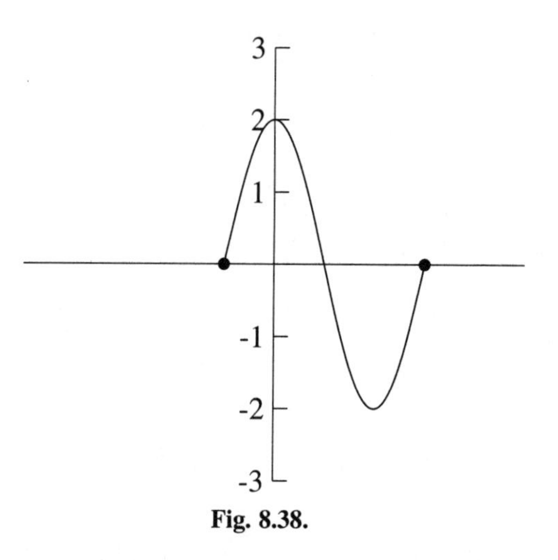

Fig. 8.38.

SOLUTIONS

1. The absolute maximum is 2, and the absolute minimum is -7.
2. The absolute maximum is 2, and the absolute minimum is -2.5.
3. The absolute maximum is 2, and the absolute minimum is -2.

We can find the absolute extrema of a function on a closed interval by finding the y-values for the endpoints of the interval as well as the y-values at any of the critical values between the endpoints. In the following problems, we will be given a function and a closed interval, $[a, b]$. We will be asked to find the absolute extrema. We will begin by putting $x = a$ and $x = b$ in $f(x)$ to find the y-values at the endpoints. After this, we will find $f'(x)$ and set it equal to zero. We only want the solution(s) that are between a and b (if there are any). We will put this number or these numbers in $f(x)$. Finally, we will observe which y-value is the largest and which is the smallest.

EXAMPLES

Find the absolute extrema for the function on the given interval.

- $f(x) = x^3 - 4x^2 - 3x + 7$ on $[0, 6]$
  We will begin by finding the y-values for $x = 0$ and $x = 6$.

$$f(0) = 0^3 - 4(0)^2 - 3(0) + 7 = 7$$

$$f(6) = 6^3 - 4(6)^2 - 3(6) + 7 = 61$$

Now we want to find any critical values for $f'(x)$ that are between $x = 0$ and $x = 6$.

$$f'(x) = 3x^2 - 8x - 3$$

$$0 = 3x^2 - 8x - 3 = (3x + 1)(x - 3)$$

$$3x + 1 = 0 \qquad x - 3 = 0$$

$$x = -\frac{1}{3} \qquad x = 3$$

Because $-\frac{1}{3}$ is not between 0 and 6, we do not need it. We do need 3.

$$f(3) = 3^3 - 4(3)^2 - 3(3) + 7 = -11$$

Of the y-values -11, 7, and 61, -11 is the smallest, and 61 is the largest. The absolute minimum value of the function on $[0, 6]$ is -11 (which occurs at $x = 3$), and the absolute maximum value is 61 (which occurs at $x = 6$).

- $f(x) = x^2 - 6x + 3$ on $[-1, 2]$

$$f(-1) = (-1)^2 - 6(-1) + 3 = 10$$

$$f(2) = (2)^2 - 6(2) + 3 = -5$$

$$f'(x) = 2x - 6$$

$$0 = 2x - 6$$

$$3 = x$$

Because 3 is outside the interval $[-1, 2]$, we cannot use it. The absolute minimum value of the function on the interval $[-1, 2]$ is -5 (which occurs at $x = 2$), and the absolute maximum value is 10 (which occurs at $x = -1$).

- $f(x) = \sqrt[3]{x^2}$ on $[-8, 1]$

$$f(-8) = \sqrt[3]{(-8)^2} = \sqrt[3]{64} = 4 \qquad f(1) = \sqrt[3]{1^2} = \sqrt[3]{1} = 1$$

$$f(x) = x^{2/3}, \text{ so, } f'(x) = \frac{2}{3}x^{-1/3} = \frac{2}{3\sqrt[3]{x}}$$

$f'(x)$ is never 0, but $f'(x)$ does not exist at $x = 0$, making 0 a critical value: $f(0) = \sqrt[3]{0^2} = 0$. The absolute minimum value of the function on $[-8, 1]$ is 0 (which occurs at $x = 0$) and the absolute maximum value is 4 (which occurs at $x = -8$).

PRACTICE

Find the absolute extrema for the function on the given interval.

1. $f(x) = 4x^3 - 21x^2 - 24x + 10$ on $[-1, 2]$
2. $f(x) = x^2 + 8x + 17$ on $[-5, 3]$
3. $f(x) = 9x - 6$ on $[3, 5]$
4. $f(x) = \frac{x}{x^2+x+1}$ at $[-2, 2]$

SOLUTIONS

1.

$$f(-1) = 4(-1)^3 - 21(-1)^2 - 24(-1) + 10 = 9$$

$$f(2) = 4(2)^3 - 21(2)^2 - 24(2) + 10 = -90$$

$$f'(x) = 12x^2 - 42x - 24$$

$$0 = 12x^2 - 42x - 24$$

$$= 6(2x^2 - 7x - 4) = 6(2x + 1)(x - 4)$$

$$2x + 1 = 0 \qquad x - 4 = 0$$

$$x = -\frac{1}{2} \qquad x = 4$$

We only need $-\frac{1}{2}$ because 4 is not between -1 and 2.

$f(-\frac{1}{2}) = 4(-\frac{1}{2})^3 - 21(-\frac{1}{2})^2 - 24(-\frac{1}{2}) + 10 = \frac{65}{4}$

The absolute minimum value of the function on the interval is -90 (which occurs at $x = 2$), and the absolute maximum value is $\frac{65}{4}$ (which occurs at $x = -\frac{1}{2}$).

2.

$$f(-5) = (-5)^2 + 8(-5) + 17 = 2 \qquad f(3) = 3^2 + 8(3) + 17 = 50$$

$$f'(x) = 2x + 8$$

$$0 = 2x + 8$$

$$-4 = x$$

$$f(-4) = (-4)^2 + 8(-4) + 17 = 1$$

The absolute minimum value of the function on the interval is 1 (which occurs at $x = -4$), and the absolute maximum value is 50 (which occurs at $x = 3$).

3.

$$f(3) = 9(3) - 6 = 21 \qquad f(5) = 9(5) - 6 = 39$$

$$f'(x) = 9$$

Because $f'(x) = 9$ for all x, there are no critical values, so the extrema occur at the endpoints. The absolute minimum value of the function on the interval is 21 (which occurs at $x = 3$), and the absolute maximum value is 39 (which occurs at $x = 5$).

4.

$$f(-2) = \frac{-2}{(-2)^2 + (-2) + 1} = -\frac{2}{3}$$

$$f(2) = \frac{2}{2^2 + 2 + 1} = \frac{2}{7}$$

$$f'(x) = \frac{1(x^2 + x + 1) - x(2x + 1)}{(x^2 + x + 1)^2}$$

$$= \frac{x^2 + x + 1 - 2x^2 - x}{(x^2 + x + 1)^2}$$

$$= \frac{-x^2 + 1}{(x^2 + x + 1)^2} = \frac{1 - x^2}{(x^2 + x + 1)^2}$$

$$1 - x^2 = 0 \qquad (x^2 + x + 1)^2 = 0$$

$$1 = x^2 \qquad x^2 + x + 1 = 0$$

$$\pm 1 = x \qquad \text{No real solution}$$

The critical values are -1 and 1.

$$f(-1) = \frac{-1}{(-1)^2 + (-1) + 1} = -1$$

$$f(1) = \frac{1}{1^2 + 1 + 1} = \frac{1}{3}$$

The absolute minimum value of the function on the interval is -1 (which occurs at $x = -1$), and the absolute maximum value is $\frac{1}{3}$ (which occurs at $x = 1$).

Why is it important that the interval be closed? The y-values can get larger and larger without ever reaching the largest y-value. For example, the function $f(x) = x$ on the open interval $(0, 1)$ has no largest y-value. The y-values include the numbers 0.9, 0.99, 0.999, 0.9999, …, and there is no largest number in the list. (The same is true for the smallest y-value.)

CHAPTER 8 REVIEW

1. What are the critical values for the function $f(x) = 2x^3 + 5x^2 - 4x + 3$?
   (a) -2 and $\frac{1}{3}$
   (b) 2 and $\frac{1}{3}$
   (c) $\frac{67}{6}$ and $\frac{53}{6}$
   (d) $\frac{67}{6}$ and $-\frac{53}{6}$

2. Is the function $f(x) = 2x^3 + 5x^2 - 4x + 3$ increasing, decreasing, or neither at $x = 0$?
   (a) Increasing
   (b) Decreasing
   (c) Neither
   (d) Cannot be determined without a graph

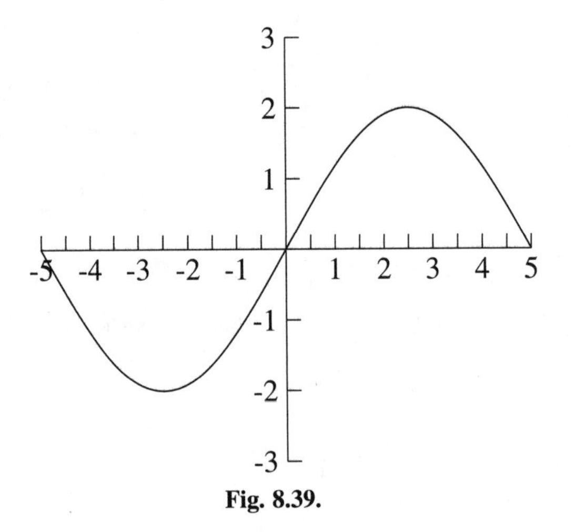

Fig. 8.39.

3. Refer to Figure 8.39. Where is this function increasing?
   (a) $(-2.5, 2)$
   (b) $(-2, 2)$
   (c) $(-\infty, -2.5)$ and $(2.5, \infty)$
   (d) $(-2.5, 2.5)$

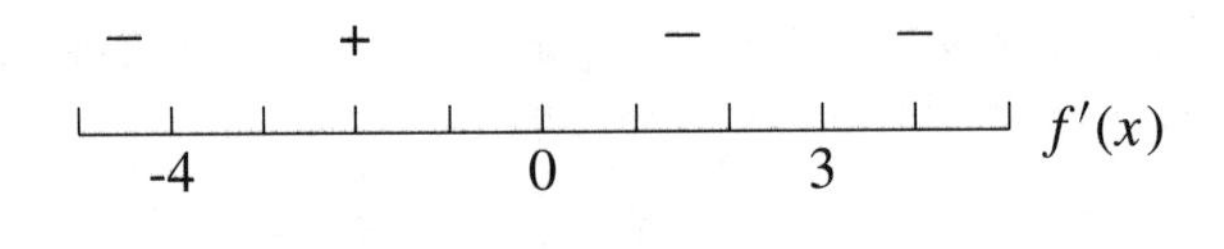

Fig. 8.40.

4. Refer to Figure 8.40, the sign graph for $f'(x)$. Which one of the following is true?
   (a) There is a relative maximum at $x = -4$ and a relative minimum at $x = 0$.

(b) There is a relative maximum at $x = -4$, a relative minimum at $x = 0$, and a relative maximum at $x = 3$.
(c) There is a relative minimum at $x = -4$ and a relative maximum at $x = 0$.
(d) There is a relative minimum at $x = -4$, a relative maximum at $x = 0$, and a relative minimum at $x = 3$.

5. Where is the function $f(x) = \sqrt[3]{(x^2 - 1)^2}$ decreasing?
(a) $(-\infty, 0)$
(b) $(0, \infty)$
(c) $(-\infty, -1)$ and $(0, 1)$
(d) $(-1, 0)$ and $(1, \infty)$

6. Which of the following is *not* true about the function $f(x) = 2x^3 - 3x^2 - 12x + 5$ on the interval $[0, 3]$.
(a) The absolute maximum is 12
(b) The absolute minimum is -15
(c) The absolute maximum is 5
(d) The absolute minimum occurs at $x = 2$

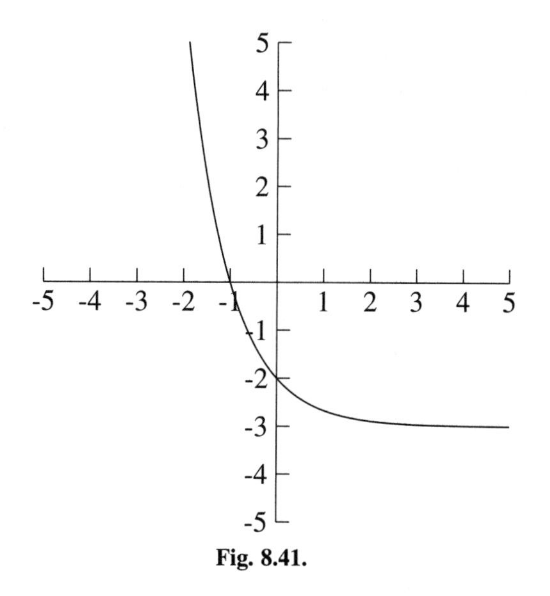

Fig. 8.41.

7. Refer to Figure 8.41. Where is this function decreasing?
(a) Everywhere
(b) Nowhere

(c) $(-1, \infty)$
(d) $(-\infty, -1)$

8. Is there a relative minimum, relative maximum, or neither at $x = 3$ for the function $f(x) = 4x^3 - 17x^2 - 6x + 10$?
(a) Relative minimum
(b) Relative maximum
(c) Neither
(d) Cannot be determined without the graph

9. In order to sketch the graph of $f(x) = 4x^3 - 3x + 5$, we should plot points for which x-values?
(a) $-2, -1, 0, 1, 2$
(b) $-1, 0, 1$
(c) $-\frac{1}{2}, \frac{1}{2}$
(d) $-1, -\frac{1}{2}, 0, \frac{1}{2}, 1$

10. What are the relative extrema for the function $f(x) = \sqrt[3]{(x^2 - 1)^2}$?
(a) There is a relative minimum at $x = 0$.
(b) There is a relative maximum at $x = 0$.
(c) There is a relative maximum at $x = -1$, a relative minimum at $x = 0$, and a relative maximum at $x = 1$.
(d) There is a relative minimum at $x = -1$, a relative maximum at $x = 0$, and a relative minimum at $x = 1$.

SOLUTIONS

1. a	2. b	3. d	4. c	5. c
6. a	7. a	8. a	9. d	10. d

CHAPTER 9

The Second Derivative and Concavity

In the same way the derivative measures how fast a function is changing, the second derivative measures how fast the derivative is changing. Suppose we have a product that sells quickly after its release but the sales taper off. While more are sold as time goes by, the sales per week are dropping off. The sales function would be increasing but the increase is diminishing. The second derivative of the sales function measures how fast sales are diminishing.

Suppose S units are sold x weeks after the product is released, where $S(x) = \frac{400x}{x+10}$, for $x = 0$ to $x = 52$ weeks. In this interval, $S'(x)$ is positive because the longer the product is available, the higher the sales. But the sales rate is slowing down, which is reflected in the fact that $S''(x)$ is negative for this interval.

$$S'(x) = \frac{400(x+10) - 400x(1)}{(x+10)^2} = \frac{4000}{(x+10)^2} = 4000(x+10)^{-2}$$

This is positive for $x = 0$ to $x = 52$.

$$S''(x) = -2(4000)(x+10)^{-3}(1) = -8000(x+10)^{-3} = -\frac{8000}{(x+10)^3}$$

This is negative for $x = 0$ to $x = 52$ (see Figure 9.1).

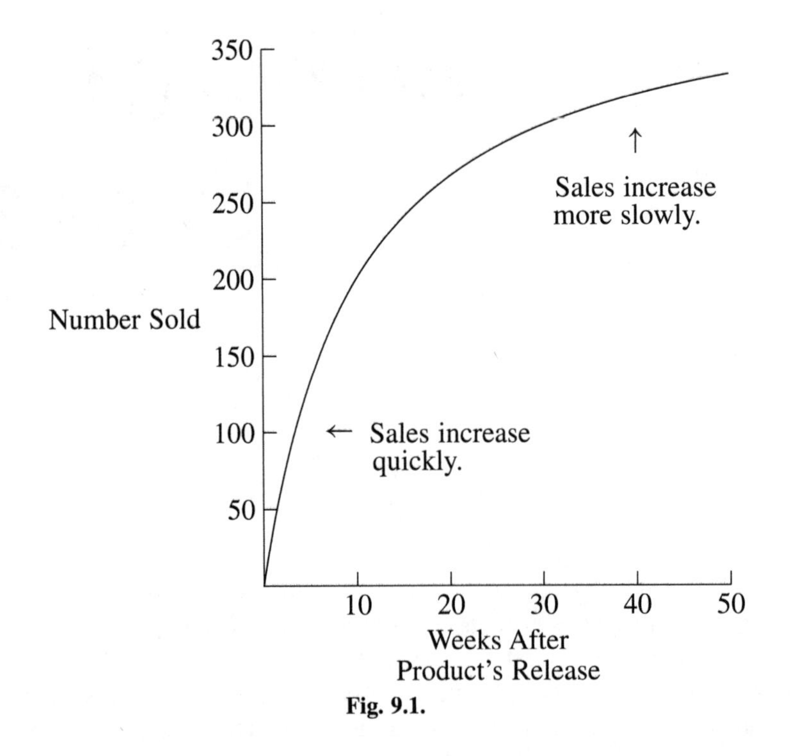

Fig. 9.1.

The relationship between increasing and decreasing intervals and the second derivative is summarized in Figures 9.2–9.5.

The second derivative of a function gives us some information on the shape of the graph of the function. Where the second derivative is positive, the graph cups upward, ⌣. Where the second derivative is negative, the graph cups

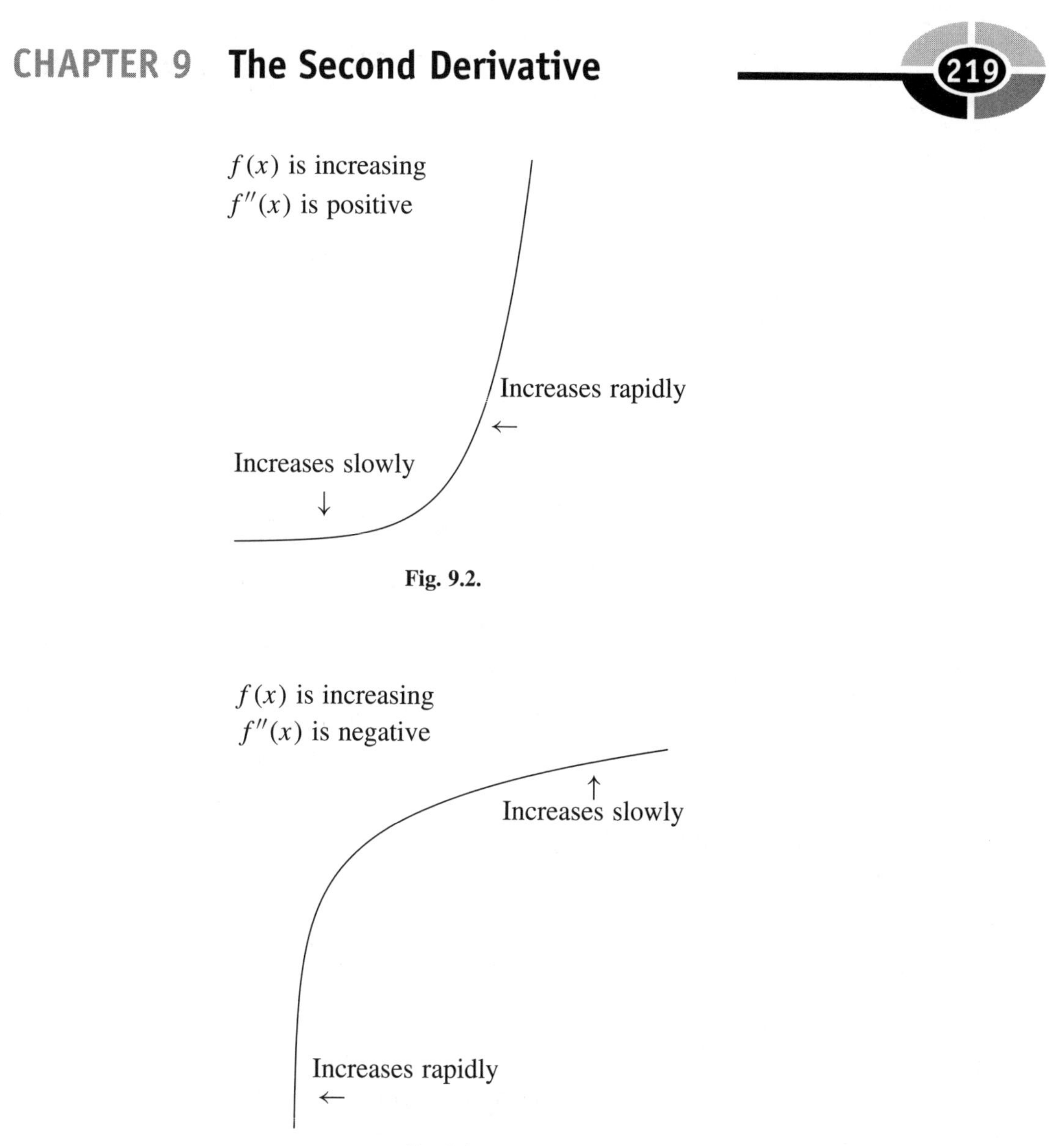

Fig. 9.2.

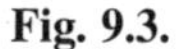

Fig. 9.3.

downward, ⌢. We say the graph is *concave up* where it cups upward and *concave down*, where it cups downward (see Figure 9.6).

The sign graph for $f''(x)$ tells us where a graph is concave up or down in the same way the sign graph for $f'(x)$ told us where a graph was increasing or decreasing. The graph of $f(x) = \frac{1}{2}x^4 - x^3 - 6x^2 + x + 15$ is shown in Figure 9.7, and the sign graph for $f''(x)$ is shown in Figure 9.8.

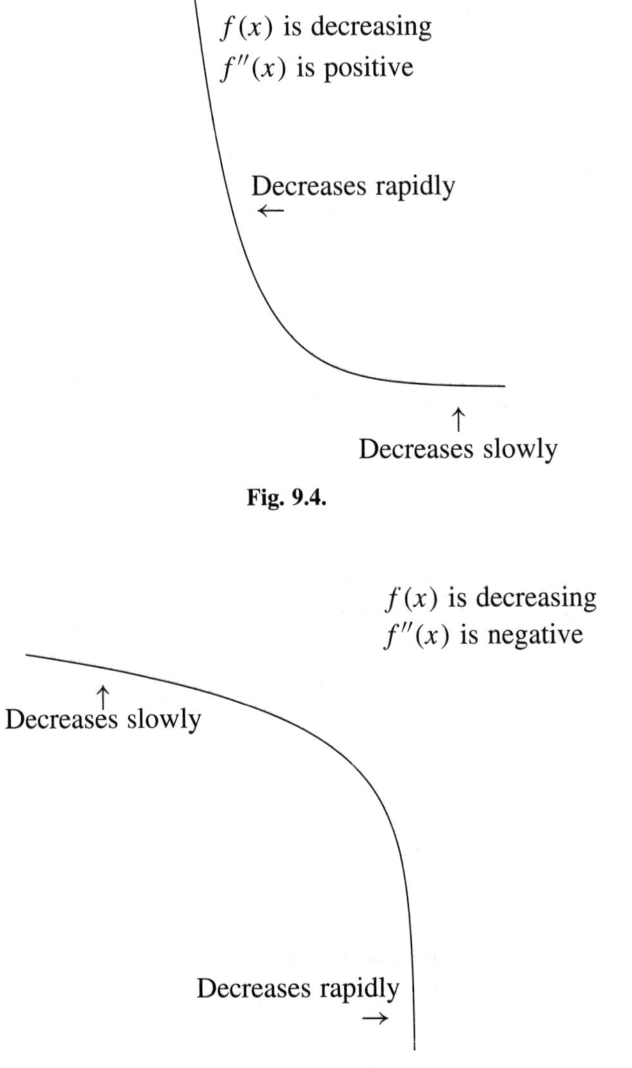

Fig. 9.4.

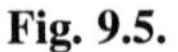

Fig. 9.5.

The sign graph for $f''(x)$ is constructed in the same way as the sign graph for $f'(x)$ is constructed. We begin by finding the first derivative, $f'(x)$, followed by the second derivative, which is the derivative of the derivative. As before, we will set this equal to zero and solve for x. These solutions, if any, are also called critical values. We will choose an x-value to the left of the smallest critical value, one between each pair of consecutive critical values (if there are more than one), and an x-value to the right of the largest critical value. We will test these numbers in the second derivative to see where the second derivative is positive (where

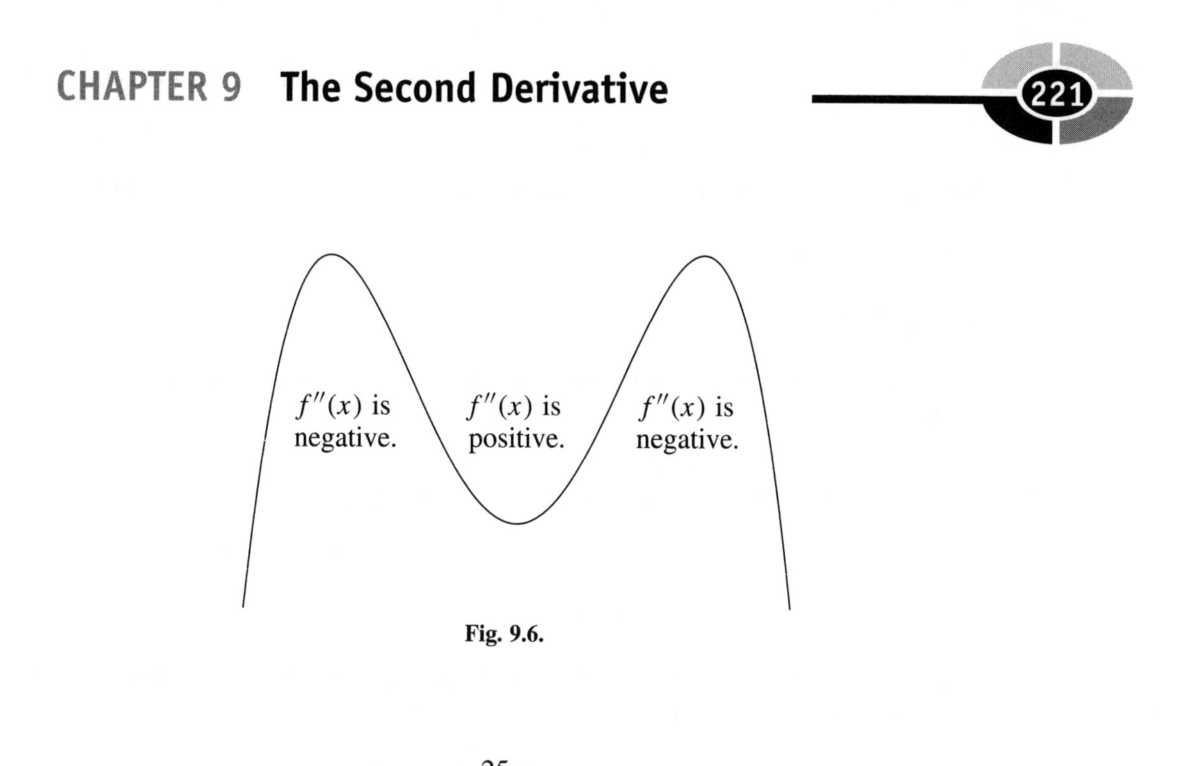

Fig. 9.6.

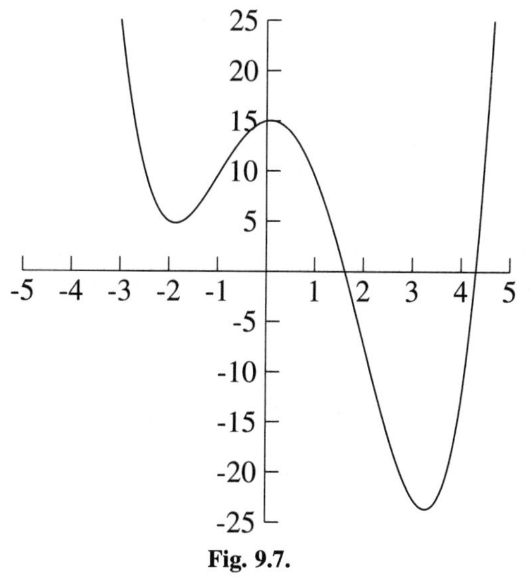

Fig. 9.7.

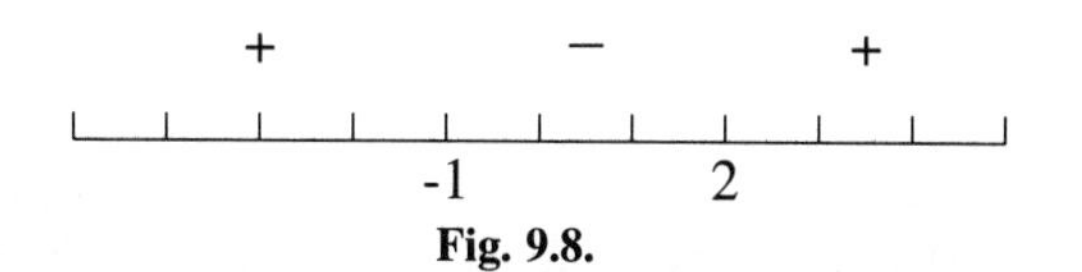

Fig. 9.8.

the graph for $f(x)$ is concave up) and where it is negative (where the graph is concave down).

EXAMPLES

Determine where the graph of each function is concave up and where it is concave down.

- $f(x) = x^4 - 8x^3 + 12x - 5$
  We will compute $f'(x)$, and then its derivative, $f''(x)$.

$$f'(x) = 4x^3 - 24x^2 + 12$$

$$f''(x) = 12x^2 - 48x$$

Now we will find the critical values by setting the second derivative equal to zero and solving for x.

$$0 = 12x^2 - 48x$$

$$0 = 12x(x - 4)$$

$$12x = 0 \qquad x - 4 = 0$$

$$x = 0 \qquad x = 4$$

The critical values are 0 and 4. We will test -1, 1 and 5 in $f''(x)$ to see where it is positive and where it is negative (Figure 9.9).

$f''(-1) = 12(-1)^2 - 48(-1) = +60$ Put a plus sign to the left of 0.

$f''(1) = 12(1)^2 - 48(1) = -36$ Put a minus sign between 0 and 4.

$f''(5) = 12(5)^2 - 48(5) = +60$ Put a plus sign to the right of 4.

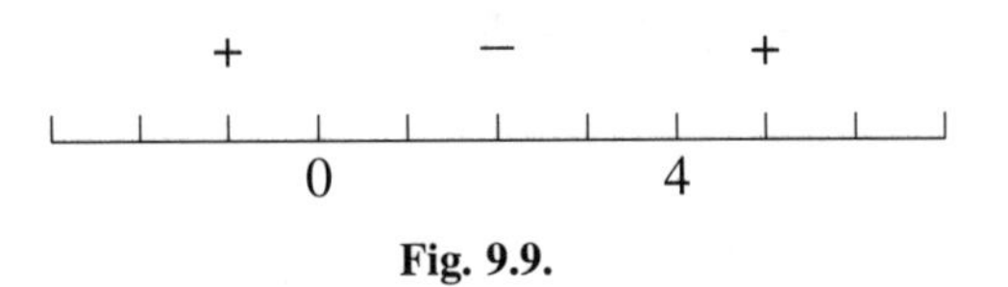

Fig. 9.9.

The graph is concave up on $(-\infty, 0)$ (to the left of 0), concave down on $(0, 4)$ (between 0 and 4), and concave up on $(4, \infty)$ (to the right of 4).

- $f(x) = 6 - x^2$

$$f'(x) = -2x$$
$$f''(x) = -2$$

Because $f''(x)$ is always -2, there are no critical values. This means that the graph never changes concavity. Because $f''(x)$ is negative, the graph is concave down everywhere.

If a function is continuous at $x = a$ and concavity changes at $x = a$, then the point $(a, f(a))$ (the point on the graph where $x = a$) is called an *inflection point*. In the first example above, there are inflection points at $x = 0$ and $x = 4$. The graph for the function in the second example has no inflection point.

PRACTICE

Determine where the graph of each function is concave up and where it is concave down. Find the inflection points, if any exist.

1. $f(x) = -x^4 + 6x^2 + 7x + 5$
2. $f(x) = x^4 - 9x^3 + 12x^2 + x - 2$
3. $f(x) = 5x^2 - 4$

SOLUTIONS

1.

$$f'(x) = -4x^3 + 12x + 7$$
$$f''(x) = -12x^2 + 12$$
$$0 = -12x^2 + 12 = 12(-x^2 + 1) = 12(1 - x^2)$$
$$0 = 12(1 - x)(1 + x)$$
$$1 - x = 0 \qquad 1 + x = 0$$
$$1 = x \qquad x = -1$$

The critical values are -1 and 1. We will test -2, 0 and 2 in $f''(x)$ (Figure 9.10).

$$f''(-2) = -12(-2)^2 + 12 = -36$$

$$f''(0) = -12(0)^2 + 12 = +12$$

$$f''(2) = -12(2)^2 + 12 = -36$$

$-$ $+$ $-$

-1 1

Fig. 9.10.

The graph is concave down on $(-\infty, -1)$, up on $(-1, 1)$ and down on $(1, \infty)$. Because concavity changes at $x = -1$ and $x = 1$, there are inflection points at $x = -1$ and $x = 1$. We will find the y-values for these points by putting -1 and 1 in the original function.

$$f(-1) = -(-1)^4 + 6(-1)^2 + 7(-1) + 5 = 3$$

$(-1, 3)$ is an inflection point.

$$f(1) = -(1^4) + 6(1^2) + 7(1) + 5 = 17$$

$(1, 17)$ is an inflection point.

2.

$$f'(x) = 4x^3 - 27x^2 + 24x + 1$$

$$f''(x) = 12x^2 - 54x + 24$$

$$0 = 12x^2 - 54x + 24 = 6(2x^2 - 9x + 4)$$

$$0 = 6(2x - 1)(x - 4)$$

$$2x - 1 = 0 \qquad x - 4 = 0$$

$$2x = 1 \qquad x = 4$$

$$x = \frac{1}{2}$$

The critical values are $\frac{1}{2}$ and 4. We will test $x = 0,\ 2$, and 5 in $f''(x)$ (Figure 9.11).

$$f''(0) = 12(0^2) - 54(0) + 24 = +24$$

$$f''(2) = 12(2^2) - 54(2) + 24 = -36$$

$$f''(5) = 12(5^2) - 54(5) + 24 = +54$$

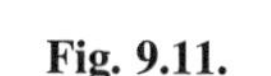

Fig. 9.11.

The graph is concave up on $(-\infty, \frac{1}{2})$, down on $(\frac{1}{2}, 4)$, and up on $(4, \infty)$. Because concavity changes at $x = \frac{1}{2}$ and 4 there inflection points at $x = \frac{1}{4}$ and 4.

$$f\left(\frac{1}{2}\right) = \left(\frac{1}{2}\right)^4 - 9\left(\frac{1}{2}\right)^3 + 12\left(\frac{1}{2}\right)^2 + \frac{1}{2} - 2 = \frac{7}{16}$$

$\left(\frac{1}{2}, \frac{7}{16}\right)$ is an inflection point.

$$f(4) = 4^4 - 9(4)^3 + 12(4)^2 + 4 - 2 = -126$$

$(4, -126)$ is an inflection point.

3.

$$f'(x) = 10x \qquad f''(x) = 10$$

$f''(x)$ is positive for all x-values, so the graph for $f(x)$ is concave up everywhere. There are no inflection points.

Concavity changes at inflection points, but it can also change at a break in the graph. For example, concavity changes at the breaks in the graph, $x = -2$ and $x = 2$, for the graph in Figure 9.12. Because of this, we need to find where the second derivative does not exist. In the above example, the second derivative (as well as the original function) does not exist at $x = -2$ and $x = 2$. For most functions in a calculus course, we can find these critical values by setting the denominator of $f''(x)$ equal to zero and solving for x.

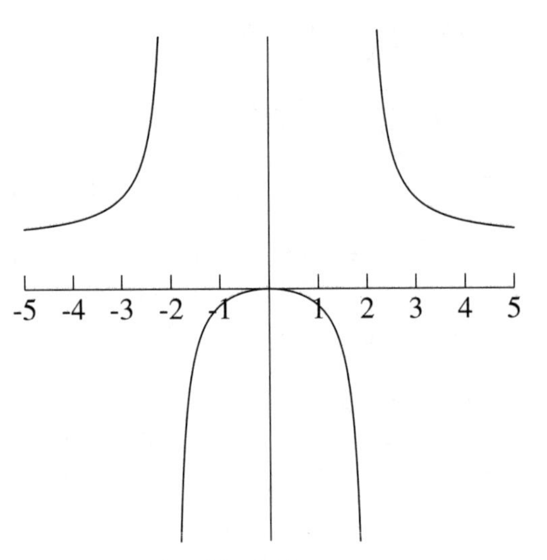

Fig. 9.12.

EXAMPLE

- $f(x) = \dfrac{2x^3}{x^2 - 9}$

$$f'(x) = \frac{6x^2(x^2 - 9) - 2x^3(2x)}{(x^2 - 9)^2} = \frac{2x^4 - 54x^2}{(x^2 - 9)^2}$$

$$f''(x) = \frac{(8x^3 - 108x)[(x^2 - 9)^2] - (2x^4 - 54x^2)(2)(x^2 - 9)(2x)}{[(x^2 - 9)^2]^2}$$

$$= \frac{36x(x^2 + 27)}{(x^2 - 9)^3}$$

We will find the critical values for $f''(x)$ by setting the numerator and denominator equal to zero.

$$36x(x^2 + 27) = 0$$

$$36x = 0 \qquad x^2 + 27 = 0$$

$$x = 0 \qquad \text{No solution}$$

One critical value is $x = 0$.

$$(x^2 - 9)^3 = 0$$

$$x^2 - 9 = 0$$

$$(x+3)(x-3)=0$$

$$x+3=0 \qquad x-3=0$$

$$x=-3 \qquad x=3$$

The other critical values are −3 and 3. We will test −4, −1, 1 and 4 in $f''(x)$ (Figure 9.13).

$$f''(-4)=\frac{36(-4)[(-4)^2+27]}{[(-4)^2-9]^3}=\frac{-144(43)}{343}=-\frac{6192}{343}$$

$$f''(-1)=\frac{36(-1)[(-1)^2+27]}{[(-1)^2-9]^3}=\frac{-36(28)}{-512}=+\frac{63}{32}$$

$$f''(1)=\frac{36(1)(1^2+27)}{(1^2-9)^3}=-\frac{63}{32}$$

$$f''(4)=\frac{36(4)(4^2+27)}{(4^2-9)^3}=+\frac{6192}{343}$$

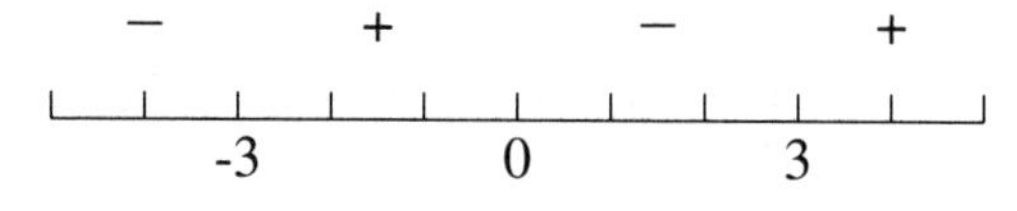

Fig. 9.13.

The graph is concave down on $(-\infty,-3)$ and $(0,3)$. The graph is concave up on $(-3,0)$ and $(3,\infty)$. The original function is not continuous at $x=-3$ and $x=3$ but is continuous at $x=0$. The inflection point is $(0,0)$ (see Figure 9.14).

The Second Derivative Test

We can use concavity to decide if a critical value for the derivative of a function gives us a relative maximum, relative minimum, or neither. Suppose we have a function $f(x)$ where $x=1$ is a critical value for $f'(x)$ (that is, when we solve $f'(x)=0$, $x=1$ is a solution). Rather than making a sign graph for $f'(x)$ and testing x-values to see where the function is increasing and where it is decreasing, we can put $x=1$ in $f''(x)$ to see if the graph is concave up or concave down or neither. If the graph is concave down at $x=1$, then we have a relative maximum at $x=1$. If the graph is concave up at $x=1$, then we have a relative minimum at $x=1$. This method is called the *second derivative test.* For a function that is

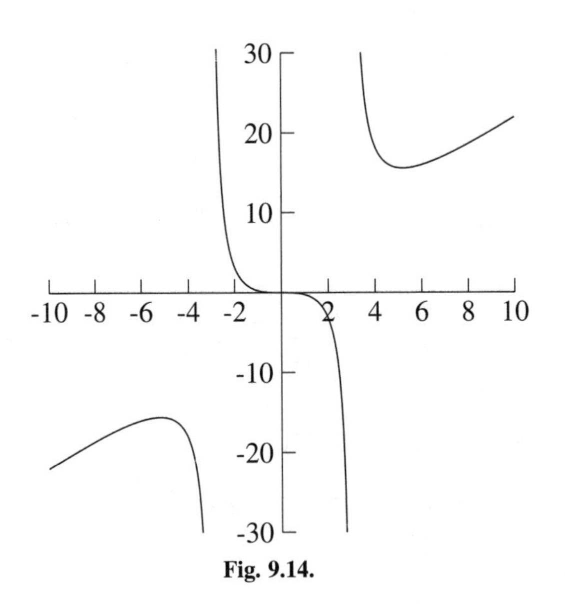

Fig. 9.14.

twice differentiable (both the derivative and the second derivative exist), we have the following fact.

> If $f'(a) = 0$ and $f''(a) < 0$, then $f(x)$ has a relative maximum at $x = a$.
> If $f'(a) = 0$ and $f''(a) > 0$, then $f(x)$ has a relative minimum at $x = a$.

Here is how we will use the second derivative test to find relative extrema.

Step 1 Compute the first derivative, $f'(x)$.
Step 2 Set the derivative equal to 0 and solve for x. The solutions to this equation are the critical values.
Step 3 Compute the second derivative, $f''(x)$.
Step 4 Put the critical values in the second derivative. If $x = a$ is a critical value and $f''(a)$ is positive, then $f(x)$ has a relative minimum at $x = a$. If $f''(a)$ is negative, then $f(x)$ has a relative maximum at $x = a$.

EXAMPLES

- $f(x) = 8x^3 - 3x^4$

  Step 1 $f'(x) = 24x^2 - 12x^3$

Step 2 $0 = 24x^2 - 12x^3 = 12x^2(2 - x)$

$$12x^2 = 0 \qquad 2 - x = 0$$

$$x = 0 \qquad 2 = x$$

Step 3 $f''(x) = 48x - 36x^2$

Step 4 Evaluate $f''(x)$ at $x = 0$ and $x = 2$.

$$f''(0) = 48(0) - 36(0^2) = 0$$

$$f''(2) = 48(2) - 36(2^2) = -48$$

$f''(0)$ is neither positive nor negative, so the function does not have a relative maximum nor a relative minimum at $x = 0$. $f''(2)$ is negative, so the function has a relative maximum at $x = 2$.

- $f(x) = \frac{1}{x^2+1}$

Step 1 $f'(x) = \frac{0(x^2+1)-1(2x)}{(x^2+1)^2} = -\frac{2x}{(x^2+1)^2}$

Step 2 $2x = 0 \qquad (x^2 + 1)^2 = 0$

$$x = 0 \qquad x^2 + 1 = 0$$

No solution

The only critical value is $x = 0$.

Step 3 $f'(x) = -2x(x^2 + 1)^{-2}$ Use the product rule.

$$f''(x) = -2(x^2 + 1)^{-2} + (-2x)(-2)(x^2 + 1)^{-3}(2x)$$

$$= \frac{-2}{(x^2 + 1)^2} + \frac{8x^2}{(x^2 + 1)^3}$$

Step 4 We will put $x = 0$ in $f''(x)$ to see if $f''(0)$ is positive or negative.

$$f''(0) = \frac{-2}{(0^2 + 1)^2} + \frac{8(0^2)}{(0^2 + 1)^3} = \frac{-2}{1} + \frac{0}{1} = -2$$

Because $f''(0)$ is negative, the function has a relative maximum at $x = 0$.

PRACTICE

Use the second derivative test to find all relative extrema.

1. $f(x) = 3x^4 + 14x^3 - 12x^2 + 10$
2. $f(x) = \sqrt[3]{x}$
3. $f(x) = \frac{x^2}{x^2+1}$

SOLUTIONS

1.
$$f'(x) = 12x^3 + 42x^2 - 24x$$
$$0 = 12x^3 + 42x^2 - 24x = 6x(2x^2 + 7x - 4)$$
$$= 6x(2x - 1)(x + 4)$$
$$6x = 0 \qquad 2x - 1 = 0 \qquad x + 4 = 0$$
$$x = 0 \qquad 2x = 1 \qquad x = -4$$
$$x = \frac{1}{2}$$
$$f''(x) = 36x^2 + 84x - 24$$
$$f''(0) = 36(0^2) + 84(0) - 24 = -24$$
$f(x)$ has a relative maximum at $x = 0$.
$$f(0) = 3(0)^4 + 14(0)^3 - 12(0)^2 + 10$$
$= 10$ is a relative maximum.
$$f''\left(\frac{1}{2}\right) = 36\left(\frac{1}{2}\right)^2 + 84\left(\frac{1}{2}\right) - 24 = +27$$
$f(x)$ has a relative minimum at $x = \frac{1}{2}$.
$$f\left(\frac{1}{2}\right) = 3\left(\frac{1}{2}\right)^4 + 14\left(\frac{1}{2}\right)^3 - 12\left(\frac{1}{2}\right)^2 + 10$$
$= \frac{143}{16}$ is a relative minimum.

$$f''(-4) = 36(-4)^2 + 84(-4) - 24 = +216$$

$f(x)$ has a relative minimum at $x = -4$.

$$f(-4) = 3(-4)^4 + 14(-4)^3 - 12(-4)^2 + 10$$

$$= -310 \text{ is a relative minimum.}$$

2. $f(x) = x^{1/3}$

$$f'(x) = \frac{1}{3}x^{-2/3} = \frac{1}{3\sqrt[3]{x^2}}$$

$$0 = \frac{1}{3\sqrt[3]{x^2}} \quad \text{This equation has no solution.}$$

$f'(x)$ does not exist at $x = 0$,

so the only critical value is $x = 0$.

$$f''(x) = \frac{1}{3} \cdot \frac{-2}{3}x^{-5/3} = \frac{-2}{9x^{5/3}} = \frac{-2}{9\sqrt[3]{x^5}}$$

$f''(0)$ does not exist, so $f(x)$ does not have a relative maximum nor minimum. There is an inflection point at $x = 0$, though.

3.

$$f'(x) = \frac{2x(x^2+1) - x^2(2x)}{(x^2+1)^2} = \frac{2x}{(x^2+1)^2}$$

$$2x = 0 \qquad (x^2+1)^2 = 0$$

$$x = 0 \qquad x^2 + 1 = 0$$

No solution

$$f'(x) = 2x(x^2+1)^{-2} \quad \text{Use the product rule}$$

$$f''(x) = 2(x^2+1)^{-2} + 2x(-2)(x^2+1)^{-3}(2x)$$

$$f''(0) = \frac{2}{(0^2+1)^2} - \frac{8(0)^2}{(0^2+1)^3} = +2$$

$f(x)$ has a relative minimum at $x = 0$.
$f(0) = \frac{0^2}{0^2+1} = 0$ is a relative minimum.

CHAPTER 9 REVIEW

1. Where is the graph of $f(x) = x^4 - 54x^2$ concave up?
   (a) $(-\infty, -\sqrt{27})$ and $(0, \sqrt{27})$
   (b) $(-\sqrt{27}, 0)$ and $(\sqrt{27}, 0)$
   (c) $(-3, 3)$
   (d) $(-\infty, -3)$ and $(3, \infty)$

2. Which of the following are inflection points for $f(x) = x^4 - 54x^2$?
   (a) $(0, 0)$, $(-\sqrt{27}, -729)$, and $(\sqrt{27} - 729)$
   (b) $(3, -405)$ and $(-3, -405)$
   (c) $(0, 0)$ only
   (d) $(0, 0)$, $(3, -405)$, and $(-3, -405)$

3. Is the graph of the function $f(x) = x^3 - 4x^2 + x - 3$ concave up, concave down, or neither at $x = -2$?
   (a) Concave up
   (b) Concave down
   (c) Neither
   (d) Cannot be determined without the graph

4. Refer to Figure 9.15 which, is the sign graph for $f''(x)$. For what interval(s) is the graph of $f(x)$ concave down?

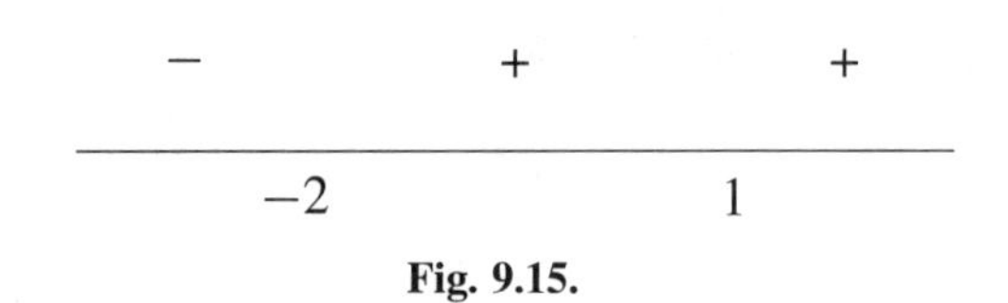

Fig. 9.15.

   (a) $(-2, 1)$
   (b) $(-2, \infty)$
   (c) $(-2, 1)$ and $(1, \infty)$
   (d) $(-\infty, -2)$

5. For some function, $f(x)$, we have $f'(10) = 0$ and $f''(10) = 3$. What does this mean?
   (a) There is a relative maximum at $x = 10$
   (b) There is a relative minimum at $x = 10$
   (c) The relative maximum is 10
   (d) The relative minimum is 10

6. For the function $f(x) = 3x^5 - 5x^4 - 200x^3$, find the critical values for $f''(x)$.
   (a) $x = 0$ only
   (b) $x = -4,\ 0$ only
   (c) $x = 5, 0$ only
   (d) $x = -4,\ 0,\ 5$ only

SOLUTIONS

1. d 2. b 3. b 4. d 5. b 6. d

Business Applications of the Derivative

We can use calculus for business applications to find the price that maximizes profit, the dimensions that minimize the cost to construct a box, and the production level that minimizes costs. Once we have a function to be optimized (maximized or minimized), we will use the same techniques we used in Chapters 8 and 9 to find the solution to our problem. We can use the first derivative test or the second derivative test on a critical value to verify that the critical value we find is the extremum we are looking for. We will skip the derivative test for most of the problems in this chapter because the problems are written so that the minimum or maximum occurs at the critical value.

EXAMPLES

- When the price for a product is p, the revenue (in \$ thousands) can be approximated by $R = -0.05p^2 + 0.98p + 18$. What price maximizes the revenue?

  We begin by finding the derivative: $R' = -0.1p + 0.98$. Now we will set R' equal to zero and solve for p.

$$-0.1p + 0.98 = 0$$

$$-0.1p = -0.98$$

$$p = \frac{-0.98}{-0.1} = 9.80$$

  Is revenue maximized for $p = 9.80$? We will use the second derivative test to verify that it is. $R'' = -0.1$. The second derivative is negative for all p, in particular for $p = 9.80$, so $p = 9.80$ gives us a maximum. Revenue is maximized when the price is \$9.80.
- The revenue for selling x thousand units of a product can be approximated by $R = x^3 - 21x^2 + 120x + 500$ (for x between 1 and 10). How many units must be sold in order to maximize revenue?

$$R' = 3x^2 - 42x + 120$$

$$0 = 3x^2 - 42x + 120 = 3(x^2 - 14x + 40)$$

$$= 3(x - 4)(x - 10)$$

$$x - 4 = 0 \qquad x - 10 = 0$$

$$x = 4 \qquad x = 10$$

  We will use the second derivative to determine which of 4 or 10 maximizes the revenue. $R'' = 6x - 42$. $R''(4) = 6(4) - 42 = -18$ and $R''(10) = 6(10) - 42 = +18$. Because R'' at $x = 4$ is negative, revenue is maximized at $x = 4$. The company should sell 4000 units to maximize revenue.
- The interest rate on an investment varied between 1990 and 2002, given by $y = -0.0866x^2 + 0.866x + 5.8$, where $x = 0$ is the year 1990 and y is the annual return, as a percent, for the year x. During what year did the investment have the highest rate of return?

The derivative is $y' = -0.1732x + 0.866$.

$$-0.1732x + 0.866 = 0$$
$$-0.1732x = -0.866$$
$$x = \frac{-0.866}{-0.1732} = 5$$

The investment had its maximum return during 1995 (the year 5).

The maximum profit often can be found by setting the marginal revenue equal to the marginal cost. This is true because the marginal profit is the difference between the marginal revenue and the marginal cost.

$$P(x) = R(x) - C(x)$$
$$\frac{d}{dx}(P(x)) = \frac{d}{dx}(R(x) - C(x))$$
$$= \frac{d}{dx}(R(x)) - \frac{d}{dx}(C(x))$$
$$P'(x) = R'(x) - C'(x)$$
$$0 = R'(x) - C'(x) \quad \text{Set the marginal profit equal to 0.}$$
$$C'(x) = R'(x) \quad \text{Add the marginal cost to both sides.}$$

If costs are higher than revenue, this method could find the maximum *loss*.

EXAMPLE

- The revenue for selling x thousand units of a product is approximated by $R(x) = -0.05x^2 + 2x + 60$, and the cost for producing x units is approximated by $C(x) = 1.5x + 20$. What level of sales maximizes profit?
  We will set the marginal revenue equal to the marginal cost.

$$R'(x) = -0.1x + 2 \qquad C'(x) = 1.5$$
$$-0.1x + 2 = 1.5$$
$$-0.1x = -0.5$$
$$x = \frac{-0.5}{-0.1} = 5$$

Maximize profit by selling 5000 units.

Calculus can also tell us the production level that minimizes the average cost per unit. Suppose one week 500 units were produced at a cost of \$8000. Then each unit costs, on average, \$8000/500 = \$16. This is different from the marginal cost for 500 units produced, which is the cost to produce one extra unit. The average cost function is the cost function divided by the number produced: $A(x) = \frac{C(x)}{x}$.

EXAMPLES

- The weekly cost to produce x units of a product is approximated by $C(x) = -0.05x^2 + 60x - 8000$ (valid for 200 to 900 units). What level of production minimizes the average cost?

  The average cost function is the total cost function divided by the production level.

$$A(x) = \frac{C(x)}{x} = \frac{-0.05x^2 + 60x - 8000}{x}$$

$$= \frac{-0.05x^2}{x} + \frac{60x}{x} - \frac{8000}{x}$$

$$= -0.05x + 60 - 8000x^{-1}$$

$$A'(x) = -0.5 - (-1)8000x^{-2} = -0.5 + \frac{8000}{x^2}$$

$$0 = -0.5 + \frac{8000}{x^2}$$

$$0.05 = \frac{8000}{x^2}$$

$$0.05x^2 = 8000$$

$$x^2 = \frac{8000}{0.05} = 160{,}000$$

$$x = 400$$

  Minimize the average cost by producing 400 units.
- The cost to produce x feet of pipe can be approximated by $C(x) = 0.02x^2 - 3x + 450$. How much pipe should be produced to minimize the average cost?

The average cost to produce x feet of pipe is

$$A(x) = \frac{C(x)}{x} = \frac{0.02x^2 - 3x + 450}{x} = \frac{0.02x^2}{x} - \frac{3x}{x} + \frac{450}{x}$$

$$= 0.02x - 3 + \frac{450}{x}.$$

$$A'(x) = 0.02 + (-1)450x^{-2} = 0.02 - \frac{450}{x^2}$$

$$0.02 - \frac{450}{x^2} = 0$$

$$0.02 = \frac{450}{x^2}$$

$$0.02x^2 = 450$$

$$x^2 = \frac{450}{0.02} = 22{,}500$$

$$x = 150$$

The manufacturer should produce 150 feet of pipe to minimize the average cost.

These average cost functions had solutions. What happens if the derivative of the average cost function has no critical value? In the simplest function, where each unit costs the same to produce and there are no fixed costs, the minimum average cost occurs for any production level because the average cost is constant. For example, say $C(x) = 5x$, where each unit costs \$5 to produce. The average cost function is $C(x) = \frac{5x}{x} = 5$. If two units are produced, the average cost per unit is \$10/2 = \$5; if 1000 units are produced, the average cost per unit is \$5000/1000 = \$5. At the other extreme, suppose we have an average cost function that is always decreasing. In this case, every time we increase the production, the average cost decreases. For example, say $C(x) = 5x + 1000$. Then $A(x) = 5 + \frac{1000}{x}$ and $A'(x) = -\frac{1000}{x^2}$. This derivative is always negative, which means that the average cost is always decreasing. We would need to determine the maximum number of units that can be produced because the maximum production level minimizes the average cost.

PRACTICE

1. The profit for selling x hundred units of a product can be approximated by $P(x) = -x^3 + 45x^2 + 1200x + 80{,}000$ (up to $x = 50$). What level of sales maximizes the profit?
2. The revenue for a product depends on the product's price. The revenue, in thousands of dollars, for the product when the price is p, is approximated by $R(p) = -0.04p^2 + 0.06p + 9.9775$. What price maximizes the revenue?
3. Over a 25-year period, the annual interest paid for on a loan can be approximated by $y = -0.00038x^3 + 0.0237x^2 - 0.296x + 5.424$, where y is the interest as a percent, and x is the number of years since 1980. During what year was the interest rate a minimum? (The critical value is a decimal number. Check the whole numbers, both smaller and larger than the critical value, in the original function to see which is the true minimum.)
4. The revenue (in $ thousand) for selling x thousand units of a product can be approximated by $R(x) = \frac{8x}{0.2x+1}$ and the cost by $C(x) = 0.5x$. How many units should be sold to maximize profit?
5. The cost to produce x units of a product can be approximated by $C(x) = 0.004x^2 - 9.6x + 7840$. How many should be produced to minimize the average cost?

SOLUTIONS

1. $P'(x) = -3x^2 + 90x + 1200$

$$0 = -3x^2 + 90x + 1200 = -3(x^2 - 30x - 400)$$

$$0 = -3(x + 10)(x - 40)$$

$$x + 10 = 0 \qquad x - 40 = 0$$

$$x = -10 \qquad x = 40$$

Because $x = -10$ cannot be a solution, the only possibility is $x = 40$. Because there are two critical values, we will use the second derivative test to see which is the maximum. The second derivative is $P''(x) = -6x + 90$, so $P''(40) = -6(40) + 90 = -150$. This is negative, so $x = 40$ leads to a maximum. Sell 4000 units (40 hundreds is 4000) to maximize the profit.

2. $R'(p) = -0.08p + 0.06.$

$$-0.08p + 0.06 = 0$$
$$-0.08p = -0.06$$
$$p = \frac{-0.06}{-0.08} = 0.75$$

Revenue is maximized when the price is \$0.75.

3. $y' = -0.00114x^2 + 0.0474x - 0.296$

$$0 = -0.00114x^2 + 0.0474x - 0.296$$
$$x = \frac{-0.0474 \pm \sqrt{(0.0474)^2 - 4(-0.00114)(-0.296)}}{2(-0.00114)}$$
$$= \frac{-0.0474 \pm \sqrt{0.00224676 - 0.00134976}}{-0.00228}$$
$$\approx 7.65,\ 33.9$$

$x = 33.9$ is outside the 0 to 25 range for the function, so we cannot consider it. Because there are two solutions, we should verify that $x = 7.65$ does give us a minimum. We will evaluate the second derivative at $x = 7.65$: $y'' = -0.00228x + 0.0474$ and $y''(7.65) = -0.00228(7.65) + 0.0474 = 0.029958$. $x = 7.65$ gives us a minimum for y. Does this mean that the minimum occurs during 1987 or 1988? We will evaluate the original function at both $x = 7$ and $x = 8$.

$$y(7) = -0.00038(7^3) + 0.0237(7^2) - 0.296(7) + 5.424 = 4.38296$$
$$y(8) = -0.00038(8^3) + 0.0237(8^2) - 0.296(8) + 5.424 = 4.37824$$

The minimum occurs during the year 1988.

4. We will compute $R'(x)$ and $C'(x)$ and set them equal to each other.

$$R'(x) = \frac{8(0.2x + 1) - 8x(0.2)}{(0.2x + 1)^2} = \frac{8}{(0.2x + 1)^2} \qquad C'(x) = 0.5$$

$$0.5 = \frac{8}{(0.2x + 1)^2}$$

$$0.5(0.2x + 1)^2 = 8$$

$$0.5[(0.2x + 1)(0.2x + 1)] = 8$$

$$0.5(0.04x^2 + 0.4x + 1) = 8$$

$$0.02x^2 + 0.2x + 0.5 = 8$$

$$0.02x^2 + 0.2x - 7.5 = 0$$

$$x = \frac{-0.2 \pm \sqrt{(0.2)^2 - 4(0.02)(-7.5)}}{2(0.02)} = \frac{-0.2 \pm \sqrt{0.64}}{0.04}$$

$$= \frac{-0.2 \pm 0.8}{0.04} = 15, \ -25 \qquad \text{Use } x = 15 \text{ only.}$$

Maximize the profit by selling 15,000 units. (You could verify that this solution gives a maximum with the second derivative test.)

5. The average cost function is

$$A(x) = \frac{C(x)}{x} = \frac{0.004x^2 - 9.6x + 7840}{x}$$

$$= \frac{0.004x^2}{x} - \frac{9.6x}{x} + \frac{7840}{x}$$

$$= 0.004x - 9.6 + \frac{7840}{x}$$

$$A'(x) = 0.004 - \frac{7840}{x^2}$$

$$0.004 - \frac{7840}{x^2} = 0$$

$$0.004 = \frac{7840}{x^2}$$

$$0.004x^2 = 7840$$

$$x^2 = \frac{7840}{0.004} = 1{,}960{,}000$$

$$x = 1400$$

Minimize the average cost by producing 1400 units.

When one variable is a function of two or more variables, sometimes increasing one variable means sacrificing another variable. For example, suppose we have 40 feet of fencing available to enclose a rectangular area. Increasing the width of the enclosed area means decreasing the length. When does increasing the width mean increasing the enclosed area, and when does increasing the width mean decreasing the enclosed area? What dimensions maximize the enclosed area? Each of the rectangles in Figure 10.1 has a perimeter of 40 feet, but their areas are very different. If we increase the width from 4 feet to 18 feet, the length is decreased from 16 feet to 2 feet, and the enclosed area decreases from 64 square feet to 36 square feet. On the other hand, if we increase the width from 4 feet to 12 feet, the length decreases from 16 feet to 8 feet, and the enclosed area increases from 64 square feet to 96 square feet.

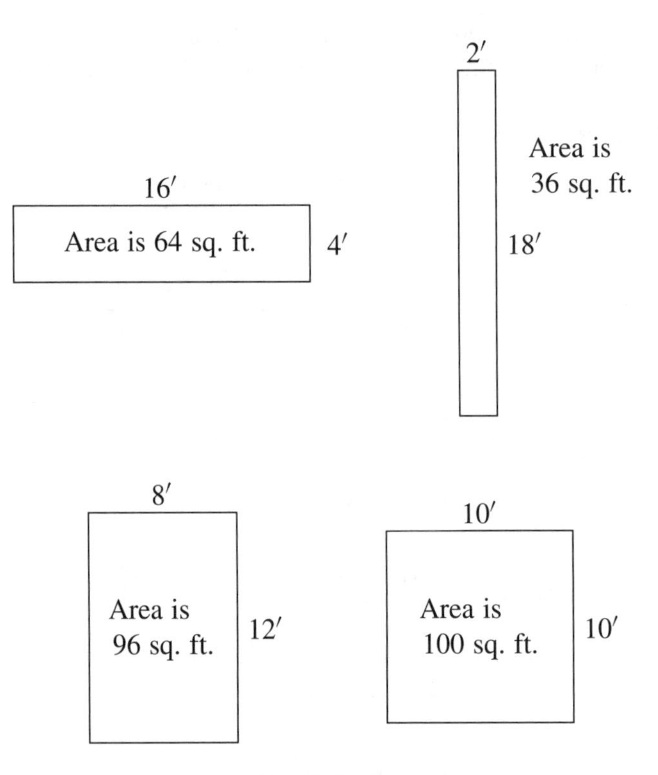

Fig. 10.1.

If we want to maximize the enclosed area with a fixed 40 feet of fencing, we can use calculus on the area formula, $A = lw$. We will solve this problem later.

Calculus can find the levels of two variables that maximizes or minimizes a third variable. One common problem involves maximizing the revenue when the price is changing. An increase in the price means more money per unit is collected but fewer units are sold. Calculus finds the price that makes the most of a price increase while minimizing the loss in sales.

In the problems that follow, the price increase (or decrease) will be given as the number of $\$a$ increases (or decreases). We will let x represent the number of times the price is increased (or decreased) by $\$a$. This makes the new price "old price + ax" (or "old price $-\ ax$"). For example, if we want to increase the price by some multiple of \$5, the new price is "old price + $5x$." The quantity sold will depend on the size of the loss (or gain) in sales from each price increase (or decrease) by $\$a$. If we lose two customers for each \$5 increase in the price, then the quantity sold is "old quantity $-\ 2x$." The revenue is the price times the quantity sold. In this example, the revenue function is "(old price $+\ 5x$)(old quantity $-\ 2x$)".

EXAMPLES

- A movie multiplex sells tickets for \$8. On Friday evenings, it averages 720 tickets sold. Each patron spends an average of \$2 on concessions. A survey shows that for each \$0.25 drop in the ticket price, 20 more people will buy tickets on Friday evenings. Assuming that they also average \$2 in concessions, what ticket price will maximize revenue?

  Let x represent the number of \$0.25 decreases in the ticket price. This makes the ticket price \$8 minus the decrease in the ticket price, which is $0.25x$: $8 - 0.25x$. The number of people buying tickets is 720 plus 20 for each x, which is represented by $20x$: $720 + 20x$. The ticket revenue is the ticket price times the number of tickets sold: $(8 - 0.25x)(720 + 20x)$. The concession revenue is \$2 times the number of tickets sold, $720 + 20x$: $2(720 + 20x)$. The total revenue is $R = (8 - 0.25x)(720 + 20x) + 2(720 + 20x) = -5x^2 + 20x + 7200$.

$$R' = -10x + 20$$

$$-10x + 20 = 0$$

$$-10x = -20$$

$$x = \frac{-20}{-10} = 2$$

  Maximize the revenue by charging $8 - 0.25(2) = \$7.50$ for tickets.

- A small farm has an apple orchard in which 250 trees are planted per acre. Each tree averages 25.8 bushels of apples. The farmer learns that for each additional tree per acre, the yield of each tree will be reduced by one-tenth of a bushel. How many trees per acre should be planted to maximize the farmer's apple harvest?

  This is another example of the increase of one variable (trees per acre) resulting in a decrease in another variable (apples per tree). We want to maximize the yield per acre, which is the number of trees per acre times the number of bushels per tree. The yield per acre is now $y = (250)(25.8) = 6450$ bushels. We will let x represent the number of trees that will be added to each acre. Then each acre will have $250 + x$ trees. For each x, we lose 0.10 bushels per tree, so 25.8 is reduced by $0.10x$. After adding x trees to each acre, each tree produces $25.8 - 0.10x$ bushels. The yield per acre becomes

$$y = \overbrace{(250 + x)}^{\text{Number of trees}} \overbrace{(25.8 - 0.10x)}^{\text{Bushels per tree}} = -0.1x^2 + 0.8x + 6450.$$

$$y' = -0.2x + 0.8$$

$$-0.2x + 0.8 = 0$$

$$-0.2x = -0.8$$

$$x = \frac{-0.8}{-0.2} = 4$$

The farmer should add 4 trees per acre for a total of 250 + 4 = 254 trees per acre in order to maximize the apple harvest.

PRACTICE

1. The manager of an office building can rent all 40 offices when the monthly rent is \$6000. The manager believes that each increase of \$1000 in the rent will result in a loss of 5 tenants with little chance of being replaced. What should be charged in rent in order to maximize revenue?
2. An athletic director of a university wants to increase the ticket price for its football games. The average attendance for home games is 3200 when the ticket price is \$10. A survey shows that for each increase of \$0.50 in the ticket price, 100 fewer fans will attend. If each fan spends an average of \$3 on concessions, what ticket price maximizes revenue?

3. A farm has a small peach orchard with 120 trees. Each tree averages 186 pounds of peaches per year. An expert has determined that each additional tree will reduce the orchard's yield by 1.5 pounds per tree. How many trees will maximize the yield?

SOLUTIONS

1. Let x represent the number of \$1000 increases in the rent. The rent is $6000 + 1000x$ and the number of tenants is $40 - 5x$. The revenue is $R = (6000 + 1000x)(40 - 5x) = -5000x^2 + 10{,}000x + 240{,}000$.

$$R' = -10{,}000x + 10{,}000$$
$$-10{,}000x + 10{,}000 = 0$$
$$-10{,}000x = -10{,}000$$
$$x = \frac{-10{,}000}{-10{,}000} = 1$$

The manager should charge $\$6000 + 1000(1) = \$7000$ rent in order to maximize revenue.

2. Let x represent the number of \$0.50 increases in the ticket price. The new ticket price is $10 + 0.50x$ and the average number attending home games is $3200 - 100x$. Ticket revenue is $(10 + 0.50x)(3200 - 100x)$ and concession revenue is \$3 for each fan: $3(3200 - 100x)$. The total revenue is $R = (10 + 0.50x)(3200 - 100x) + 3(3200 - 100x) = -50x^2 + 300x + 41{,}600$.

$$R' = -100x + 300$$
$$-100x + 300 = 0$$
$$-100x = -300$$
$$x = \frac{-300}{-100} = 3$$

In order to maximize revenue, the ticket price should be $\$10 + 3(0.50) = \$11.50$.

3. Let x represent the number of extra trees planted. The total number of trees is $120 + x$. Each tree's yield is reduced by 1.5 pounds for each extra tree, so the yield is decreased by $1.5x$. The yield per tree is $186 - 1.5x$.

The total yield is $y = (120 + x)(186 - 1.5x) = -1.5x^2 + 6x + 22320$.

$$y' = -3x + 6$$
$$-3x + 6 = 0$$
$$-3x = -6$$
$$x = 2$$

The farmer will maximize the peach harvest by adding two peach trees, for a total of 122 trees.

Recall from the previous section the problem of using 40 feet of fencing to enclose a rectangular area. If we want to maximize the enclosed area, we can use calculus to maximize the area formula, $A = lw$. This function has no maximum unless we restrict the variables l and w. This is where we use the fact that 40 feet of fencing are available. This forces the perimeter of the enclosed area to be 40. We will use the perimeter formula, $P = 2l + 2w$, substituting 40 for P: $40 = 2l + 2w$. Using this equation, we can replace one of l or w in $A = lw$, forming an equation to fit our conditions. We will solve for l.

$$2l + 2w = 40$$
$$2l = 40 - 2w$$
$$l = \frac{40 - 2w}{2} = 20 - w$$

Now we will replace l with $20 - w$ in $A = lw$: $A = (20 - w)w = 20w - w^2$. This equation gives us the area of any rectangular region whose perimeter is 40 feet.

$$A = 20w - w^2$$
$$A' = 20 - 2w$$
$$0 = 20 - 2w$$
$$2w = 20$$
$$w = \frac{20}{2} = 10 \qquad l = 20 - w = 20 - 10 = 10$$

The enclosed area is maximized when the width is 10 feet and the length is also 10 feet.

If we want to optimize a geometric formula or one that is based on a geometric formula, we will use information given in the problem to eliminate one of the variables. Sometimes we can simply replace a variable directly with a number. Usually we will have to use a relationship between the variables to write one of the variables in terms of the other (like we did above with $2l + 2w = 40$). After making a substitution, we will have an equation with two variables that we will differentiate.

EXAMPLES

- A thin piece of metal, 20" × 16", will be used to construct an open-topped box. A square will be cut from each corner (see Figure 10.2). After the corners are removed, the sides will be folded up (see Figure 10.3). What size corner should be cut so that the box's volume is maximized?

  We want to maximize the volume of a rectangular box, so we will begin with the formula $V = lwh$. When x inches are removed from each corner, the length is reduced to $20 - x - x = 20 - 2x$, and the width is reduced to $16 - x - x = 16 - 2x$. We can replace l with $20 - 2x$ and w with $16 - 2x$. The formula $V = lwh$ becomes $V = (20 - 2x)(16 - 2x)h$. The height of the box is the size of the corner, so we can replace h with x. We now have $V = (20 - 2x)(16 - 2x)x$.

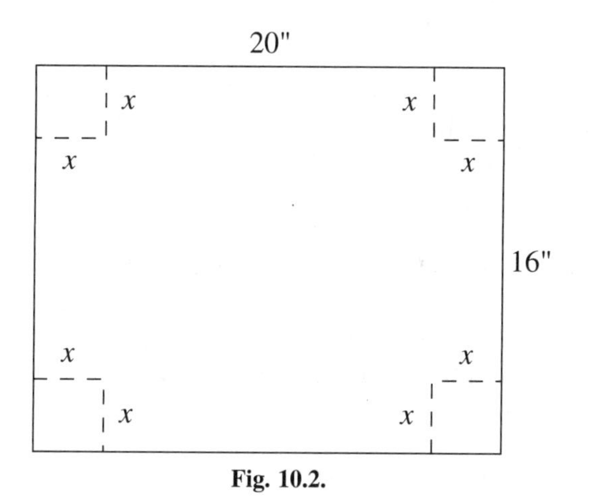

Fig. 10.2.

$$V = [(20 - 2x)(16 - 2x)]x = (320 - 72x + 4x^2)x$$
$$= 4x^3 - 72x^2 + 320x$$
$$V' = 12x^2 - 144x + 320$$

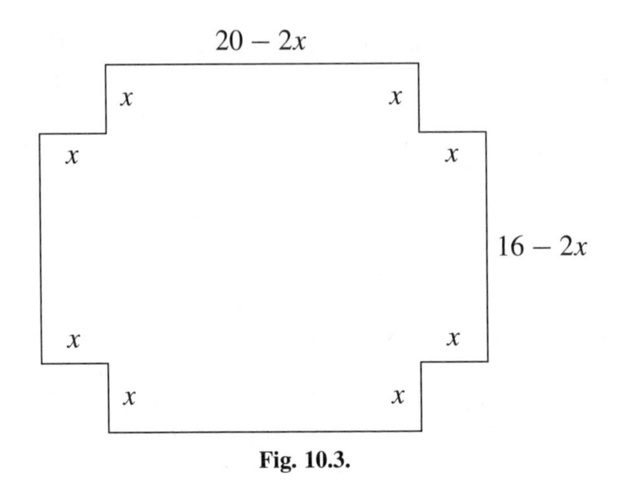

Fig. 10.3.

$$0 = 12x^2 - 144x + 320$$

$$x = \frac{-(-144) \pm \sqrt{(-144)^2 - 4(12)(320)}}{2(12)} = \frac{144 \pm \sqrt{5376}}{24}$$

$$\approx 2.94,\ 9.06$$

We cannot consider $x = 9.06$ because we would need a piece of metal that is more than 18 inches on each side so that 9" could be cut from each corner. We will use the second derivative test to verify that 2.94 leads to a maximum.

$$V'' = 24x - 144 \qquad V''(2.94) = 24(2.94) - 144 = -73.44$$

The second derivative is negative at $x = 2.94$, so we have a maximum at $x = 2.94$. The volume of the box is maximized when about 2.94" is cut from each corner.

The next two problems are other common fencing problems in which a fixed amount of fencing is available and we want to find the dimensions that maximize the enclosed area. In the first problem, only three sides of the rectangular area are to be fenced because the fourth side is some other boundary. In the second, a rectangular area is subdivided into two or more areas. The equation to be maximized in each case is $A = lw$. As before, we will use information about the available fencing to eliminate either l or w in $A = lw$. Once we have A written in terms of l only or w only, we can maximize the area.

EXAMPLES

- The manager of a large retail store wants to enclose an area behind the store. There are 80 feet of fencing material available. The side against the building does not need to be fenced. What dimensions will maximize the enclosed area? (see Figure 10.4)

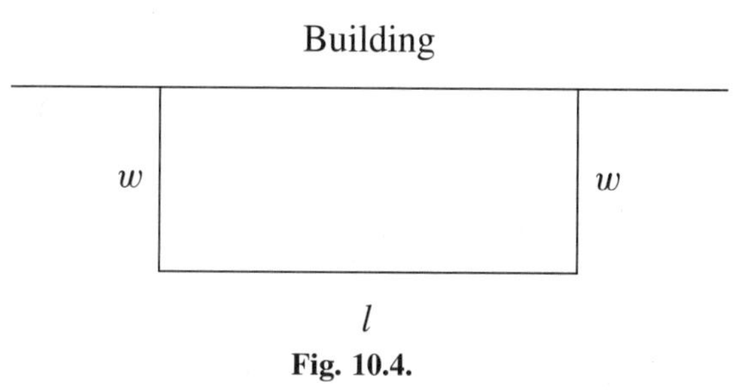

Fig. 10.4.

From the figure, we see that $w + w + l$ must equal 80: $2w + l = 80$. We will solve this equation for l: $l = 80 - 2w$. We could solve the equation for w but this would involve using a fraction. Now we will replace l with $80 - 2w$ in $A = lw$.

$$A = (80 - 2w)w = 80w - 2w^2$$

$$A' = 80 - 4w$$

$$0 = 80 - 4w$$

$$4w = 80$$

$$w = \frac{80}{4} = 20 \qquad l = 80 - 2w = 80 - 2(20) = 40$$

Maximize the enclosed area with a width of 20 feet and a length of 40 feet.

- A rancher wants to enclose a rectangular area divided into two pens (see Figure 10.5). If there is 900 feet of fencing available, what dimensions will maximize the enclosed area?

  Because 900 feet of fencing is available, we must have $l + w + w + w + l$ be 900: $2l + 3w = 900$. We will solve this equation for l (solving for w works, too).

$$2l + 3w = 900$$

$$2l = 900 - 3w$$

$$l = \frac{900 - 3w}{2} = 450 - \frac{3}{2}w$$

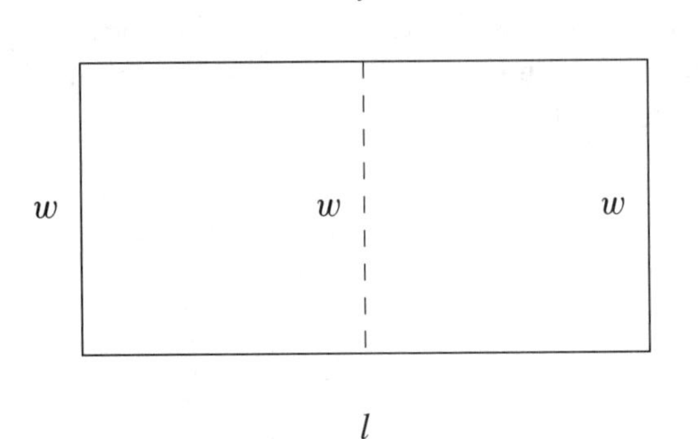

Fig. 10.5.

We will replace l with $450 - \frac{3}{2}w$ in $A = lw$.

$$A = \left(450 - \frac{3}{2}w\right)w = 450w - \frac{3}{2}w^2$$

$$A' = 450 - 3w$$

$$0 = 450 - 3w$$

$$3w = 450$$

$$w = \frac{450}{3} = 150 \qquad l = 450 - \frac{3}{2}w = 450 - \frac{3}{2}(150) = 225$$

Maximize the enclosed area with a width of 150 feet and a length of 225 feet.

PRACTICE

1. An open-topped box is to be constructed from a 12" × 16" piece of cardboard by cutting a square from each corner and folding up the sides. What size corner should be cut out so that the box's volume is maximum?
2. A school administrator wants to enclose a practice field. One side of the property is already fenced. There are 400 meters of fencing material available. If only three sides of the area needs to be fenced, what dimensions will maximize the area? (see Figure 10.6).
3. The owner of a kennel has 90 feet of fencing available to enclose three pens (see Figure 10.7). What dimensions maximize the enclosed area?

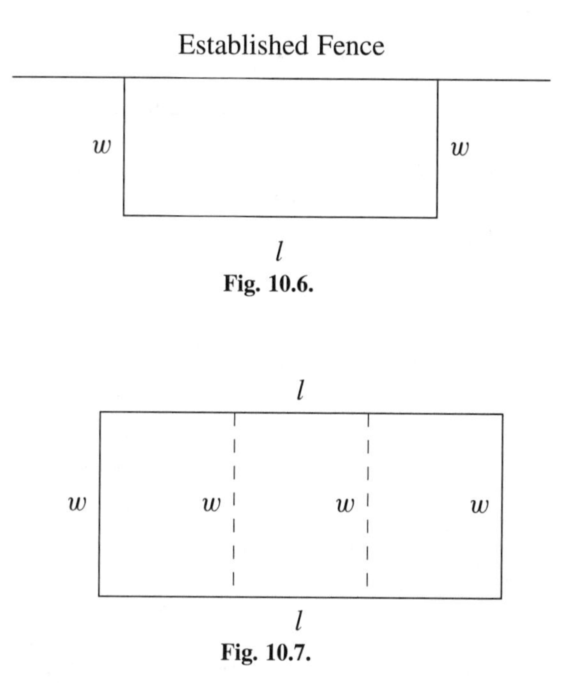

Fig. 10.6.

Fig. 10.7.

SOLUTIONS

1. Let x represent the length, in inches, to be cut from each side. Then after the cut, the lengths of the sides are $12 - 2x$ and $16 - 2x$ inches. The height of the box is x inches. The volume $V = lwh$ becomes $V = (12 - 2x)(16 - 2x)x$.

$$V = [(12 - 2x)(16 - 2x)]x = (192 - 56x + 4x^2)x$$

$$= 4x^3 - 56x^2 + 192x$$

$$V' = 12x^2 - 112x + 192 = 4(3x^2 - 28x + 48)$$

$$0 = 4(3x^2 - 28x + 48)$$

$$0 = 3x^2 - 28x + 48$$

$$x = \frac{-(-28) \pm \sqrt{(-28)^2 - 4(3)(48)}}{2(3)} = \frac{28 \pm \sqrt{208}}{6} \approx 2.26,\ 7.07$$

Two sides of the cardboard are only 12 inches, so we cannot cut 7.07 inches from each corner. The only possibility is 2.26. Because there are two solutions, we will make sure that 2.26 leads to a maximum.

$$V'' = 24x - 112 \qquad V''(2.26) = 24(2.26) - 112 = -57.76$$

Because the second derivative is negative, $x = 2.26$ leads to a maximum. The volume is maximized when about 2.26 inches is cut from each corner.

2. Because 400 meters of fencing material is available, we have $2w + l = 400$. We will solve for l, giving us $l = 400 - 2w$. We will substitute $400 - 2w$ for l in $A = lw$.

$$A = (400 - 2w)w = 400w - 2w^2$$

$$A' = 400 - 4w$$

$$0 = 400 - 4w$$

$$4w = 400$$

$$w = \frac{400}{4} = 100 \qquad l = 400 - 2w = 400 - 2(100) = 200$$

Maximize the area with a width of 100 meters and a length of 200 meters.

3. Using the fact that 90 feet of fencing is available, we have $4w + 2l = 90$. We will solve for l.

$$4w + 2l = 90$$

$$2l = 90 - 4w$$

$$l = \frac{90 - 4w}{2} = 45 - 2w$$

Substituting $45 - 2w$ for l in $A = lw$ gives us $A = (45 - 2w)w = 45w - 2w^2$.

$$A' = 45 - 4w$$

$$0 = 45 - 4w$$

$$4w = 45$$

$$w = \frac{45}{4} = 11.25 \qquad l = 45 - 2w = 45 - 2(11.25) = 22.5$$

Maximize the area with a width of 11.25 feet and a length of 22.5 feet.

Calculus can help optimize geometric problems in which some parts are weighted more heavily than other parts. In this book, parts will be weighted more if they cost more money to construct. We will begin with fencing problems where one side of the fence costs more or less than the other sides. There will be two versions of each problem—one in which the budget is fixed and we want to maximize the area, and the other in which the area is fixed and we want to minimize the cost.

EXAMPLES

- Refer to Figure 10.8.

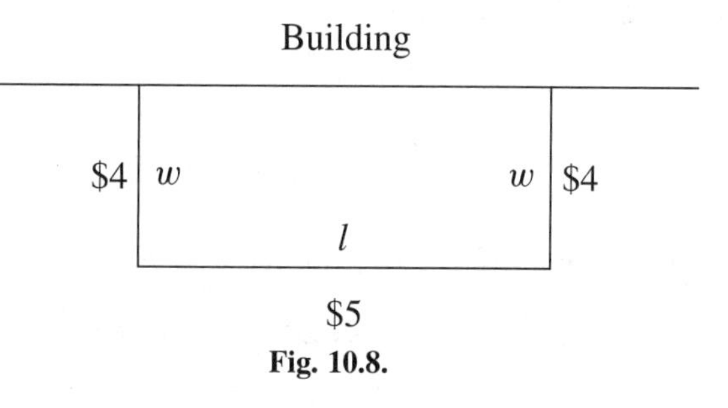

Fig. 10.8.

1. Minimize the cost if the area must be 800 square feet.
2. Maximize the area if there is \$100 available to spend on fencing.

Each side that makes up the width costs \$4 per foot and the side that makes up the length costs \$5 per foot. Two sides cost $\$4w$ each and one side costs $\$5l$, where the total cost is $C = 4w + 4w + 5l = 8w + 5l$.

1. The area is a fixed 800 square feet, so $A = lw$ becomes $800 = lw$. Solving for l gives us $l = \frac{800}{w}$. Now we can substitute $\frac{800}{w}$ for l in the cost function, $C = 8w + 5l$.

$$C = 8w + 5\left(\frac{800}{w}\right) = 8w + \frac{4000}{w}$$

$$C' = 8 - \frac{4000}{w^2}$$

$$0 = 8 - \frac{4000}{w^2}$$

$$8 = \frac{4000}{w^2}$$

$$8w^2 = 4000$$

$$w^2 = \frac{4000}{8} = 500$$

$$w \approx 22.4 \qquad l = \frac{800}{w} \approx \frac{800}{22.4} \approx 35.7$$

Minimize the cost by letting the width be about 22.4 feet and the length be about 35.7 feet.

2. This time we want to maximize $A = lw$ and will use the cost function to eliminate l. Because \$100 is available to spend on the fence, the cost function $C = 8w + 5l$ becomes $8w + 5l = 100$. Solving for l gives us

$$5l = 100 - 8w$$

$$l = \frac{100 - 8w}{5} = 20 - 1.6w$$

We will substitute $20 - 1.6w$ for l in $A = lw$.

$$A = (20 - 1.6w)w = 20w - 1.6w^2$$

$$A' = 20 - 3.2w$$

$$0 = 20 - 3.2w$$

$$3.2w = 20$$

$$w = \frac{20}{3.2} = 6.25 \qquad l = 20 - 1.6w = 20 - 1.6(6.25) = 10$$

Maximize the area by letting the width be 6.25 feet and the length be 10 feet.

- Refer to Figure 10.9.

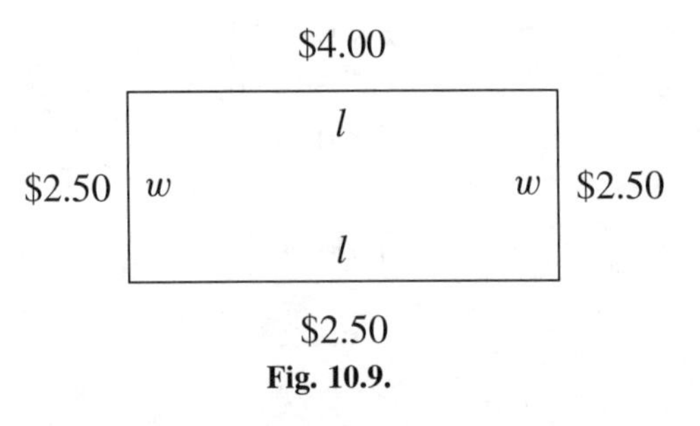

Fig. 10.9.

1. Minimize the cost if the area is to be 4000 square feet.
2. Maximize the area if the fence budget is \$650.

The width costs $2.50w + 2.50w = 5w$ and the length costs $4l + 2.50l = 6.50l$. This makes the total cost $C = 5w + 6.50l$.

1. The area is 4000 square feet, so $A = lw$ becomes $4000 = lw$. We will solve for l: $l = \frac{4000}{w}$ and substitute this for l in the cost function.

$$C = 5w + 6.50\left(\frac{4000}{w}\right) = 5w + \frac{26{,}000}{w}$$

$$C' = 5 - \frac{26{,}000}{w^2}$$

$$0 = 5 - \frac{26{,}000}{w^2}$$

$$5 = \frac{26{,}000}{w^2}$$

$$5w^2 = 26{,}000$$

$$w^2 = \frac{26{,}000}{5} = 5200$$

$$w \approx 72.1 \qquad l = \frac{4000}{w} \approx \frac{4000}{72.1} \approx 55.5$$

Minimize the cost by letting the width be about 72.1 feet and the length be about 55.5 feet.

2. Because the fence budget is \$650, the cost function becomes $5w + 6.50l = 650$. We will solve for w.

$$5w = 650 - 6.50l$$

$$w = \frac{650 - 6.50l}{5} = 130 - 1.30l$$

We are ready to substitute $130 - 1.30l$ for w in the area function.

$$A = l(130 - 1.30l) = 130l - 1.30l^2$$

$$A' = 130 - 2.60l$$

$$0 = 130 - 2.60l$$

$$2.60l = 130$$

$$l = \frac{130}{2.60} = 50 \qquad w = 130 - 1.30l = 130 - 1.30(50) = 65$$

Maximize the area by letting the length be 50 feet and the width be 65 feet.

- Refer to Figure 10.10.

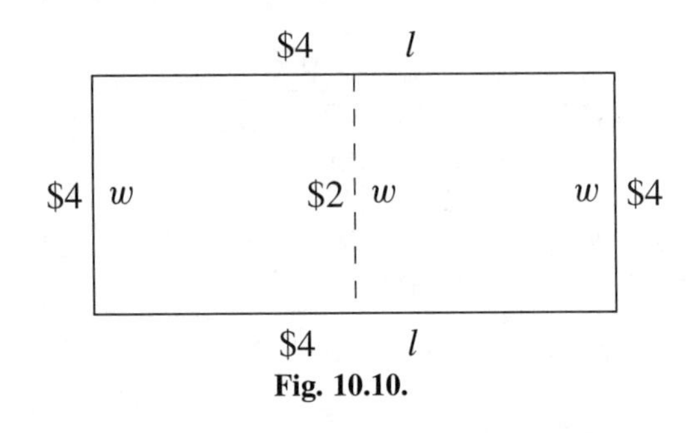

Fig. 10.10.

1. The enclosed area is to be 5000 square feet. What dimensions minimize the cost?
2. \$300 is available for fencing materials. What dimensions maximize the enclosed area?

The width costs $4w + 2w + 4w = 10w$, and the length costs $4l + 4l = 8l$. The total cost function is $C = 10w + 8l$.

1. The area is 5000 square feet, so $5000 = lw$. Solving for l gives us $l = \frac{5000}{w}$. With this substitution, the cost function becomes

$$C = 10w + 8\left(\frac{5000}{w}\right) = 10w + \frac{40{,}000}{w}$$

$$C' = 10 - \frac{40{,}000}{w^2}$$

$$0 = 10 - \frac{40{,}000}{w^2}$$

$$10 = \frac{40{,}000}{w^2}$$

$$10w^2 = 40{,}000$$

$$w^2 = \frac{40{,}000}{10} = 4000$$

$$w \approx 63.2 \qquad l = \frac{5000}{w} \approx \frac{5000}{63.2} \approx 79.1$$

Minimize the cost by letting the width be about 63.2 feet and the width be about 79.1 feet.

2. The cost is \$300, so the cost function becomes $10w + 8l = 300$. We will solve this for w.

$$10w = 300 - 8l$$

$$w = \frac{300 - 8l}{10} = 30 - 0.8l$$

Substituting $30 - 0.8l$ for w in $A = lw$ gives us $A = l(30 - 0.8l) = 30l - 0.8l^2$.

$$A' = 30 - 1.6l$$

$$0 = 30 - 1.6l$$

$$1.6l = 30$$

$$l = \frac{30}{1.6} = 18.75 \qquad w = 30 - 0.8l = 30 - 0.8(18.75) = 15$$

Maximize the area by letting the length be 18.75 feet and the width be 15 feet.

PRACTICE

1. Refer to Figure 10.11.

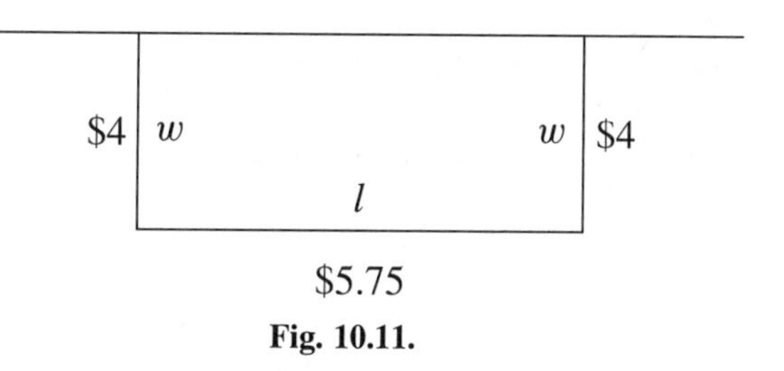

Fig. 10.11.

(a) What dimensions minimize the cost if the area is to be 1600 square feet?

(b) The fence budget is \$690. What dimensions maximize the area?

2. Refer to Figure 10.12.

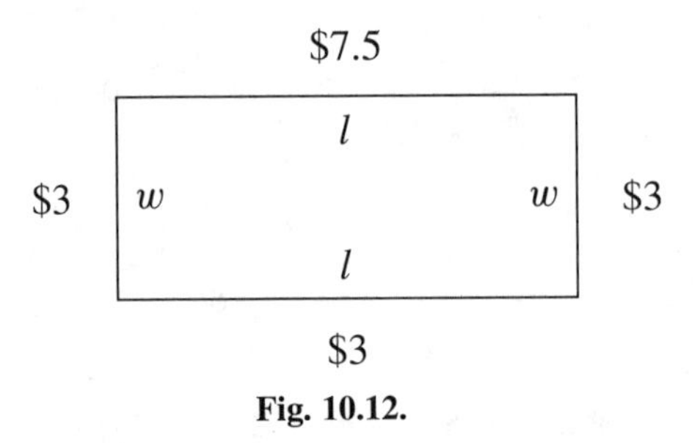

Fig. 10.12.

(a) What dimensions minimize the cost if the area is to be 1225 square feet?
(b) The fence budget is $150. What dimensions maximize the area?

3. Refer to Figure 10.13.

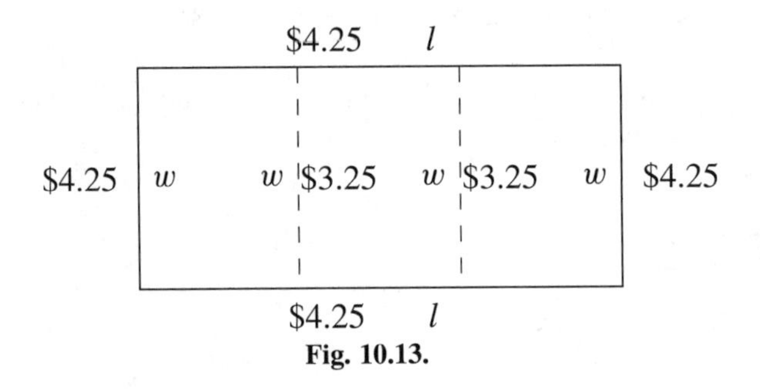

Fig. 10.13.

(a) What dimensions minimize the cost if the area is to be 1500 square feet?
(b) The fence budget is $1020. What dimensions maximize the area?

SOLUTIONS

1. The cost function is $C = 4w + 4w + 5.75l = 8w + 5.75l$.

(a) $A = lw$ becomes $1600 = lw$. Solving for l gives us $l = \frac{1600}{w}$. The cost function becomes

$$C = 8w + 5.75\left(\frac{1600}{w}\right) = 8w + \frac{9200}{w}$$

$$C' = 8 - \frac{9200}{w^2}$$

$$0 = 8 - \frac{9200}{w^2}$$

$$8 = \frac{9200}{w^2}$$

$$8w^2 = 9200$$

$$w^2 = \frac{9200}{8} = 1150$$

$$w \approx 33.9 \qquad l = \frac{1600}{w} \approx \frac{1600}{33.9} \approx 47.2$$

Minimize the cost by letting the width be about 33.9 feet and the length be about 47.2 feet.

(b) The cost function is $8w + 5.75l = 690$. We will solve for w.

$$8w = 690 - 5.75l$$

$$w = \frac{690 - 5.75l}{8} = 86.25 - 0.71875l$$

Substituting $86.25 - 0.71875l$ for w in the area function gives us $A = l(86.25 - 0.71875l) = 86.25l - 0.71875l^2$.

$$A' = 86.25 - 1.4375l$$

$$0 = 86.25 - 1.4375l$$

$$1.4375l = 86.25$$

$$l = \frac{86.25}{1.4375} = 60 \qquad w = 86.25 - 0.71875l$$

$$= 86.25 - 0.71875(60) = 43.125$$

Maximize the area by letting the length be 60 feet and the width be 43.125 feet.

2. The cost is $3w + 7.5l + 3w + 3l = 6w + 10.5l$

(a) The area formula is $1225 = lw$. We will solve for w: $w = \frac{1225}{l}$.

$$C = 6w + 10.5l = 6\left(\frac{1225}{l}\right) + 10.5l = \frac{7350}{l} + 10.5l$$

$$C' = -\frac{7350}{l^2} + 10.5$$

$$0 = -\frac{7350}{l^2} + 10.5$$

$$10.5 = \frac{7350}{l^2}$$

$$10.5l^2 = 7350$$

$$l^2 = \frac{7350}{10.5} = 700$$

$$l \approx 26.5 \qquad w = \frac{1225}{l} \approx \frac{1225}{26.5} \approx 46.2$$

Minimize the cost by letting the length be about 26.5 feet and the width be about 46.2 feet.

(b) The cost function is $6w + 10.5l = 150$. We will solve for w.

$$6w = 150 - 10.5l$$

$$w = \frac{150 - 10.5l}{6} = 25 - 1.75l$$

We will substitute for w in the area function.

$$A = l(25 - 1.75l) = 25l - 1.75l^2$$

$$A' = 25 - 3.5l$$

$$0 = 25 - 3.5l$$

$$3.5l = 25$$

$$l = \frac{25}{3.5} \approx 7.14 \quad w = 25 - 1.75l \approx 25 - 1.75(7.14) \approx 12.5$$

Maximize the area by letting the length be about 7.14 feet and the width be about 12.5 feet.

3. The cost function is $C = 4.25w + 4.25l + 4.25w + 4.25l + 3.25w + 3.25w = 15w + 8.50l$.

   (a) The area function is $1500 = lw$. We will solve for l: $l = \frac{1500}{w}$. This gives us the cost function

$$C = 15w + 8.50\left(\frac{1500}{w}\right) = 15w + \frac{12{,}750}{w}$$

$$C' = 15 - \frac{12{,}750}{w^2}$$

$$0 = 15 - \frac{12{,}750}{w^2}$$

$$15 = \frac{12{,}750}{w^2}$$

$$15w^2 = 12{,}750$$

$$w^2 = \frac{12{,}750}{15} = 850$$

$$w \approx 29.2 \quad l = \frac{1500}{w} \approx \frac{1500}{29.2} \approx 51.4$$

   Minimize the cost by letting the width be about 29.2 feet and the length be about 51.4 feet.

   (b) The cost function becomes $15w + 8.50l = 1020$. We will solve for w.

$$15w = 1020 - 8.50l$$

$$w = \frac{1020 - 8.50l}{15} = 68 - \frac{17}{30}l$$

   With this substitution, the area function becomes

$$A = l\left(68 - \frac{17}{30}l\right) = 68l - \frac{17}{30}l^2$$

$$A' = 68 - 2 \cdot \frac{17}{30}l = 68 - \frac{17}{15}l$$

$$0 = 68 - \frac{17}{15}l$$

$$\frac{17}{15}l = 68$$

$$l = 68 \cdot \frac{15}{17} = 60 \qquad w = 68 - \frac{17}{30}l$$

$$= 68 - \frac{17}{30}(60) = 34$$

Maximize the area by letting the length be 60 feet and the width be 34 feet.

We can optimize similar problems with many other shapes. Next, we will work with three-dimensional shapes—the rectangular box and the right-circular cylinder, which is the shape of a can. In the first problems, we will be given a fixed volume and will be asked to find the dimensions that minimize the surface area. Later, we will minimize the cost when different surfaces (top, bottom, sides) have different costs to construct. In the same way we used the area information in the problems above, we will use the volume information to eliminate one of the variables. Then we will make the substitution in the cost function to minimize the cost.

EXAMPLES

- A box with a square bottom is to have a volume of 64 cubic inches. What dimensions will minimize the surface area? (see Figure 10.14.)

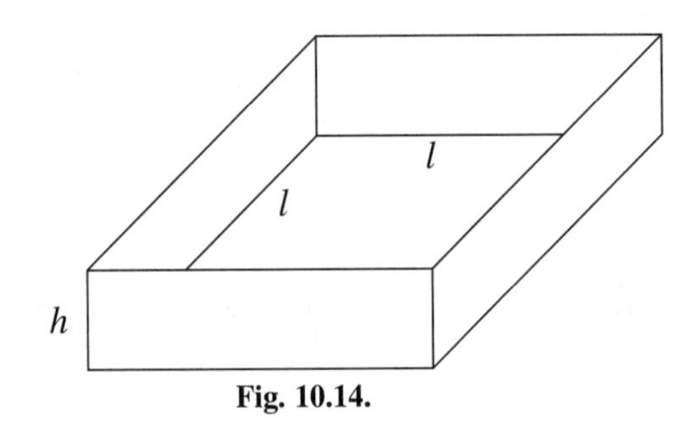

Fig. 10.14.

The volume of a rectangular box is $V = lwh$. Because the bottom of the box is square, we can replace w with l (replacing l with w works, also). $V = lwh$ becomes $V = l \cdot lh = l^2h$. Because the volume is 64, we have $64 = l^2h$. The surface of the box consists of six parts: the top, bottom, and four sides. The area of the top (and bottom) is l^2. The area of each of the four sides is lh. The total area is

$$SA = \overbrace{l^2}^{\text{Top}} + \overbrace{l^2}^{\text{Bottom}} + \overbrace{4lh}^{\text{4 Sides}} = 2l^2 + 4lh.$$

We will eliminate h by solving for h in $64 = l^2h$. Solving for l would require taking square roots, making the formula a little more complicated. From $64 = l^2h$, we have $h = \frac{64}{l^2}$. The surface area becomes

$$SA = 2l^2 + 4l\left(\frac{64}{l^2}\right) = 2l^2 + \frac{256}{l}$$

This is what we want to minimize.

$$SA' = 4l - \frac{256}{l^2}$$

$$0 = 4l - \frac{256}{l^2}$$

$$4l = \frac{256}{l^2}$$

$$4l^3 = 256$$

$$l^3 = \frac{256}{4} = 64$$

$$l = 4 \qquad h = \frac{64}{l^2} = \frac{64}{4^2} = 4$$

The surface area is minimized when the length, width, and height are each 4 inches.

- A can, in the shape of a right-circular cylinder, is to be constructed with a volume of 45 cubic inches. What dimensions will minimize the surface area? (see Figure 10.15).

  The volume formula for a right circular cylinder is $V = \pi r^2 h$. Because the volume is 45, the formula becomes $45 = \pi r^2 h$. The surface of the can comes in three parts: the top, bottom, and sides. The area of the top (and the bottom) is the area of a circle with radius r, πr^2. The area of the sides is $2\pi rh$. The total surface area is

$$SA = \overbrace{\pi r^2}^{\text{Top}} + \overbrace{\pi r^2}^{\text{Bottom}} + \overbrace{2\pi rh}^{\text{Sides}} = 2\pi r^2 + 2\pi rh.$$

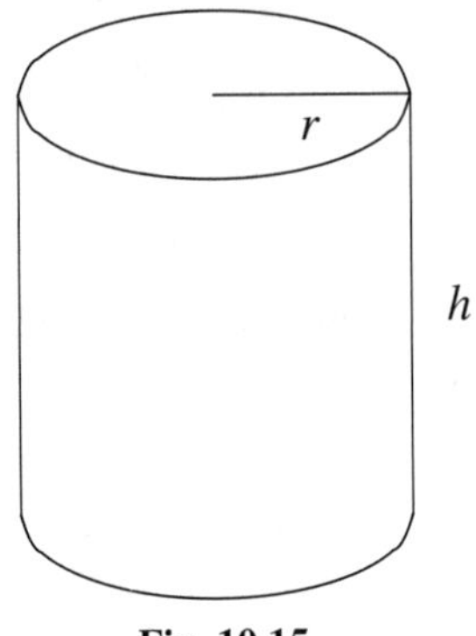

Fig. 10.15.

We will use the volume information to eliminate h in the surface area formula: $h = \frac{45}{\pi r^2}$. After we make the substutition, we can minimize the surface area equation.

$$SA = 2\pi r^2 + 2\pi r\left(\frac{45}{\pi r^2}\right) = 2\pi r^2 + \frac{90}{r}$$

$$SA' = 4\pi r - \frac{90}{r^2}$$

$$0 = 4\pi r - \frac{90}{r^2}$$

$$4\pi r = \frac{90}{r^2}$$

$$4\pi r^3 = 90$$

$$r^3 = \frac{90}{4\pi}$$

$$r \approx 1.93 \qquad h = \frac{45}{\pi r^2} \approx \frac{45}{\pi(1.93)^2} \approx 3.85$$

The surface area is minimized when the radius of the can is about 1.93 inches and the height about 3.85 inches.

- The volume of a box is to be 12 cubic feet. The width will be two-thirds the length. What dimensions minimize the surface area of the box?

  From the information given in the problem, we can eliminate two variables in the volume formula, $V = lwh$. The volume is 12 cubic feet, allowing us to replace V with 12. From the fact that the width is two-thirds the length, we can replace w with $\frac{2}{3}l$. The volume formula becomes $12 = l \cdot \frac{2}{3}lh = \frac{2}{3}l^2h$. We will solve this equation for h.

$$\frac{2}{3}l^2h = 12$$

$$l^2h = \frac{3}{2} \cdot 12 = 18$$

$$h = \frac{18}{l^2}$$

The area of the top (and the bottom) is lw or $l \cdot \frac{2}{3}l = \frac{2}{3}l^2$. The area of each of four sides is lh. The surface area is $SA = \frac{2}{3}l^2 + \frac{2}{3}l^2 + 4lh = \frac{4}{3}l^2 + 4lh$. Once we substitute $\frac{18}{l^2}$ for h, we have

$$SA = \frac{4}{3}l^2 + 4l\left(\frac{18}{l^2}\right) = \frac{4}{3}l^2 + \frac{72}{l}$$

We are ready to minimize this function.

$$SA' = 2\left(\frac{4}{3}l\right) - \frac{72}{l^2} = \frac{8}{3}l - \frac{72}{l^2}$$

$$0 = \frac{8}{3}l - \frac{72}{l^2}$$

$$\frac{8}{3}l = \frac{72}{l^2}$$

$$\frac{8}{3}l^3 = 72$$

$$l^3 = \frac{3}{8} \cdot 72 = 27$$

$$l = 3 \qquad w = \frac{2}{3}l = \frac{2}{3} \cdot 3 = 2 \qquad h = \frac{18}{l^2} = \frac{18}{3^2} = 2$$

The surface area is minimized when the length is 3 feet and each of the width and height is 2 feet.

PRACTICE

1. The volume of a box is to be 40.5 cubic inches. The width must be three-fourths the length. What dimensions minimize the surface area?
2. A barrel is to be constructed in the shape of a right-circular cylinder having a volume of 20 cubic feet. What dimensions minimize the surface area?

SOLUTIONS

1. We can eliminate two variables in the volume formula. We can replace w with $\frac{3}{4}l$ because the width is three-fourths the length. We can replace V with 40.5 because the volume is 40.5 cubic inches. The volume formula, $V = lwh$, becomes $40.5 = l(\frac{3}{4}l)h = \frac{3}{4}l^2h$. We will solve this equation for h.

$$\frac{3}{4}l^2h = 40.5$$

$$l^2h = \frac{4}{3} \cdot 40.5 = 54$$

$$h = \frac{54}{l^2}$$

The area of the top of the box (and the bottom) is $wl = \frac{3}{4}l \cdot l = \frac{3}{4}l^2$. The area of each of the four sides is lh. The surface area of the box is $SA = \frac{3}{4}l^2 + \frac{3}{4}l^2 + 4lh = 2 \cdot \frac{3}{4}l^2 + 4lh = \frac{3}{2}l^2 + 4lh$.

$$SA = \frac{3}{2}l^2 + 4lh$$

$$SA = \frac{3}{2}l^2 + 4l\left(\frac{54}{l^2}\right) \quad \text{Replace } h \text{ with } \frac{54}{l^2}$$

$$SA = \frac{3}{2}l^2 + \frac{216}{l}$$

$$SA' = 2\left(\frac{3}{2}l\right) - \frac{216}{l^2} = 3l - \frac{216}{l^2}$$

$$0 = 3l - \frac{216}{l^2}$$

$$3l = \frac{216}{l^2}$$

$$3l^3 = 216$$

$$l^3 = \frac{216}{3} = 72$$

$$l \approx 4.16 \qquad w = \frac{3}{4}l \approx \frac{3}{4}(4.16) \approx 3.12$$

$$h = \frac{54}{l^2} \approx \frac{54}{4.16^2} \approx 3.12$$

Minimize the surface area by letting the length be about 4.16 inches, and each of the height and width be about 3.12 inches.

2. Because the volume is 20, we can replace V in the volume formula, $V = \pi r^2 h$: $20 = \pi r^2 h$. Solving this for h gives us $h = \frac{20}{\pi r^2}$. The surface area is

$$SA = \overbrace{\pi r^2}^{\text{Top}} + \overbrace{\pi r^2}^{\text{Bottom}} + \overbrace{2\pi r h}^{\text{Sides}} = 2\pi r^2 + 2\pi r h = 2\pi r^2 + 2\pi r\left(\frac{20}{\pi r^2}\right)$$

$$= 2\pi r^2 + \frac{40}{r}.$$

$$SA = 2\pi r^2 + \frac{40}{r}$$

$$SA' = 4\pi r - \frac{40}{r^2}$$

$$0 = 4\pi r - \frac{40}{r^2}$$

$$4\pi r = \frac{40}{r^2}$$

$$4\pi r^3 = 40$$

$$r^3 = \frac{40}{4\pi}$$

$$r \approx 1.47 \qquad h = \frac{20}{\pi r^2} =\approx \frac{20}{\pi(1.47^2)} \approx 2.94$$

Minimize the surface area by letting the radius be about 1.47 feet and the height be about 2.94 feet.

Instead of minimizing the surface area in the next set of problems, we will minimize the cost of constructing the box or cylinder (we will ignore material that is scrapped). Our containers will use different materials for the top, bottom, and sides, giving us different costs. We will compute how much each part costs and will minimize their sum. As we did in the previous problems, we will use the fixed volume to eliminate one of the variables in the cost function.

EXAMPLES

- An open-topped tank in the shape of a right-circular cylinder is to be constructed having a volume of 15 cubic feet. The material for the bottom costs \$1.25 per square foot, and the material for the sides cost \$0.90 per square foot. What dimensions will minimize the cost of the tank?

  The volume is 15, allowing us to replace V with 15 in the volume formula $V = \pi r^2 h$: $15 = \pi r^2 h$. Now we have $h = \frac{15}{\pi r^2}$. The bottom costs \$1.25 per square foot, and there are πr^2 square feet in the bottom. This makes the bottom cost $1.25\pi r^2$. The sides cost \$0.90 per square foot, and there are $2\pi rh$ square feet in the sides. This makes the sides cost $0.90(2\pi rh) = 1.80\pi rh$. The total cost is

$$
\begin{aligned}
C &= 1.25\pi r^2 + 1.80\pi rh \\
&= 1.25\pi r^2 + 1.80\pi r\left(\frac{15}{\pi r^2}\right) && \text{Substitute } \frac{15}{\pi r^2} \text{ for } h. \\
&= 1.25\pi r^2 + \frac{27}{r}. \\
C' &= 2.5\pi r - \frac{27}{r^2} \\
0 &= 2.5\pi r - \frac{27}{r^2} \\
2.5\pi r &= \frac{27}{r^2} \\
2.5\pi r^3 &= 27 \\
r^3 &= \frac{27}{2.5\pi} \\
r &\approx 1.51 \qquad h = \frac{15}{\pi r^2} \approx \frac{15}{\pi(1.51^2)} \approx 2.1
\end{aligned}
$$

Minimize the cost with a radius of about 1.51 feet and a height of about 2.1 feet.

- A box is being constructed. It will have a square bottom and needs to have a volume of 5 cubic feet. Material for the top costs \$0.20 per square foot; the bottom, \$0.40 per square foot; and the sides, \$0.35 per square foot. What dimensions minimize the material cost?

  The box has a square bottom, so the length and width are the same. This lets us replace w with l in the volume formula and the area formula. The volume is 5 cubic feet, so we can replace V with 5 in the volume formula, $V = lwh$: $5 = lwh = l \cdot lh = l^2h$. Solving for h gives us $h = \frac{5}{l^2}$. The top costs \$0.20 per square foot, and there are $lw = l \cdot l = l^2$ square feet. The top costs a total of $0.20l^2$. The bottom costs \$0.40 per square foot, and there are l^2 square feet. The bottom costs a total of $0.40l^2$. Each of the four sides costs \$0.35 per square foot. Each side has lh square feet. Each side costs $0.35lh$ and all four sides cost $4(0.35lh) = 1.4lh$. Replacing h with $\frac{5}{l^2}$ gives us $1.4l(\frac{5}{l^2}) = \frac{7}{l}$. The total cost for the box material is $C = 0.2l^2 + 0.4l^2 + \frac{7}{l} = 0.6l^2 + \frac{7}{l}$.

$$C = 0.6l^2 + \frac{7}{l}$$

$$C' = 2(0.6)l - \frac{7}{l^2} = 1.2l - \frac{7}{l^2}$$

$$0 = 1.2l - \frac{7}{l^2}$$

$$1.2l = \frac{7}{l^2}$$

$$1.2l^3 = 7$$

$$l^3 = \frac{7}{1.2}$$

$$l \approx 1.8 \qquad h = \frac{5}{l^2} \approx \frac{5}{1.8^2} \approx 1.54$$

Minimize the cost with a length and width of about 1.8 feet and a height of about 1.54 feet.

PRACTICE

1. Minimize the cost of a box that is to have a volume of 10 cubic feet and square bottom. Material for the bottom costs \$0.30 per square foot; the top, \$0.25 per square foot; and the sides, \$0.20 per square foot.
2. A container in the shape of a right-circular cylinder will be made so that it has a volume of 120 cubic inches. Material for the top costs \$0.25 per square inch; the bottom, \$0.60 per square inch; and the sides, \$0.40 per square inch. What dimensions minimize the cost?

SOLUTIONS

1. The top and bottom are square, so $l = w$. The volume is 10 cubic feet. Now we can write the volume formula, $V = lwh$ as $10 = l \cdot lh = l^2h$. Now we have $h = \frac{10}{l^2}$. The top costs $0.25l^2$; the bottom, $0.30l^2$; and each of the sides, $0.20lh$. The total cost function is $C = 0.25l^2 + 0.30l^2 + 4(0.20lh) = 0.55l^2 + 0.80lh$. Replacing h with $\frac{10}{l^2}$ gives us $C = 0.55l^2 + 0.80l(\frac{10}{l^2}) = 0.55l^2 + \frac{8}{l}$.

$$C' = 2(0.55l) - \frac{8}{l^2} = 1.10l - \frac{8}{l^2}$$

$$0 = 1.10l - \frac{8}{l^2}$$

$$1.10l = \frac{8}{l^2}$$

$$1.10l^3 = 8$$

$$l^3 = \frac{8}{1.10}$$

$$l \approx 1.94 \qquad h = \frac{10}{l^2} \approx \frac{10}{1.94^2} \approx 2.66$$

Minimize the cost of the box with a length and width of about 1.94 feet and a height of about 2.66 feet.

2. The volume is 120 cubic inches, so the volume formula $V = \pi r^2 h$ becomes $120 = \pi r^2 h$. This gives us $h = \frac{120}{\pi r^2}$. The area of the top is πr^2, so it costs $0.25\pi r^2$. The area of the bottom is πr^2, so it costs is $0.60\pi r^2$. The area of the sides is $2\pi rh$, so it costs $0.40(2\pi rh) = 0.80\pi rh$. Replacing h with $\frac{120}{\pi r^2}$ gives us $0.80\pi r(\frac{120}{\pi r^2}) = \frac{96}{r}$. The total

cost is $C = 0.25\pi r^2 + 0.60\pi r^2 + \frac{96}{r} = 0.85\pi r^2 + \frac{96}{r}$.

$$C' = 2(0.85\pi r) - \frac{96}{r^2} = 1.7\pi r - \frac{96}{r^2}$$

$$0 = 1.7\pi r - \frac{96}{r^2}$$

$$1.7\pi r = \frac{96}{r^2}$$

$$1.7\pi r^3 = 96$$

$$r^3 = \frac{96}{1.7\pi}$$

$$r \approx 2.62 \qquad h = \frac{120}{\pi r^2} \approx \frac{120}{\pi(2.62^2)} \approx 5.56$$

Minimize the cost with a radius of about 2.62 inches and a height of about 5.56 inches.

Economic Lot Size

Some products that are kept in inventory have two costs associated with them—storage costs and order costs. If storing a product is expensive, it might be cheaper to order it in small quantities. If ordering the product is expensive, it might be cheaper to order the product in large quantities. Calculus can determine how many to order at a time to minimize both costs. This quantity is called the *economic lot size*. We will only consider products whose use is spread evenly throughout the year instead of products that are used at some times more than others.

In our next set of problems, we will be given information on how much one unit costs to store for a year and how much each order costs to place. We will compute the annual storage and annual order costs. The sum of these costs is called the annual inventory cost. The storage cost is computed by multiplying the cost to store one unit for one year by the average number in storage. The order cost is computed by multiplying the number of orders per year by the cost for each order.

We will let x represent the number of units per order. To find the number of orders per year, we will divide the total needed per year by the number of units per order. For example, if we need 200 units per year and order 10 units per order, there will be 200/10 = 20 orders per year. Because we assume that these products are used uniformly through the year, $\frac{x}{2}$ is the average number of units in storage.

Why is this so? Suppose we operate five days per week and use 6 units each day, for a total of 30 per week. At the beginning of the day on Monday, there are 30 units in stock. At the end of the day on Monday, there are $30-6=24$; at the end of the day on Tuesday, $24-6=18$; Wednesday, 12; Thursday, 6; and Friday, 0. The average of 30, 24, 18, 12, 6, 0 is 15 (= 30/2).

$$\frac{30+24+18+12+6+0}{6}=15$$

Strictly speaking, when storage costs are very expensive and order costs are very inexpensive, the minimum inventory cost could occur when one unit is ordered at a time. On the other hand, if storage costs are very inexpensive and order costs are very expensive, the minimum inventory cost could occur when one order is placed per year. In either of these cases, the minimum cost might not occur at the critical value for the derivative of the cost function. We would have to compute these costs separately. This is done in the first example only.

EXAMPLES

- An office supply distributor sells 4500 cases of paper each year. It costs \$1.50 to store one case of paper for one year. Each order costs \$0.60. How many cases should be ordered at a time? How many orders should be placed in a year?

  Let x represent the number of cases of paper per order. The average number in storage is $\frac{x}{2}$. One case costs \$1.50 to store for one year, so annual storage costs are $1.50(\frac{x}{2})=0.75x$. The number of orders per year is $\frac{4500}{x}$. This makes the annual order cost $0.60(\frac{4500}{x})=\frac{2700}{x}$. The annual inventory cost is

$$C=0.75x+\frac{2700}{x}$$

  We will minimize this function.

$$C'=0.75-\frac{2700}{x^2}$$

$$0=0.75-\frac{2700}{x^2}$$

$$0.75=\frac{2700}{x^2}$$

$$0.75x^2=2700$$

$$x^2 = \frac{2700}{0.75} = 3600$$

$$x = 60$$

Minimize annual inventory costs by ordering 60 cases at a time, 4500/60 = 75 times per year.

Let us take a moment to compare the cost of ordering 45 times per year (60 cases) with ordering once per year (4500 cases). We will let $x = 60$ and $x = 4500$ in the inventory cost function.

$$C(x) = 0.75x + \frac{2700}{x}$$

$$C(60) = 0.75(60) + \frac{2700}{60} = 90$$

$$C(4500) = 0.75(4500) + \frac{2700}{4500} = 3375.60$$

At 75 orders per year, the cost is \$90. At one order per year, the cost is \$3375.60. Compare these costs to ordering one case at a time: $C(1) = 0.75(1) + \frac{2700}{1} = \$2700.75$.

- A convenience store sells 980 cases of milk per year. Each case of milk costs \$4 to store one year, and each order costs \$2.50. How many cases of milk should the store manager order at a time?

  The average number of cases of milk in storage is $\frac{x}{2}$, and there are $\frac{980}{x}$ orders per year. The storage costs are $4(\frac{x}{2}) = 2x$, and the annual order costs are $2.50(\frac{980}{x}) = \frac{2450}{x}$. The annual inventory costs are $C = 2x + \frac{2450}{x}$.

$$C' = 2 - \frac{2450}{x^2}$$

$$0 = 2 - \frac{2450}{x^2}$$

$$2 = \frac{2450}{x^2}$$

$$2x^2 = 2450$$

$$x^2 = \frac{2450}{2} = 1225$$

$$x = 35$$

The manager should order 35 cases of milk at a time.

PRACTICE

1. A car repair shop expects to use 1620 boxes of spark plugs per year. Each box costs \$3 to store for one year, and each order costs \$7.50. How many boxes should be ordered each year? How many orders per year will there be?
2. A furniture store sells 90 designer lamps per year. It costs \$10 to store one lamp for one year. Each order costs \$8. How many lamps should be ordered at a time?
3. A home improvement store sells 5000 batteries per year. Each battery costs \$0.80 to store for one year, and each order costs \$5. How many orders should be placed each year?

SOLUTIONS

1. The average number of boxes in storage is $\frac{x}{2}$, so the annual storage costs are $3(\frac{x}{2}) = 1.50x$. There will be $\frac{1620}{x}$ orders per year, so the annual order costs are $7.50(\frac{1620}{x}) = \frac{12{,}150}{x}$. Annual inventory costs are $C = 1.50x + \frac{12{,}150}{x}$.

$$C' = 1.50 - \frac{12,150}{x^2}$$

$$0 = 1.50 - \frac{12,150}{x^2}$$

$$1.50 = \frac{12,150}{x^2}$$

$$1.50x^2 = 12,150$$

$$x^2 = \frac{12,150}{1.50} = 8100$$

$$x = 90$$

Minimize the annual inventory costs by ordering 90 boxes of spark plugs at a time, 1620/90 = 18 times per year.

2. The average number of lamps in inventory is $\frac{x}{2}$, so the annual storage costs are $10(\frac{x}{2}) = 5x$. There are $\frac{90}{x}$ orders each year, so the annual order

costs are $8(\frac{90}{x}) = \frac{720}{x}$. Annual inventory costs are $C = 5x + \frac{720}{x}$.

$$C' = 5 - \frac{720}{x^2}$$

$$0 = 5 - \frac{720}{x^2}$$

$$5 = \frac{720}{x^2}$$

$$5x^2 = 720$$

$$x^2 = \frac{720}{5} = 144$$

$$x = 12$$

The store manager should order 12 lamps at a time.

3. The average number of batteries in inventory is $\frac{x}{2}$, so the annual storage costs are $0.80(\frac{x}{2}) = 0.40x$. There are $\frac{5000}{x}$ orders each year, so the annual order costs are $5(\frac{5000}{x}) = \frac{25{,}000}{x}$. Annual inventory costs are $C = 0.40x + \frac{25{,}000}{x}$.

$$C' = 0.40 - \frac{25{,}000}{x^2}$$

$$0 = 0.40 - \frac{25{,}000}{x^2}$$

$$0.40 = \frac{25{,}000}{x^2}$$

$$0.40x^2 = 25{,}000$$

$$x^2 = \frac{25{,}000}{0.40} = 62,500$$

$$x = 250$$

The store manager should make 5000/250 = 20 orders per year.

CHAPTER 10 REVIEW

1. When a builders' supplier sells x tons of a material, the revenue can be approximated by $R = 1000x^3 - 28,125x^2 + 174{,}500x + 14{,}000$ and the cost by $C = 6000x + 10{,}000$, valid up to seven tons. How many tons should be sold to maximize the profit?
   (a) 2.25
   (b) 2.85
   (c) 3.35
   (d) 3.75

2. A storage area is to be enclosed with an area of 4200 square feet. Material and labor cost \$6 per foot for three sides and the side next to a road costs \$8 per foot (Figure 10.16). What is the least the fence can cost?

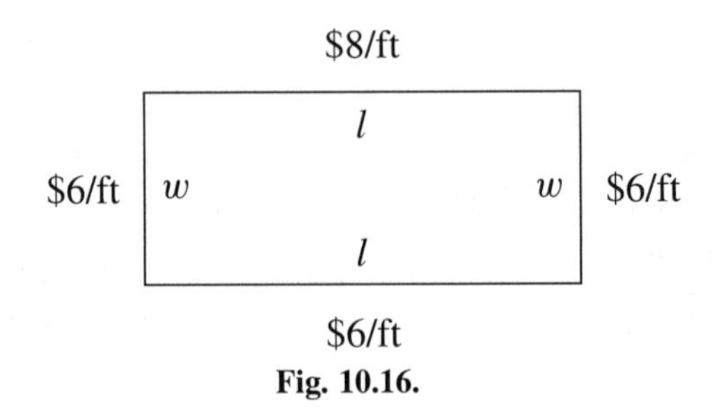

Fig. 10.16.

   (a) \$1420
   (b) \$1680
   (c) \$1710
   (d) \$1860

3. A farm and ranch store sells 100 pallets of a mineral mix per year. Each pallet of mix costs \$3 to store for one year. Each order costs \$6. How many orders per year will minimize the annual inventory cost?
   (a) 5
   (b) 10
   (c) 15
   (d) 20

4. When the price for a product is p, the revenue can be approximated by $R = -12p^2 + 432p + 2112$ (valid for prices up to \$40). What price maximizes revenue?

(a) \$18
(b) \$6
(c) \$40
(d) \$24

5. A piece of cardboard, 21" × 18", will be used to construct an open-topped box. A square will be cut from each corner and the sides folded up. How much should be cut from each corner in order to maximize the volume?
(a) About 3.2"
(b) About 3.8"
(c) About 4.3"
(d) About 4.6"

6. The number of customers served by a company from 1990 to 1996 can be approximated by $y = -85x^3 + 918x^2 - 2624x + 11219$, where $x = 0$ corresponds to the year 1990. During what year were there the fewest customers? (Round x up. For example, $x = 2.85$ would be during the year 1993.)
(a) 1991
(b) 1992
(c) 1994
(d) 1995

7. A box with a square bottom and a volume of eight cubic feet is to be constructed with cardboard. The cardboard that is used for the bottom costs \$0.24 per square foot. Cardboard used for the top costs \$0.16 per square foot; and for the sides, \$0.20 per square foot. What is the height of the box that minimizes the cost?
(a) 4 feet
(b) $\sqrt{2}$ feet
(c) 1 foot
(d) 2 feet

8. The revenue for selling x units of a product can be approximated by $R = -0.003x^2 + 30x - 63{,}000$. How many units must be sold in order to maximize revenue?
(a) 2000
(b) 3500
(c) 4800
(d) 5000

9. A video store owner rents videos for \$2 per night. Each week, an average of 1400 videos are rented. A consultant informed the owner that for every increase of \$0.20 in the price, there would be a loss of 100 rentals each week. What should she charge for the videos in order to maximize revenue?
   (a) \$2.10
   (b) \$2.30
   (c) \$2.40
   (d) \$2.50

10. A rancher wants to enclose a rectangular pasture next to a creek. The side along the creek does not need to be fenced. The rancher has 600 feet of fencing materials available. What is the maximum area that can be enclosed?
    (a) 40,000 square feet
    (b) 45,000 square feet
    (c) 50,000 square feet
    (d) 55,000 square feet

11. The cost to produce x thousand feet of wire can be approximated by $C = 0.035x^2 - 0.297x + 4.586$. How much wire should be produced to minimize the average cost?
    (a) About 4.2 thousand feet
    (b) About 4.6 thousand feet
    (c) About 11.4 thousand feet
    (d) About 41.3 thousand feet

SOLUTIONS

1. d	2. b	3. a	4. a	5. a	6. b	7. d
8. d	9. c	10. b	11. c			

CHAPTER 11

Exponential and Logarithmic Functions

When a quantity's rate of change is a fixed percentage over time, then it is changing *exponentially*. An exponential function has a variable in the exponent: $y = 5^x$, $y = 2^{x-1}$, $f(x) = 4 - 7^x$, and $R(x) = 3000(0.9)^x$. For example, the value of a \$1000 investment that grows 10% per year can be found using the exponential function $y = 1000(1.10)^x$, where the investment is worth y dollars after x years. If the investment is five years old, then it is worth $y = 1000(1.10)^5 = 1000(1.61051) =$ \$1610.51.

Many quantities in the natural and social sciences change exponentially. A city that grows at the rate of 3% per year is growing exponentially, the number of bacteria growing on food increases exponentially, the value of equipment that is depreciated 10% per year decreases exponentially, and the radioactivity of plutonium decreases exponentially.

For an exponential function that is increasing, the rate of change increases, too. Suppose \$1000 is invested at 10% interest for ten years. If the interest is not compounded, that is, the interest does not earn interest, then \$100 is earned each of the ten years. If the interest is compounded, then 10% of the previous year's balance earns interest, too. The \$100 earned in the first year earns interest for nine years, the \$100 earned in the second year earns interest for eight years, and so on (see Table 11.1).

Table 11.1

Year	Compounded interest	Noncompounded (simple) interest	Difference
1	\$100.00	\$100	\$0
2	\$110.00	\$100	\$10.00
3	\$121.00	\$100	\$21.00
4	\$133.10	\$100	\$33.10
5	\$146.41	\$100	\$46.41
6	\$161.05	\$100	\$61.05
7	\$177.16	\$100	\$77.16
8	\$194.87	\$100	\$94.87
9	\$214.36	\$100	\$114.36
10	\$235.80	\$100	\$135.80
Total Interest	\$1593.75	\$1000	

The difference between interest that is compounded and not compounded gets larger and larger each year. This effect gets more dramatic over time. The graphs in Figure 11.1 show the value of these investments over twenty years. The line shows the value of the account with simple (noncompounded) interest, and the curve shows the value of the account with compounded interest. Notice how the distance between the curve and the line gets larger each year.

The graph for an increasing exponential function (of the form $y = a^x$) looks like the graph in Figure 11.2, and the graph for a decreasing exponential function (of the form $y = a^x$) looks like the graph in Figure 11.3.

Because the derivative of a function tells us where the function is increasing and where it is decreasing, the derivative of an increasing exponential function, of the form $y = a^x$, is always positive. The derivative of a decreasing exponential function, of the form $y = a^x$, is always negative. Both graphs cup upward, so the second derivative is always positive. Remember that the second derivative

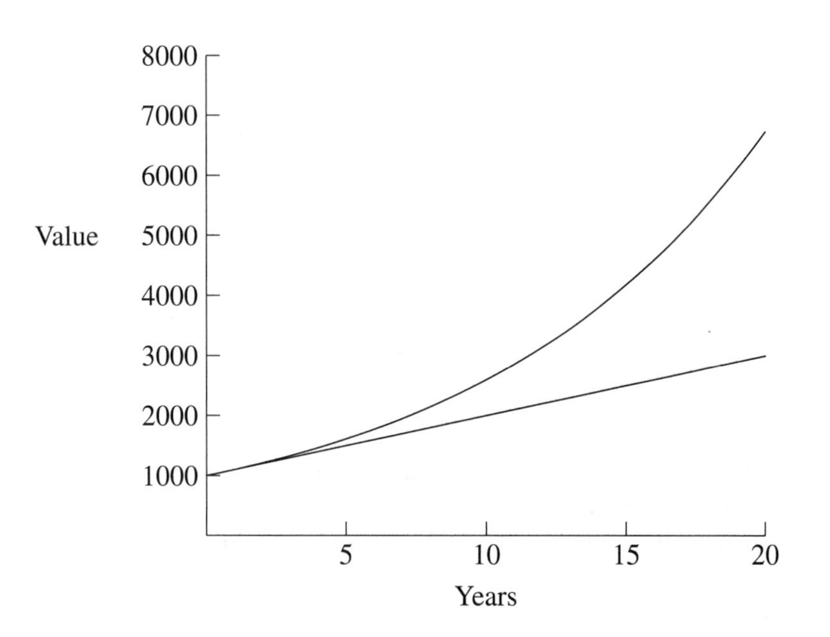

Fig. 11.1.

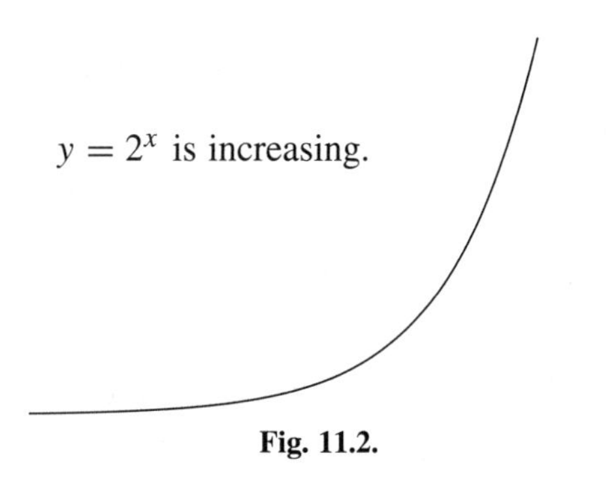

Fig. 11.2.

describes how fast the rate of change is changing. On an increasing exponential function, this means that the function is increasing faster and faster. This fact is illustrated in Table 11.1, which shows the compound interest (the *change* in the value of the investment) growing more each year.

Before we work with the derivative of exponential functions, we will review some algebraic properties of exponents. The number being raised to a power is the *base*.

- The base for the function $y = 1000(1.10)^x$ is 1.10.
- The base for the function $y = 5^x$ is 5.

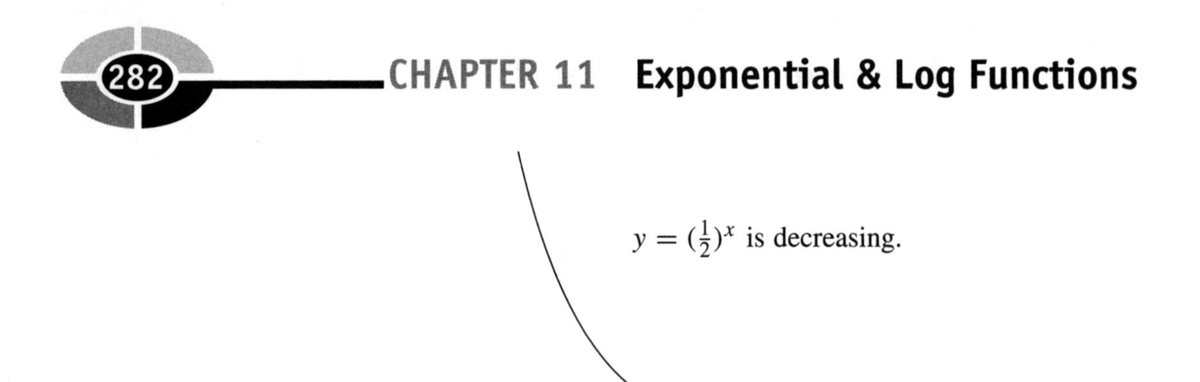

Fig. 11.3.

The base of an increasing function is larger than one. The base of a decreasing function is positive but less than one.

Let a be a positive number and m and n be any real numbers.

1. $a^m \cdot a^n = a^{m+n}$
2. $\dfrac{a^m}{a^n} = a^{m-n}$
3. $(a^m)^n = a^{mn}$
4. $a^0 = 1$
5. $a^{-n} = \dfrac{1}{a^n}$
6. $\sqrt[n]{a^m} = a^{m/n}$

EXAMPLES

Use Properties 1–3 to simplify the expression.

- $3^4 \cdot 3^2 = 3^{4+2} = 3^6$ Property 1
- $7 \cdot 7^3 = 7^1 \cdot 7^3 = 7^{1+3} = 7^4$ Property 1
- $2^x \cdot 2^3 = 2^{x+3}$ Property 1
- $\dfrac{6^5}{6^2} = 6^{5-2} = 6^3$ Property 2
- $\dfrac{6}{6^3} = \dfrac{6^1}{6^3} = 6^{1-3} = 6^{-2}$ Property 2
- $\dfrac{a^2}{a^x} = a^{2-x}$ Property 2
- $(5^2)^3 = 5^{2 \cdot 3} = 5^6$ Property 3

- $(e^x)^2 = e^{x \cdot 2} = e^{2x}$ Property 3
- $(2^{-3})^5 = 2^{(-3)(5)} = 2^{-15}$ Property 3

PRACTICE

Use Properties 1–3 to simplify the expression.

1. $4 \cdot 4^x$
2. $\dfrac{9^3}{9^x}$
3. $(a^3)^x$
4. $7^2 \cdot 7^5$
5. $\dfrac{a^9}{a^4}$
6. $(2^3)^4$

SOLUTIONS

1. $4 \cdot 4^x = 4^1 \cdot 4^x = 4^{1+x}$
2. $\dfrac{9^3}{9^x} = 9^{3-x}$
3. $(a^3)^x = a^{3x}$
4. $7^2 \cdot 7^5 = 7^{2+5} = 7^7$
5. $\dfrac{a^9}{a^4} = a^{9-4} = a^5$
6. $(2^3)^4 = 2^{3 \cdot 4} = 2^{12}$

An important base for exponential functions is the number e, named in honor of Leonard Euler. It is the limit, as $n \to \infty$, of $(1+\frac{1}{n})^n$. Its decimal approximation is 2.718281828.... Because e is larger than 1, $y = e^x$ is an increasing exponential function. It is used when an exponential quantity is continuously changing. Large populations of people continuously change, a radioactive substance continuously decays, a hot cup of coffee continuously cools. The number e is used in all of these applications. This function has the remarkable property that it is its own derivative. It is the only function (other than $y = 0$) that is its own rate of change. That is, the derivative of e^x is e^x.

The derivative of $y = e^x$ is $y' = e^x$.

By the chain rule, the derivative of $y = e^{f(x)}$ is $y' = f'(x)e^{f(x)}$. The derivative is the original function multiplied by the power's derivative. (We will see why this works later.)

If $f(x)$ is a differentiable function and $y = e^{f(x)}$, then $y' = f'(x)e^{f(x)}$.

EXAMPLES

Find y'.

- $y = e^{3x+1}$

  The derivative of the power, $3x + 1$, is 3, so $y' = 3e^{3x+1}$.

- $y = e^{x+5}$

  The derivative of $x + 5$ is 1, so $y' = 1 \cdot e^{x+5} = e^{x+5}$.

- $y = e^{x^2-2x+3}$

  The derivative of $x^2 - 2x + 3$ is $2x - 2$. This makes $y' = (2x - 2)e^{x^2-2x+3}$.

- Find the tangent line to $y = e^{2x-4}$ at (2, 1).

The slope of the tangent line is y' evaluated at $x = 2$.

$$y' = 2e^{2x-4}$$

$$m = 2e^{2(2)-4} = 2e^0 = 2 \cdot 1 = 2 \qquad \text{By Property 4, } e^0 = 1.$$

With $x = 2$, $y = 1$, and $m = 2$, $y = mx + b$ becomes $1 = 2 \cdot 2 + b$, which gives us $b = -3$. The tangent line is $y = 2x - 3$.

PRACTICE

Find y' for problems 1–6.

1. $y = e^{6x}$
2. $y = e^{x^3-1}$
3. $y = e^{x^2+4x+2}$
4. $y = e^{x-10}$
5. $y = e^{\sqrt{x}}$
6. $y = e^{\sqrt{3x+1}}$
7. Find the tangent line to $y = e^{x^2-1}$ at $(-1, 1)$.

SOLUTIONS

1. $y' = 6e^{6x}$
2. $y' = 3x^2 e^{x^3-1}$
3. $y' = (2x+4)e^{x^2+4x+2}$
4. $y' = 1 \cdot e^{x-10} = e^{x-10}$
5. $y = e^{\sqrt{x}} = e^{x^{1/2}}$

$$y' = \frac{1}{2}x^{-1/2}e^{\sqrt{x}} \text{ or } \frac{1}{2x^{1/2}}e^{\sqrt{x}} = \frac{e^{\sqrt{x}}}{2\sqrt{x}}$$

6. $y = e^{\sqrt{3x+1}} = e^{(3x+1)^{1/2}}$

$$y' = \frac{1}{2}(3x+1)^{-1/2}(3)e^{\sqrt{3x+1}} \text{ or } \frac{3}{2} \cdot \frac{1}{(3x+1)^{1/2}}e^{\sqrt{3x+1}} = \frac{3e^{\sqrt{3x+1}}}{2\sqrt{3x+1}}$$

7. The slope of the tangent line is the derivative evaluated at $x = -1$.

$$y' = 2xe^{x^2-1}$$

$$m = 2(-1)e^{(-1)^2-1} = -2e^0 = -2(1) = -2$$

With $x = -1$, $y = 1$, and $m = -2$, $y = mx + b$ becomes $1 = -2(-1) + b$. This gives us $b = -1$. The tangent line is $y = -2x - 1$.

The derivative of $y = e^{f(x)}$ is $y = f'(x)e^{f(x)}$ because of the chain rule, $\frac{dy}{dx} = \frac{dy}{du} \cdot \frac{du}{dx}$. When $u = f(x)$, $\frac{du}{dx} = f'(x)$. This allows us to write $y = e^{f(x)}$ as $y = e^u$, which means that $y' = e^u$.

$$\frac{dy}{dx} = \frac{dy}{du} \cdot \frac{du}{dx}$$

$$= e^u \frac{du}{dx} \qquad \text{Replace } e^u \text{ with } e^{f(x)} \text{ and } \frac{du}{dx} \text{ with } f'(x).$$

$$= e^{f(x)} f'(x)$$

Now that we can differentiate $e^{f(x)}$, we can differentiate products and quotients containing exponential functions. The power rule is not necessary as we will see in the next example.

EXAMPLES

Find y'.

- $y = (e^{3x-2})^4$

  Because of the exponent rule $(a^m)^n = a^{mn}$, we can rewrite $(e^{3x-2})^4$ as $e^{4(3x-2)} = e^{12x-8}$, which is a little easier to differentiate: $y' = 12e^{12x-8}$. Compare this method to using the power rule on $(e^{3x-2})^4$.

$$y' = 4(e^{3x-2})^3(3e^{3x-2}) = 12(e^{3x-2})^3(e^{3x-2})^1 = 12(e^{3x-2})^{3+1}$$

$$= 12(e^{3x-2})^4 \text{ or } 12e^{4(3x-2)} = 12e^{12x-8}$$

- $y = xe^{6x}$

  xe^{6x} is the product of $F(x) = x$ and $G(x) = e^{6x}$. We will use the product rule.

$$y' = \overbrace{1}^{F'} \cdot \overbrace{e^{6x}}^{G} + \overbrace{x}^{F} \cdot \overbrace{6e^{6x}}^{G'} = e^{6x} + 6xe^{6x}$$

- $y = x^2 - 3x + x^2e^{-x}$

$$y' = 2x - 3 + \overbrace{2x}^{F'}\ \overbrace{e^{-x}}^{G} + \overbrace{x^2}^{F}\ \overbrace{(-1)e^{-x}}^{G'} = 2x - 3 + 2xe^{-x} - x^2e^{-x}$$

- $y = \frac{e^{5x}}{e^{x^2}}$

  Rather than use the quotient rule, we can use the exponent property

$\frac{a^m}{a^n} = a^{m-n}$ to rewrite this as $y = e^{5x-x^2}$. This is much easier to differentiate: $y' = (5-2x)e^{5x-x^2}$.

- $y = \frac{e^{4x+3}}{x^2+1}$

  We have the quotient of $F(x) = e^{4x+3}$ and $G(x) = x^2 + 1$. By the quotient rule,

$$y' = \frac{4e^{4x+3}(x^2+1) - e^{4x+3}(2x)}{(x^2+1)^2} \quad \text{or} \quad \frac{2e^{4x+3}(2(x^2+1)-x)}{(x^2+1)^2}$$

$$= \frac{2e^{4x+3}(2x^2+2-x)}{(x^2+1)^2}$$

PRACTICE

Find y'.

1. $y = 4xe^{3x}$
2. $y = 2x^3e^{5x^2}$
3. $y = (e^{x^3})^6$
4. $y = 9x^4 - e^{5x}$
5. $$y = \frac{e^{x^2+3}}{e^{x-4}}$$
6. $$y = \frac{e^{x^2+x}}{4x^2-1}$$

SOLUTIONS

1. $y' = 4e^{3x} + 4x(3)e^{3x} = 4e^{3x} + 12xe^{3x}$
2. $y' = 6x^2e^{5x^2} + 2x^3(10x)e^{5x^2} = 6x^2e^{5x^2} + 20x^4e^{5x^2}$
3. Instead of using the power rule, we will rewrite the function as $y = e^{x^3 \cdot 6} = e^{6x^3}$. This gives us $y' = 18x^2e^{6x^3}$.
4. $y' = 36x^3 - 5e^{5x}$

5. We will use the exponent rule $\frac{a^m}{a^n} = a^{m-n}$ to simplify the function before differentiating.

$$y = \frac{e^{x^2+3}}{e^{x-4}} = e^{(x^2+3)-(x-4)} = e^{x^2-x+7}$$

$$y' = (2x-1)e^{x^2-x+7}$$

6.

$$y' = \frac{(2x+1)e^{x^2+x}(4x^2-1) - e^{x^2+x}(8x)}{(4x^2-1)^2}$$

$$\text{or } \frac{e^{x^2+x}((2x+1)(4x^2-1)-8x)}{(4x^2-1)^2} = \frac{e^{x^2+x}(8x^3+4x^2-10x-1)}{(4x^2-1)^2}$$

The *logistic function* is related to the exponential function. The logistic function describes some quantities that increase for a time and then level off. The basic logistic function has the form

$$y = \frac{a}{1+be^{-cx}}$$

where a, b, and c are fixed numbers. The graph of $y = \frac{10}{1+5e^{-0.15x}}$ is in Figure 11.4.

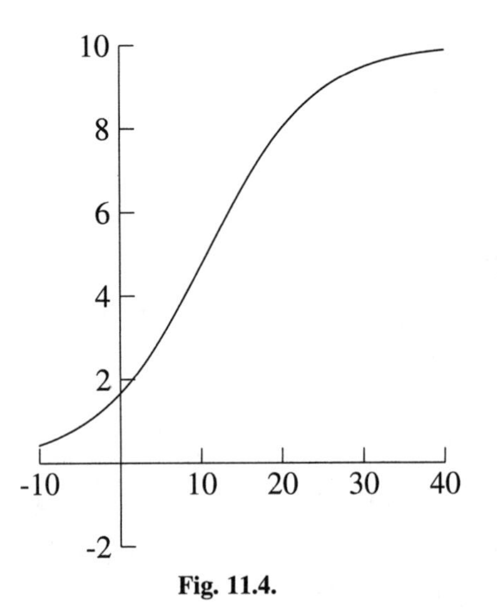

Fig. 11.4.

We will use the quotient rule to differentiate logistic functions.

EXAMPLE

- $y = \frac{100}{50+e^{-0.1x}}$

$$y' = \frac{0(50+e^{-0.1x}) - 100(-0.1e^{-0.1x})}{(50+e^{-0.1x})^2}$$
$$= \frac{10e^{-0.1x}}{(50+e^{-0.1x})^2}$$

Logarithms

Quantities that begin rising (or falling) quickly at first and then more slowly can sometimes be approximated by logarithmic functions. The logarithmic equation $y = \log_a x$ is simply a different way of writing the exponential equation $a^y = x$. The number a is the base for both the logarithm and the exponent. The graphs of logarithmic functions are very similar to the graphs of exponential functions. The dashed graph in Figure 11.5 is the graph of $y = 2^x$. The solid graph is the graph of $y = \log_2 x$.

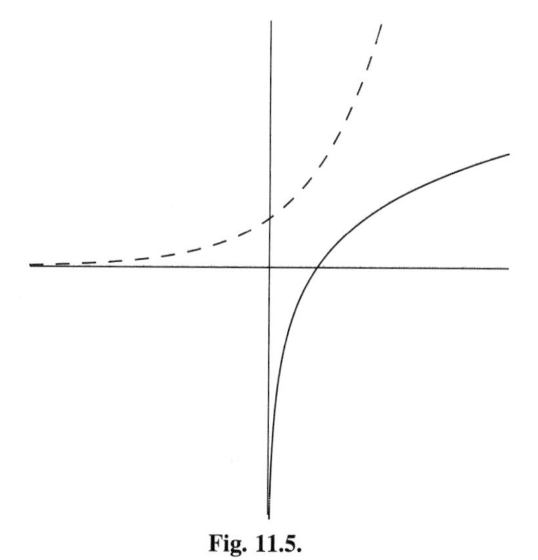

Fig. 11.5.

Logarithms are used to solve exponential equations. For example, suppose \$1000 is an investment that earns 10% annual interest, compounded annually. Recall that the investment is worth y dollars after x years, where $y = 1000(1.10)^x$. Suppose we want to know how long it would take for this investment to grow to \$2000. In other words, we want to solve the equation $2000 = 1000(1.10)^x$. We will be able to solve this equation later.

Before getting to the derivative of logarithmic functions, we will practice rewriting exponential and logarithmic equations as well as cover some logarithm properties.

EXAMPLES

Using the fact that $y = \log_a x$ means $a^y = x$, rewrite the logarithmic equations as exponential equations and the exponential equations as logarithmic equations.

- $6^x = 10$ becomes $\log_6 10 = x$
- $3^2 = 9$ becomes $\log_3 9 = 2$
- $10^{-2} = 0.01$ becomes $\log_{10} 0.01 = -2$
- $4^{3/2} = 8$ becomes $\log_4 8 = \frac{3}{2}$
- $\log_2 x = 6$ becomes $2^6 = x$
- $\log_6 \frac{1}{36} = -2$ becomes $6^{-2} = \frac{1}{36}$
- $\log_{10} 1000 = 3$ becomes $10^3 = 1000$
- $\log_a m = n$ becomes $a^n = m$

The logarithm with base e has its own notation, $\ln x$ means $\log_e x$. "ln" is called the *natural logarithm.*

- $\ln x = 10$ means $e^{10} = x$
- $e^2 = 7.389$ means $\ln 7.389 = 2$

PRACTICE

Rewrite the logarithmic equations as exponential equations and the exponential equations as logarithmic equations.

1. $5^2 = 25$
2. $3^x = 29$
3. $\log_4 9 = x$
4. $\ln 4 = 1.3863$
5. $\log_{10} 0.1 = -1$

6. $e^{3x} = 16$
7. $1.10^x = 2$
8. $\log_{49} 7 = \frac{1}{2}$

SOLUTIONS

1. $\log_5 25 = 2$
2. $\log_3 29 = x$
3. $4^x = 9$
4. $e^{1.3863} = 4$
5. $10^{-1} = 0.1$
6. $\ln 16 = 3x$
7. $\log_{1.10} 2 = x$
8. $49^{1/2} = 7$

There are two cancelation properties—one in which logarithms cancel exponents, and the other in which exponents cancel logarithms. Both come from rewriting exponential and logarithmic equations.

$$a^{\log_a x} = x \quad \text{and} \quad \log_a a^x = x$$

When the base is e,

$$e^{\ln x} = x \quad \text{and} \quad \ln e^x = x$$

If we rewrite the exponential equation $a^{\log_a x} = x$ as a logarithmic equation, where $\log_a x$ is the exponent, we have $\log_a x = \log_a x$. When we rewrite the logarithmic equation, $\log_a a^x = x$ as an exponential equation, we have $a^x = a^x$. We will use these properties later when we find the derivatives of $y = a^x$ and $y = \log_a x$.

EXAMPLES

- $5^{\log_5 7} = 7$
- $\ln e^{16} = 16$
- $9^{\log_9 4} = 4$
- $\log_{10} 10^x = x$
- $e^{\ln 24} = 24$
- $\log_{20} 20^t = t$

We will use other logarithm properities to make differentiation a little easier.

1. $\log_a mn = \log_a m + \log_a n$
2. $\log_a \frac{m}{n} = \log_a m - \log_a n$
3. $\log_a m^n = n \log_a m$

EXAMPLES

- $\log_{10} 6x = \log_{10} 6 + \log_{10} x$ Property 1
- $\log_4 9 - \log_4 5 = \log_4 \dfrac{9}{5}$ Property 2
- $2\log_3 x = \log_3 x^2$ Property 3
- $\ln \dfrac{x}{10} = \ln x - \ln 10$ Property 2
- $\log_5 7^3 = 3\log_5 7$ Property 3
- $\ln 15y = \ln 15 + \ln y$ Property 1

We can find the derivative of $y = \ln x$ from the fact that the derivative of $e^{f(x)}$ is $f'(x)e^{f(x)}$. We will begin by rewriting $y = \ln x$ as an exponential equation: $e^y = x$. Now we will differentiate each side of this equation implicitly, with respect to x.

$$\frac{d}{dx}(e^y) = \frac{d}{dx}(x)$$

$$e^y \frac{dy}{dx} = 1$$ The derivative of e^y is e^y times the derivative of y.

$$\frac{dy}{dx} = \frac{1}{e^y}$$ Divide both sides of the equation by e^y.

$$\frac{dy}{dx} = \frac{1}{x}$$ We know that e^y is x.

The derivative of $y = \ln x$ is $y' = \frac{1}{x}$.

Using implicit differentiation and the fact that the derivative of $\ln x$ is $\frac{1}{x}$, we can find the derivative of $y = \ln f(x)$. Again, we will rewrite the logarithmic equation as an exponential equation: $e^y = f(x)$ and use implicit differentiation.

$$\frac{d}{dx}(e^y) = \frac{d}{dx}(f(x))$$

$$e^y \frac{dy}{dx} = f'(x)$$

$$\frac{dy}{dx} = \frac{f'(x)}{e^y}$$

$$\frac{dy}{dx} = \frac{f'(x)}{f(x)} \qquad \text{We know that } e^y \text{ is } f(x).$$

The derivative of the natural logarithm of $f(x)$ is a fraction whose numerator is $f'(x)$ and whose denominator is $f(x)$.

If $f(x)$ is a differentiable function and $y = \ln f(x)$, then $y' = \frac{f'(x)}{f(x)}$.

EXAMPLES

Find y'.

- $y = \ln(x^2 - 9)$

  The derivative of $x^2 - 9$ is $2x$.

  $$y' = \frac{2x}{x^2 - 9}$$

- $y = \ln(4x^2 + 3x - 1)$

  The derivative of $4x^2 + 3x - 1$ is $8x + 3$.

  $$y' = \frac{8x + 3}{4x^2 + 3x - 1}$$

- $y = \ln(x^{-2} - x^{-3})$

  The derivative of $x^{-2} - x^{-3}$ is $-2x^{-3} - (-3)x^{-4} = -2x^{-3} + 3x^{-4}$.

  $$y' = \frac{-2x^{-3} + 3x^{-4}}{x^{-2} - x^{-3}}$$

At times, the logarithm properties can save us from having to use the product rule, quotient rule, and/or power rule. For example, the property $\log_a m^n = n \log_a m$ allows us to rewrite $y = \ln(x^2 + 10)^4$ as $y = 4\ln(x^2 + 10)$ and avoid using the power rule. Let us take a moment to compare differentiating this function with and without using this property. If we do not use the property and differentiate using the power rule, we have

$$y' = \frac{4(x^2+10)^3(2x)}{(x^2+10)^4} = \frac{8x}{x^2+10}.$$

If we use the property and differentiate $y = 4\ln(x^2+10)$, we have

$$y' = 4\left(\frac{2x}{x^2+10}\right) = \frac{8x}{x^2+10}.$$

EXAMPLES

- $y = \ln \frac{x^2}{x+5}$

  We will use the property $\log_a \frac{m}{n} = \log_a m - \log_a n$ to rewrite the function as $y = \ln x^2 - \ln(x+5)$.

  $$y' = \frac{2x}{x^2} - \frac{1}{x+5} = \frac{2}{x} - \frac{1}{x+5}$$

  (We could also rewrite $\ln x^2$ as $2\ln x$.) Compare this to differentiating without using the logarithm property.

  $$y' = \frac{\frac{2x(x+5)-x^2(1)}{(x+5)^2}}{\frac{x^2}{x+5}}$$

- $y = \ln[(x^2-4)(2x+5)]$

  We will use the logarithm property $\log_a mn = \log_a m + \log_a n$ to rewrite the function as $y = \ln(x^2-4) + \ln(2x+5)$.

  $$y' = \frac{2x}{x^2-4} + \frac{2}{2x+5}$$

- $y = \ln \sqrt[3]{x+1}$

  Because $\sqrt[3]{x+1} = (x+1)^{1/3}$, we have $y = \ln(x+1)^{1/3}$. By the third logarithm property, this is equal to $y = \frac{1}{3}\ln(x+1)$.

  $$y' = \frac{1}{3}\left(\frac{1}{x+1}\right) = \frac{1}{3(x+1)} = \frac{1}{3x+3}$$

We have to be careful not to evaluate the derivative at an x-value that is not allowed in the original function. The graph in Figure 11.6 is the graph of $y =$

$\ln(x-3)$. The derivative of this function is $y' = \frac{1}{x-3}$. Although we can evaluate $\frac{1}{x-3}$ at $x = 1$, we cannot let $x = 1$ in the original function. As you can see in the graph, the function does not exist to the left of $x = 3$.

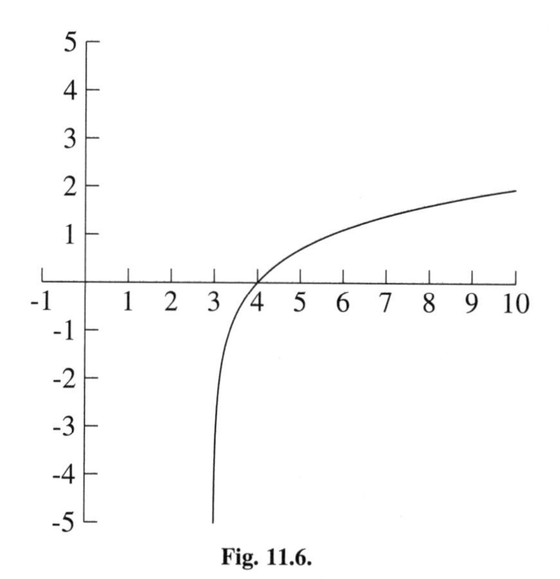

Fig. 11.6.

PRACTICE

Find y'.

1. $y = \ln(12x - 7)$
2. $y = \ln(x^2 + 3)$
3. $y = \ln(5x)$
4. $y = \ln[(6x^2 - x)(x + 4)]$
5. $y = \ln\sqrt{x^2 - 6x + 10}$
6. $y = \ln\frac{x+1}{x-1}$
7. $y = \ln(14x^2 + x - 3)^5$
8. $y = \ln[(x + 3)(x - 6)(x^2 + 2)]$

SOLUTIONS

1.

$$y' = \frac{12}{12x - 7}$$

2.

$$y' = \frac{2x}{x^2 + 3}$$

3.

$$y' = \frac{5}{5x} = \frac{1}{x}$$

4. $y = \ln[(6x^2 - x)(x + 4)] = \ln(6x^2 - x) + \ln(x + 4)$

$$y' = \frac{12x - 1}{6x^2 - x} + \frac{1}{x + 4}$$

5. $y = \ln\sqrt{x^2 - 6x + 10} = \ln(x^2 - 6x + 10)^{1/2} = \frac{1}{2}\ln(x^2 - 6x + 10)$

$$y' = \frac{1}{2}\left(\frac{2x - 6}{x^2 - 6x + 10}\right) = \frac{1}{2} \cdot \frac{2(x - 3)}{x^2 - 6x + 10} = \frac{x - 3}{x^2 - 6x + 10}$$

6. $y = \ln\frac{x+1}{x-1} = \ln(x + 1) - \ln(x - 1)$

$$y' = \frac{1}{x + 1} - \frac{1}{x - 1}$$

7. $y = \ln(14x^2 + x - 3)^5 = 5\ln(14x^2 + x - 3)$

$$y' = 5\left(\frac{28x + 1}{14x^2 + x - 3}\right) = \frac{5(28x + 1)}{14x^2 + x - 3} = \frac{140x + 5}{14x^2 + x - 3}$$

8. $y = \ln[(x + 3)(x - 6)(x^2 + 2)] = \ln(x + 3) + \ln(x - 6) + \ln(x^2 + 2)$

$$y' = \frac{1}{x + 3} + \frac{1}{x - 6} + \frac{2x}{x^2 + 2}$$

Now that we can differentiate $\ln f(x)$, we will differentiate products, quotients, and powers of functions involving logarithms.

EXAMPLES

- $y = \sqrt{x+5}\ln(4x-3)$

  We have the product of $F(x) = \sqrt{x+5} = (x+5)^{1/2}$ and $G(x) = \ln(4x-3)$.

$$y' = \overbrace{\frac{1}{2}(x+5)^{-1/2}(1)}^{F'}\overbrace{\ln(4x-3)}^{G} + \overbrace{(x+5)^{1/2}}^{F}\overbrace{\frac{4}{4x-3}}^{G'}$$

$$\text{or } \frac{1}{2\sqrt{x+5}}\ln(4x-3) + \sqrt{x+5}\frac{4}{4x-3} = \frac{\ln(4x-3)}{2\sqrt{x+5}} + \frac{4\sqrt{x+5}}{4x-3}$$

- $y = [\ln(5x^2+7)]^3$

  We will use the power rule, $y' = n(f(x))^{n-1}f'(x)$, where $f(x) = \ln(5x^2+7)$, and $f'(x) = \frac{10x}{5x^2+7}$.

$$y' = 3[\ln(5x^2+7)]^2\left(\frac{10x}{5x^2+7}\right)$$

  The logarithm property $\log_a m^n = n\log_a m$ does not apply to $y = [\ln(5x^2+7)]^3$, which is $y = [\ln(5x^2+7)]\cdot[\ln(5x^2+7)]\cdot[\ln(5x^2+7)]$.

- $y = \frac{\ln(x^2-6)}{2x+4}$

  By the quotient rule,

$$y' = \frac{\frac{2x}{x^2-6}(2x+4) - [\ln(x^2-6)](2)}{(2x+4)^2}.$$

PRACTICE

Find y'.

1. $y = (16x^2+1)\cdot\ln(x-3)$
2. $y = \sqrt{\ln(3x^5-2x^3+x)}$
3. $y = \frac{\ln 10x}{x+4}$
4. $y = e^{10x}\ln(12x^2-9)$

SOLUTIONS

1.

$$y' = 32x\ln(x-3) + (16x^2+1)\cdot\frac{1}{x-3}$$

2. $y = [\ln(3x^5 - 2x^3 + x)]^{1/2}$

$$y' = \frac{1}{2}[\ln(3x^5 - 2x^3 + x)]^{-1/2}\cdot\frac{15x^4 - 6x^2 + 1}{3x^5 - 2x^3 + x}$$

3.

$$y' = \frac{\frac{10}{10x}(x+4) - (\ln 10x)(1)}{(x+4)^2}$$

4.

$$y' = 10e^{10x}\ln(12x^2-9) + e^{10x}\cdot\frac{24x}{12x^2-9}$$

Logarithms can greatly simplify the differentiation of complicated functions. For example, if we want to find the derivative of

$$y = \left(\frac{(x+3)(x^2+10)}{2x-1}\right)^4$$

we could use the power, quotient, and product rules. Using all of these formulas would be very messy, and the derivative would need tedious simplification. Instead, we will take the natural logarithm of both sides of the equation and use logarithm properties to simplify the right side of the equation, finally implicitly differentiating both sides of the equation.

$$\ln y = \ln\left(\frac{(x+3)(x^2+10)}{2x-1}\right)^4$$

$$\ln y = 4\ln\left[\frac{(x+3)(x^2+10)}{2x-1}\right] \qquad \text{Property 3}$$

$$\ln y = 4[\ln[(x+3)(x^2+10)] - \ln(2x-1)] \qquad \text{Property 2}$$

$$\ln y = 4[\ln(x+3) + \ln(x^2+10) - \ln(2x-1)] \qquad \text{Property 1}$$

We are ready to implicitly differentiate both sides of the equation.

$$\frac{\frac{dy}{dx}}{y} = 4\left(\frac{1}{x+3} + \frac{2x}{x^2+1} - \frac{2}{2x-1}\right) \qquad \text{The derivative of } \ln y \text{ is } \frac{\frac{dy}{dx}}{y}.$$

$$\frac{\frac{dy}{dx}}{y} = 4\left(\frac{1}{x+3}\right) + 4\left(\frac{2x}{x^2+1}\right) - 4\left(\frac{2}{2x-1}\right)$$

$$\frac{\frac{dy}{dx}}{y} = \frac{4}{x+3} + \frac{8x}{x^2+1} - \frac{8}{2x-1}$$

We will clear the fraction on the left side of the equation by multiplying both sides of the equation by y.

$$\frac{dy}{dx} = y\left(\frac{4}{x+3} + \frac{8x}{x^2+1} - \frac{8}{2x-1}\right)$$

Finally, we will replace y with $(\frac{(x+3)(x^2+10)}{2x-1})^4$ from the original equation.

$$\frac{dy}{dx} = \left(\frac{(x+3)(x^2+10)}{2x-1}\right)^4 \left(\frac{4}{x+3} + \frac{8x}{x^2+1} - \frac{8}{2x-1}\right)$$

One of the cancelation properties of logarithms and exponents allows us to differentiate a function of the form $y = a^x$, when a is any positive number that is not 1. From the cancelation property $t = b^{\log_b t}$, we have $a = e^{\ln a} (= e^{\log_e a})$. We will raise both sides of this equation to the x power.

$$a^x = (e^{\ln a})^x$$

By the exponent property $(a^m)^n = a^{mn}$, the above can be rewritten as $a^x = e^{x \ln a}$. Now we know that $y = a^x$ and $y = e^{x \ln a}$ are the same function. Recall that the derivative of $e^{f(x)}$ is $f'(x)e^{f(x)}$. Here, $f(x) = x \ln a$, x times the fixed number $\ln a$. This means that $f'(x) = \ln a$. The derivative of $y = e^{x \ln a}$ is $y' = (\ln a)e^{x \ln a}$. We can replace $e^{x \ln a}$ with a^x, so the derivative becomes $y' = (\ln a)(a^x)$.

The derivative of $y = a^x$ is $y' = (\ln a)(a^x)$.

EXAMPLES

Find y'.

- $y = 3^x$ $\qquad y' = (\ln 3)(3^x)$
- $y = 10^x$ $\qquad y' = (\ln 10)(10^x)$
- $y = 28^x$ $\qquad y' = (\ln 28)(28^x)$
- $y = \left(\frac{2}{3}\right)^x$ $\qquad y' = \left(\ln \frac{2}{3}\right)\left(\frac{2}{3}\right)^x$

Using the chain rule, we can find the derivative of $y = a^{f(x)}$. To fit the chain rule formula $\frac{dy}{dx} = \frac{dy}{du} \cdot \frac{du}{dx}$, we will write $y = a^u$, where $u = f(x)$ and $\frac{du}{dx} = f'(x)$.

$$\begin{aligned} \frac{dy}{dx} &= \frac{dy}{du} \cdot \frac{du}{dx} \\ &= (\ln a)a^u \cdot \frac{du}{dx} \\ &= (\ln a)a^{f(x)} \cdot f'(x) \end{aligned}$$

> If $f(x)$ is a differentiable function and $y = a^{f(x)}$,
> then $y' = f'(x)(\ln a)a^{f(x)}$.

EXAMPLES

Find y'.

- $y = 2^{x^3}$

  The derivative of the exponent is $3x^2$.

  $$y' = 3x^2(\ln 2)(2^{x^3})$$

- $y = 8^{2x+5}$

  The derivative of the exponent is 2.

  $$y' = 2(\ln 8)(8^{2x+5})$$

- $5^{\sqrt{x}}$

  The derivative of the exponent is $\frac{1}{2}x^{-1/2}$.

$$y' = \frac{1}{2}x^{-1/2}(\ln 5)(5^{\sqrt{x}})$$

PRACTICE

Find y'.

1. $y = 6^x$
2. $y = 15^x$
3. $y = (\frac{1}{2})^x$
4. $y = 10^{4x^3+6x+2}$
5. $y = 9^{\sqrt{x}+5x}$
6. $y = 21^{\ln 6x}$

SOLUTIONS

1. $y' = (\ln 6)(6^x)$
2. $y' = (\ln 15)(15^x)$
3. $y' = [\ln(\frac{1}{2})](\frac{1}{2})^x$
4. $y' = (12x^2 + 6)(\ln 10)(10^{4x^3+6x+2})$
5. $$y' = \left(\frac{1}{2}x^{-1/2} + 5\right)(\ln 9)(9^{\sqrt{x}+5x}) \text{ or } \left(\frac{1}{2\sqrt{x}} + 5\right)(\ln 9)(9^{\sqrt{x}+5x})$$
6. The derivative of $\ln 6x$ is $\frac{6}{6x} = \frac{1}{x}$.

$$y' = \frac{1}{x}(\ln 21)(21^{\ln 6x})$$

With the change of base formula, we can rewrite any logarithm as a natural logarithm. This allows us to differentiate functions of the form $y = \log_a f(x)$.

When we want to rewrite a logarithm in old base b to new base a (usually base 10 or base e), we can use the formula

$$\log_b m = \frac{\log_a m}{\log_a b}.$$

For now, we will practice using the formula. Later we will see where it comes from and how to use it to differentiate $y = \log_a f(x)$.

EXAMPLES

Rewrite the logarithm using the indicated base.

- $\log_3 20$, base 8

  The new base is $a = 8$, the old base is $b = 3$ and $m = 20$.

  $$\log_3 20 = \frac{\log_8 20}{\log_8 3}$$

- $\log_{15}(x^2 + 10)$, base 4

  $a = 4,\ b = 15$, and $m = x^2 + 10$

  $$\log_{15}(x^2 + 10) = \frac{\log_4(x^2 + 10)}{\log_4 15}$$

- $\log_{1.8} 2$, base 10

  $a = 10$, $b = 1.8$, and $m = 2$

  $$\log_{1.8} 2 = \frac{\log_{10} 2}{\log_{10} 1.8}$$

- $\log_{0.9} 6$, as a natural logarithm

  The base of the natural logarithm is e, so $a = e$, $b = 0.9$, and $m = 6$.

  $$\log_{0.9} 6 = \frac{\log_e 6}{\log_e 0.9} = \frac{\ln 6}{\ln 0.9}$$

PRACTICE

Rewrite the logarithm using the indicated base.

1. $\log_{20} x$, base 4
2. $\log_{1.3}(x - 9)$, base 5

3. $\log_{10} 2$, base e
4. $\log_{18} 15$, base 10
5. $\ln 5t$, base 1.4

SOLUTIONS

1.

$$\log_{20} x = \frac{\log_4 x}{\log_4 20}$$

2.

$$\log_{1.3}(x - 9) = \frac{\log_5(x - 9)}{\log_5 1.3}$$

3.

$$\log_{10} 2 = \frac{\log_e 2}{\log_e 10} = \frac{\ln 2}{\ln 10}$$

4.

$$\log_{18} 15 = \frac{\log_{10} 15}{\log_{10} 18}$$

5.

$$\ln 5t = \log_e 5t = \frac{\log_{1.4} 5t}{\log_{1.4} e}$$

To see where the change of base formula comes from, let us try to find a decimal approximation for $\log_5 28$ using only what we have learned about logarithms and exponents. We begin with the equation $x = \log_5 28$. Rewriting this equation as an exponential equation gives us $5^x = 28$. We will take the natural logarithm of both sides of this equation and solve for x. The reason we want the natural logarithm is that most calculators have a natural logarithm key, marked "LN."

$$5^x = 28$$

$$\ln 5^x = \ln 28$$

$$x \ln 5 = \ln 28 \qquad \text{Logarithm Property 3}$$

$$x = \frac{\ln 28}{\ln 5} \qquad \text{Divide both sides of the equation by } \ln 5.$$

$$x \approx \frac{3.33220451}{1.609437912} \approx 2.0704$$

These same steps allow us to write $\log_b m$ as $\frac{\log_a m}{\log_a b}$.

$$y = \log_b m$$

$$b^y = m \qquad \text{Rewrite as an exponential equation.}$$

$$\log_a b^y = \log_a m \qquad \text{Take logarithms of both sides.}$$

$$y \log_a b = \log_a m \qquad \text{Logarithm Property 3}$$

$$y = \frac{\log_a m}{\log_a b}$$

$$\log_b m = \frac{\log_a m}{\log_a b} \qquad \text{Replace } y \text{ with } \log_b m.$$

Recall the problem in which \$1000 is invested for x years earning 10% interest, compounded annually: $y = 1000(1.10)^x$. We want to know how long it would take for this investment to grow to \$2000. To do this, we need to solve the equation $2000 = 1000(1.10)^x$.

$$2000 = 1000(1.10)^x$$

$$2 = 1.10^x \qquad \text{Divide both sides by 1000.}$$

$$x = \log_{1.10} 2 \qquad \text{Rewrite the equation as a logarithmic equation.}$$

$$x = \frac{\ln 2}{\ln 1.10} \qquad \text{Use the change of base formula.}$$

$$x \approx 7.27$$

It would take 7.27 years (or 8 years if interest is not paid until the year is over) for the investment to grow to \$2000.

By the change of base formula, we can write $y = \log_a f(x)$ as

$$y = \frac{\ln f(x)}{\ln a} = \frac{1}{\ln a} \ln f(x).$$

The derivative of this function is $y' = \frac{1}{\ln a} \frac{f'(x)}{f(x)}$ or $y' = \frac{f'(x)}{(\ln a) f(x)}$.

If $f(x)$ is a differentiable function and $y = \log_a f(x)$, then $y' = \frac{f'(x)}{(\ln a) f(x)}$.

EXAMPLES

Find y'.

- $y = \log_5(10x + 3)$

$$y' = \frac{10}{(\ln 5)(10x + 3)}$$

- $y = \log_{10} 16x$

$$y' = \frac{16}{(\ln 10)16x} = \frac{1}{(\ln 10)x}$$

- $y = \log_4(x^2 - 6)$

$$y' = \frac{2x}{(\ln 4)(x^2 - 6)}$$

PRACTICE

Find y'.

1. $y = \log_9(14x + 15)$
2. $y = \log_{10} x^2$
3. $y = \log_{2.4}(5x^3 + 2x^2 - 3)$
4. $y = \log_{34} \sqrt{x}$

SOLUTIONS

1.

$$y' = \frac{14}{(\ln 9)(14x + 15)}$$

2.

$$y' = \frac{2x}{(\ln 10)x^2} = \frac{2}{(\ln 10)x}$$

3.

$$y' = \frac{15x^2 + 4x}{(\ln 2.4)(5x^3 + 2x^2 - 3)}$$

4. $y = \log_{34} \sqrt{x} = \log_{34} x^{1/2} = \frac{1}{2} \log_{34} x$

$$y' = \frac{1}{2} \cdot \frac{1}{(\ln 34)x} = \frac{1}{2(\ln 34)x}$$

Applications

We will finish the chapter with some applications of exponential and logarithmic functions. Because these business and science applications use one of the forms $y = a^x$, $y = \log_a x$, and $y = \frac{a}{b+ce^{dx}}$, there are no relative extrema. Any maximum or minimum occurs at an endpoint (the smallest and largest x-values allowed). The derivative gives us the rate of change of these functions, however.

EXAMPLES

- \$1000 is invested at 10% interest, compounded annually. The value of the account after x years can be found with $y = 1000(1.10)^x$. How fast is the account growing at 9 years?

  The derivative of $y' = 1000(1.10)^x$ approximates the interest earned in the year x.

$$y' = 1000(\ln 1.10)1.10^x$$

At 9 years, $y = 1000(\ln 1.10)1.10^9 \approx 224.74$, which means the account is growing at the rate of \$224.74 per year at 9 years.

- The temperature of a cup of coffee x minutes after sitting on a table can be approximated by $y = 80 + 120e^{-0.03x}$. How fast is the temperature cooling at 4 minutes? At 8 minutes?

  We will evaluate the derivative at $x = 4$ and $x = 8$.

$$y' = 120(-0.03)e^{-0.03x} = -3.6e^{-0.03x}$$

$$y' = -3.6e^{-0.03(4)} = -3.6e^{-0.12} \approx -3.2$$

$$y' = -3.6e^{-0.03(8)} = -3.6e^{-0.24} \approx -2.8$$

  At 4 minutes, the coffee is cooling at the rate of 3.2 degrees per minute. At 8 minutes, the coffee is cooling at the rate of 2.8 degrees per minute.
- A human resources manager has determined that when x thousand dollars is spent on a training program, productivity can be approximated by $y = 19.97 + 7.05 \ln x$ (between $x = 0.5$ and $x = 7$). Productivity is measured in y thousand units per day. How fast is productivity increasing when $1000 is spent on the program? When $4000 is spent?

$$y' = 7.05\left(\frac{1}{x}\right)$$

  At $x = 1$, $y' = \frac{7.05}{1} = 7.05$. Productivity is increasing at the rate of about 7.1 thousand units per day. At $t = 4$, $y' = \frac{7.05}{4} = 1.7625$. Productivity is increasing at the rate of about 1.8 thousand per day.
- After x weeks of a product's release, y thousand units are sold, where y can be approximated by

$$y = \frac{10.3}{1 + 1.77e^{-0.29x}}.$$

  How fast are sales increasing at 2 weeks? 10 weeks?

$$y' = \frac{(-10.3)(1.77)(-0.29)e^{-0.29x}}{(1 + 1.77e^{-0.29x})^2} = \frac{5.28699e^{-0.29x}}{(1 + 1.77e^{-0.29x})^2}$$

$$x = 2 \quad y' = \frac{5.28699e^{-0.29(2)}}{(1 + 1.77e^{-0.29(2)})^2} \approx 0.7467$$

$$x = 10 \quad y' = \frac{5.28699e^{-0.29(10)}}{(1 + 1.77e^{-0.29(10)})^2} \approx 0.24156$$

  The product is selling at the rate of about 750 per week after two weeks and 240 per week after 10 weeks.

PRACTICE

1. For the years 1970 to 2000, a city's population was y thousand, x years after 1970, where $y = 150e^{0.04x}$. How fast was the population increasing in the year 1975? 1995?

2. The value of a piece of equipment x years after it is purchased is $y = 150{,}000(0.9)^x$. How fast is it losing its value 5 years after purchase?

3. A sample of 100 mg of a radioactive substance has y mg of the substance remaining after x days, where $y = 100e^{-0.005x}$. How fast is the substance decaying after 10 days? 20 days?

4. The purchasing power of the dollar between 1975 and 2000 can be approximated by $y = 1.841 - 0.3937 \ln x$, x years after the year 1975. How fast is the value of the dollar dropping in the year 1980? 1990?
(The equation is based on data obtained from the *Statistical Abstract of the United States, 2004–05*, Table 697.)

5. The sales level y of a product, after x thousand dollars is spent on advertising, can be approximated by

$$y = \frac{22}{1 + 30e^{-0.7x}}.$$

How fast are sales increasing when \$5000 is spent on advertising? \$15,000?

6. A mathematics teacher collected data on how many minutes of class time was spent covering a topic and the average class score on a test. When x minutes are spent on the concept, the class average is y, where

$$y = \frac{69.13}{1 + 6.3e^{-0.139x}}.$$

How fast is the class average increasing when 15 minutes is spent covering the concept? 45 minutes?

SOLUTIONS

1. $y' = 150(0.04)e^{0.04x} = 6e^{0.04x}$. At $x = 5$, $y' = 6e^{0.04(5)} \approx 7.3$. In the year 1975, the population is increasing at the rate of about 7.3 thousand per year. At $x = 25$, $y' = 6e^{0.04(25)} \approx 16.3$. In the year 1995, the population is increasing at the rate of about 16.3 thousand per year.

2. $y' = 150{,}000\ (\ln 0.9)(0.9)^x$. At $x = 5$, $y' = 150{,}000(\ln 0.9)(0.9)^5 \approx -9332.15$. The equipment is losing value at the rate of about \$9330 per year at 5 years.

3. $y' = 100(-0.005)e^{-0.005x} = -0.5e^{-0.005x}$. At $x = 10$, $y' = -0.5e^{-0.005(10)} \approx -0.48$. At 10 days, the radioactive substance is decaying at the rate of about 0.48 mg per day. At $x = 20$, $y' = -0.5e^{-0.005(20)} \approx -0.45$. At 20 days, the radioactive substance is decaying at the rate of about 0.45 mg per day.

4.

$$y' = \frac{-0.3937}{x}$$

At $x = 5$, $y' = \frac{-0.3937}{5} = -0.07874$. In the year 1980, the dollar was losing value at the rate of about 7.9 cents per year. At $x = 15$, $y' = \frac{-0.3937}{15} = -0.026$. In the year 1990, the dollar was losing value at the rate of about 2.6 cents per year.

5.

$$y' = \frac{-22(-21e^{-0.7x})}{(1+30e^{-0.7x})^2} = \frac{462e^{-0.7x}}{(1+30e^{-0.7x})^2}$$

$$x = 5 \quad y' = \frac{462e^{-0.7(5)}}{(1+30e^{-0.7(5)})^2} \approx 3.841$$

$$x = 15 \quad y' = \frac{462e^{-0.7(15)}}{(1+30e^{-0.7(15)})^2} \approx 0.013$$

When \$5000 is spent on advertising, sales are increasing at the rate of 3.8 per thousand spent on advertising. When \$15,000 is spent, sales are increasing at the rate of 0.013 units per thousand spent on advertising.

6.

$$y' = \frac{-69.13(-0.139)(6.3)e^{-0.139x}}{(1+6.3e^{-0.139x})^2} = \frac{60.537141e^{-0.139x}}{(1+6.3e^{-0.139x})^2}$$

$$x = 15 \quad y' = \frac{60.537141e^{-0.139(15)}}{(1+6.3e^{-0.139(15)})^2} \approx 2.3667$$

$$x = 45 \quad y' = \frac{60.537141e^{-0.139(45)}}{(1+6.3e^{-0.139(45)})^2} \approx 0.1135$$

When 15 minutes is spent on the concept, the class average is increasing at the rate of about 2.4 points per minute spent on the concept. When

45 minutes is spent on the concept, the class average is increasing at the rate of about 0.1 points per minute spent on the concept.

CHAPTER 11 REVIEW

1. $y = e^{5x^4 - x^2}$
   (a) $y' = (20x^3 - 2x)e^{20x^3 - 2x}$
   (b) $y' = (20x^3 - 2x)e^{5x^4 - x^2}$
   (c) $y' = (5x^4 - x^4)e^{20x^3 - 2x}$
   (d) $y' = (5x^4 - x^2)e^{5x^4 - x^2 - 1}$

2. $4^x \cdot 4^9 =$
   (a) $16x^{9x}$
   (b) 16^{x+9}
   (c) 4^{9x}
   (d) 4^{x+9}

3. $y = \ln(8x^3 + 7)$
   (a) $y' = 24x^2 \ln(8x^3 + 7)$
   (b) $y' = \frac{8x^3+7}{24x^2}$
   (c) $y' = \frac{24x^2}{8x^3+7}$
   (d) $y' = \ln \frac{24x^2}{8x^3+7}$

4. $y = 7^x$
   (a) $y' = x7^{x-1}$
   (b) $y' = (\ln 7)(7^x)$
   (c) $y' = \frac{1}{x}$
   (d) $y' = e^7 \cdot 7^x$

5. $\log_6[(x + 10)(x - 4)] =$
   (a) $[\log_6(x + 10)][\log_6(x - 4)]$
   (b) $\log_6(x + 10) - \log_6(x - 4)$
   (c) $\log_6(x + 10) + \log_6(x - 4)$
   (d) $(x + 10)\log_6(x - 4)$

6. $y = \ln(4x^2 + x)^2$

(a)

$$y' = \frac{16x + 2}{4x^2 + x}$$

(b) $y' = 2\ln(4x^2 + x)$

(c)

$$y' = \left(\frac{8x + 1}{4x^2 + x}\right)^2$$

(d)

$$y' = \frac{8x + 1}{4x^2 + x}$$

7. $y = 4^{x+6}$
   (a) $y' = (x + 6)(\ln 4)(4^{x+6})$
   (b) $y' = (x + 5)(\ln 4)(4^{x+6})$
   (c) $y' = (x + 6)(4^{x+6})$
   (d) $y' = (\ln 4)(4^{x+6})$

8. The number of bacteria in a culture x hours after 1:00 can be approximated by $y = 600e^{0.40x}$. How fast is the number of bacteria increasing at 4:00?
   (a) About 600 per hour
   (b) About 800 per hour
   (c) About 1000 per hour
   (d) About 1200 per hour

9. $y = (e^{3x+4})^2$
   (a) $y' = 6e^{6x+8}$
   (b) $y' = 2(e^{3x+4})$
   (c) $y' = (3e^{3x+4})^2$
   (d) $y' = 12e^{3x+4}$

10. $\log_7(1 - 2x) =$
   (a) $\frac{\ln(1-2x)}{\ln 7}$
   (b) $\frac{\ln 7}{\ln(1-2x)}$
   (c) $\frac{\ln 1 - \ln 2x}{\ln 7}$
   (d) $\frac{\ln 7}{\ln 1 - \ln 2x}$

11. $y = \frac{x^3-4x^2+x}{e^{5x}}$
(a) $y' = \frac{(3x^2-8x+1)(e^{5x})-(x^3-4x^2+x)(5e^{5x})}{(e^{5x})^2}$
(b) $y' = 3x^2 - 8x + 1 - 5e^{5x}$
(c) $y' = (3x^2 - 8x + 1)e^{5x} + (x^3 - 4x^2 + x)(5e^{5x})$
(d) $y' = \frac{3x^2-8x+1}{5e^{5x}}$

12. \$750 is invested for x years, earning 8% interest, compounded annually. It is worth $y = 750(1.08)^x$ after x years. How fast is the account growing at the end of ten years?
(a) About \$115 per year
(b) About \$125 per year
(c) About \$135 per year
(d) About \$145 per year

13. $y = \ln \frac{14x}{3x^2+1}$
(a) $y' = \frac{7}{3x}$
(b) $y' = \ln \frac{7}{3x}$
(c) $y' = \frac{1}{x} - \frac{6x}{3x^2+1}$
(d) $y' = \ln(14x) - \ln(3x^2 + 1)$

14. $y = \sqrt{e^{x^2-1}}$
(a) $y' = \frac{1}{2}(e^{x^2-1})^{-1/2}(2xe^{x^2-1})$
(b) $y' = \sqrt{2xe^{x^2-1}}$
(c) $y' = 2x\sqrt{e^{x^2-1}}$
(d) $y' = \sqrt{2x}e^{x^2-1}$

15. $y = 10^{1-x^2}$
(a) $y' = -x^2 10^{1-x^2}$
(b) $y' = -x^2(\ln 10)(10^{1-x^2})$
(c) $y' = -2x(\ln 10)(10^{1-x^2})$
(d) $y' = (\ln 10)(10^{1-x^2})$

SOLUTIONS

1. b	2. d	3. c	4. b	5. c	6. a	7. d	8. b
9. a	10. a	11. a	12. b	13. c	14. a	15. c	

CHAPTER 12

Elasticity of Demand

When the price of many products increases, demand decreases. The combination of the increase in price and decrease in demand can affect revenue—the revenue increases, decreases, or even remains the same. If the decrease in demand is small enough, the revenue will increase because the price increase makes up for the loss in sales. If the decrease in demand is large enough, the revenue will decrease because the increase in price will not make up for the loss in sales. Economists measure the sensitivity of demand to price increases with a number called the *elasticity of demand.* Elasticity of demand is a ratio of the percent change in demand and the percent change in the price. The Greek letter η (eta) is used to represent this number.

$$\eta = \frac{\text{Percent change in demand}}{\text{Percent change in price}}$$

EXAMPLE

- The demand for a product sold by the pound is $D(p) = 100 - 5p$, where D represents the number of pounds demanded when the price per pound is p. Find η for a 10% price increase when the prices are \$6, \$8, and \$12.

For each of $p = 6$, $p = 8$, and $p = 12$, we will compute the demand before and after the price increase so that we can find the percent decrease in demand. And for comparison, we will compute the change in revenue.

When the price is \$6 per pound, there are $D(6) = 100-5(6) = 70$ pounds demanded. The revenue is (\$6)(70) = \$420. When the price increases 10%, or \$0.60, demand has decreased to $D(6.60) = 100 - 5(6.60) = 67$, and the revenue has increased to (\$6.60)(67) = \$442.20. The percent decrease in demand is $\frac{70-67}{70} = \frac{3}{70} \approx 0.043 = 4.3\%$. At \$6, a 10% increase in the price results in a 4.3% decrease in demand and an increase in revenue.

$$\eta = \frac{4.3\%}{10\%} = 0.43$$

When the price is \$8 per pound, demand is $D(8) = 100 - 5(8) = 60$, and the revenue is (\$8)(60) = \$480. When the price increases 10%, or \$0.80, demand has decreased to $D(8.80) = 100 - 5(8.80) = 56$, and revenue has increased to (\$8.80)(56) = \$492.80. The percent decrease in demand is $\frac{60-56}{60} = \frac{4}{60} \approx 0.067 = 6.7\%$. At \$8, a 10% increase in the price results in a 6.7% decrease in the demand and an increase in revenue.

$$\eta = \frac{6.7\%}{10\%} = 0.67$$

When the price is \$12 per pound, demand is $D(12) = 100 - 5(12) = 40$, and the revenue is (\$12)(40) = \$480. When the price increases 10%, or \$1.20, demand has decreased to $D(13.20) = 100 - 5(13.20) = 34$, and revenue has decreased to (\$13.20)(34) = \$448.80. As a percent, demand has dropped by $\frac{40-34}{40} = \frac{6}{40} = 0.15 = 15\%$. At \$12, a 10% increase in the price results in a 15% decrease in the demand and a decrease in revenue.

$$\eta = \frac{15\%}{10\%} = 1.5$$

A 10% increase in the price at each of \$6 and \$8 per pound resulted in an *increase* in revenue while revenue *decreased* at \$12 per pound. When a price increase causes revenue to increase, demand is *inelastic*. When a price increase causes revenue to decrease, demand is *elastic*. When η is smaller than 1, demand is inelastic, and when it is larger than 1, demand is elastic.

When $\eta < 1$, demand is inelastic.
When $\eta > 1$, demand is elastic.

When the demand function is differentiable, there is a simple formula for η. We will call this function $E(p)$.

$$E(p) = \frac{-pD'(p)}{D(p)} \times 100\%$$

To see where it comes from, let h represent the price increase. In the previous example, $h = 0.60$, $h = 0.80$, and $h = 1.20$. The percent increase in the price is $\frac{h}{p} \times 100\%$. For example, $\frac{h}{p} \times 100\% = \frac{0.60}{6} = 10\%$. The percent decrease in demand is

$$\frac{D(p+h) - D(p)}{D(p)}$$

For example, $\frac{56-60}{60} \approx -6.7\%$. This number is negative because the demand function is decreasing. With this notation, we can represent elasticity at price p as

$$\eta = \frac{\text{Percent change in demand}}{\text{Percent change in price}} = \frac{\frac{D(p+h)-D(p)}{D(p)}}{\frac{h}{p}}$$

We will use algebra to rewrite this expression in a way that allows us to use calculus.

$$\begin{aligned}
\eta &= \frac{\frac{D(p+h)-D(p)}{D(p)}}{\frac{h}{p}} \\
&= \frac{D(p+h) - D(p)}{D(p)} \div \frac{h}{p} = \frac{D(p+h) - D(p)}{D(p)} \cdot \frac{p}{h} \\
&= \frac{p[D(p+h) - D(p)]}{D(p) \cdot h} \\
&= \frac{p}{D(p)} \cdot \frac{D(p+h) - D(p)}{h}
\end{aligned}$$

If $D(p)$ is a differentiable function, we can take the limit of $\frac{D(p+h)-D(p)}{h}$ as $h \to 0$, which we know is $D'(p)$.

$$D'(p) = \lim_{h\to 0} \frac{D(p+h) - D(p)}{h}$$

This allows us to use $D'(p)$ in place of $\frac{D(p+h)-D(p)}{h}$ for η and $E(p)$. $E(p) = \frac{-pD'(p)}{D(p)}$ gives us the elasticity of demand for a small increase in the price. Because demand is usually decreasing, $D'(p)$ is usually negative. For the sake of convenience, we want η to be positive, so we use $-p$ in the formula instead of simply p.

EXAMPLES

- Use $E(p)$ to find the elasticity of demand for the demand function and prices in the previous example.

  The demand function is $D(p) = 100 - 5p$, so $D'(p) = -5$.

$$E(p) = \frac{-p(-5)}{100 - 5p} = \frac{5p}{5(20 - p)} = \frac{p}{20 - p}$$

$$\text{At \$6, } E(6) = \frac{6}{20 - 6} = 0.43$$

$$\text{At \$8, } E(8) = \frac{8}{20 - 8} = 0.67$$

$$\text{At \$12, } E(12) = \frac{12}{20 - 12} = 1.5$$

- Find the elasticity of demand for $D(p) = 1000 - 6p$ at $p = 70$ and $p = 90$. Determine if demand is elastic or inelastic.
  $D'(p) = -6$

$$E(p) = \frac{-pD'(p)}{D(p)} = \frac{-p(-6)}{1000 - 6p} = \frac{6p}{1000 - 6p}$$

$$= \frac{6p}{2(500 - 3p)} = \frac{3p}{500 - 3p}$$

$$\eta = E(70) = \frac{3(70)}{500 - 3(70)} = \frac{210}{290} \approx 0.72$$

$$\eta = E(90) = \frac{3(90)}{500 - 3(90)} = \frac{270}{230} \approx 1.17$$

Because 0.72 is smaller than 1, demand is inelastic at \$70. Because 1.17 is larger than 1, demand is elastic at \$90.

- Find the elasticity of demand for $D(p) = \frac{100}{p-10}$ at $p = 20$ and $p = 25$. Determine if demand is elastic or inelastic.

$$D'(p) = \frac{0(p-10) - 100(1)}{(p-10)^2} = \frac{-100}{(p-10)^2}$$

$$E(p) = \frac{-p \cdot \frac{-100}{(p-10)^2}}{\frac{100}{p-10}}$$

$$= \frac{\frac{100p}{(p-10)^2}}{\frac{100}{p-10}} = \frac{100p}{(p-10)^2} \div \frac{100}{p-10}$$

$$= \frac{100p}{(p-10)^2} \cdot \frac{p-10}{100} = \frac{p}{p-10}$$

$$\eta = E(20) = \frac{20}{20-10} = 2 \qquad \eta = E(25) = \frac{25}{25-10} \approx 1.67$$

Because 2 and 1.67 are larger than 1, demand is elastic for both \$20 and \$25.

- Find the elasticity of demand for $D(p) = 100(0.9^p)$ at $p = 5$ and $p = 8$. Determine if demand is elastic or inelastic.

$$D'(p) = 100(\ln 0.9)(0.9^p) \approx -10.5(0.9^p)$$

$$E(p) = \frac{-p(-10.5)(0.9^p)}{100(0.9^p)} = \frac{10.5p}{100}$$

$$\eta = E(5) = \frac{10.5(5)}{100} = 0.525 \qquad \eta = E(8) = \frac{10.5(8)}{100} = 0.84$$

Both 0.525 and 0.84 are less than 1, so demand is inelastic at \$5 and \$8.

PRACTICE

Find the elasticity of demand for the given function and price. Determine if demand is elastic or inelastic.

1. $D(p) = 125 - 4p; \quad p = 15$
2. $D(p) = 8 - 0.3p; \quad p = 18$
3. $D(p) = 75(0.85^x); \quad p = 9$
4. $D(p) = \frac{60}{p-8}; \quad p = 20$
5. $D(p) = \frac{50}{\sqrt{p+1}}; \quad p = 3$

SOLUTIONS

1. $D'(p) = -4$

$$E(p) = \frac{-p(-4)}{125 - 4p} = \frac{4p}{125 - 4p}$$

$$\eta = E(15) = \frac{4(15)}{125 - 4(15)} \approx 0.92$$

Demand is inelastic at \$15.

2. $D'(p) = -0.3$

$$E(p) = \frac{-p(-0.3)}{8 - 0.3p} = \frac{0.3p}{8 - 0.3p}$$

$$\eta = E(18) = \frac{0.3(18)}{8 - 0.3(18)} \approx 2.08$$

Demand is elastic at \$18.

3. $D'(p) = 75(\ln 0.85)(0.85^p)$

$$E(p) = \frac{-p[75(\ln 0.85)(0.85^p)]}{75(0.85^p)} = -p(\ln 0.85)$$

$$\eta = E(9) = -9(\ln 0.85) \approx 1.46$$

Demand is elastic at \$9.

4.

$$D'(p) = \frac{0(p-8) - 60(1)}{(p-8)^2} = \frac{-60}{(p-8)^2}$$

$$E(p) = \frac{-p \cdot \frac{-60}{(p-8)^2}}{\frac{60}{p-8}} = \frac{\frac{60p}{(p-8)^2}}{\frac{60}{p-8}}$$

$$E(p) = \frac{60p}{(p-8)^2} \div \frac{60}{p-8} = \frac{60p}{(p-8)^2} \cdot \frac{p-8}{60}$$

$$E(p) = \frac{p}{p-8}$$

$$\eta = E(20) = \frac{20}{20-8} \approx 1.67$$

Demand is elastic at \$20.

5. $D(p) = 50(p+1)^{-1/2}$

$$D'(p) = 50\left(-\frac{1}{2}\right)(p+1)^{-3/2} = \frac{-25}{(p+1)^{3/2}} = \frac{-25}{\sqrt{(p+1)^3}}$$

$$E(p) = \frac{-p \cdot \frac{-25}{\sqrt{(p+1)^3}}}{\frac{50}{\sqrt{p+1}}}$$

$$E(p) = \frac{\frac{25p}{\sqrt{(p+1)^3}}}{\frac{50}{\sqrt{p+1}}} = \frac{25p}{\sqrt{(p+1)^3}} \cdot \frac{\sqrt{p+1}}{50}$$

$$E(p) = \frac{p\sqrt{p+1}}{2\sqrt{(p+1)^3}}$$

$$\eta = E(3) = \frac{3\sqrt{3+1}}{2\sqrt{(3+1)^3}} = \frac{3(2)}{2\sqrt{4^3}} = \frac{6}{2(8)} = 0.375$$

Demand is inelastic at \$3.

We can use η as a measure of how fast revenue is increasing or decreasing. If η is larger than 1, a small increase in the price results in a decrease in revenue.

The larger η is, the faster the revenue is decreasing. If $\eta = 2.9$, revenue is falling more sharply than if $\eta = 1.3$. If η is between 0 and 1, revenue is increasing. The closer to 0 η is, the faster the increase (see Figure 12.1).

Revenue is increasing faster.	Revenue is increasing.	Revenue is decreasing.	Revenue is decreasing faster.
$\eta = 0$		$\eta = 1$	

Fig. 12.1.

When $\eta = 1$, demand is *unit elastic* or has *unit elasticity*. We will see later the importance of unit elasticity. For now, we will find the price, if it exists, for which demand is unit elastic.

EXAMPLES

Find the price for which demand is unit elastic.

- $D(p) = 240 - 15p$

  We will find $E(p)$ and solve the equation $E(p) = 1$.

$$E(p) = \frac{-p(-15)}{240 - 15p} = \frac{15p}{240 - 15p}$$

$$\frac{15p}{240 - 15p} = 1$$

$$15p = 240 - 15p \qquad \text{Multiply by sides by } 240 - 15p.$$

$$30p = 240$$

$$p = \frac{240}{30} = 8$$

Demand is unit elastic when the price is \$8.

- $D(p) = 5e^{-0.4p}$

$$D'(p) = 5(-0.4)e^{-0.4p} = -2e^{-0.4p}$$

$$E(p) = \frac{-p(-2e^{-0.4p})}{5e^{-0.4p}} = \frac{2pe^{-0.4p}}{5e^{-0.4p}} = \frac{2p}{5} = 0.4p$$

$$E(p) = 1$$

$$0.4p = 1$$

$$p = \frac{1}{0.4} = 2.5$$

Demand is unit elastic when the price is \$2.50.

When any continuous function changes from increasing to decreasing, it passes through a maximum. This is what happens to the revenue function when elasticity changes from increasing (when η is less than 1) to decreasing (when η is greater than 1). Revenue is maximized for the price at which demand is unit elastic. Let us see what happens when we find the critical values for the derivative of the revenue function. Revenue is found by taking the product of the price and the number of units sold. The price is p, and the number of units sold is the demand, $D(p)$, so the revenue function is $R = p \cdot D(p)$. Using the product rule gives us $R' = 1 \cdot D(p) + p \cdot D'(p)$.

$$R' = D(p) + pD'(p) \qquad \text{Set } R' = 0$$

$$0 = D(p) + pD'(p)$$

$$-pD'(p) = D(p)$$

$$\frac{-pD'(p)}{D(p)} = \frac{D(p)}{D(p)}$$

$$\frac{-pD'(p)}{D(p)} = 1$$

$$E(p) = 1 \qquad \text{Substitute } \frac{-pD'(p)}{D(p)} \text{ for } E(p).$$

PRACTICE

Find the price that maximizes revenue by finding the price at which demand is unit elastic.

1. $D(p) = 975 - 39p$
2. $D(p) = 60e^{-0.1p}$
3. $D(p) = -0.01p^2 - 0.03p + 600$ (Please give your answer to the nearest dollar.)

SOLUTIONS

1.

$$D'(p) = -39 \qquad E(p) = \frac{-p(-39)}{975 - 39p} = \frac{39p}{975 - 39p}$$

$$\frac{39p}{975 - 39p} = 1$$

$$39p = 975 - 39p$$

$$78p = 975$$

$$p = \frac{975}{78} = 12.5$$

Revenue is maximized when the price is \$12.50.

2.

$$D'(p) = 60(-0.1)e^{-0.1p} = -6e^{-0.1p}$$

$$E(p) = \frac{-p(-6e^{-0.1p})}{60e^{-0.1p}} = \frac{6pe^{-0.1p}}{60e^{-0.1p}} = \frac{p}{10}$$

$$\frac{p}{10} = 1$$

$$p = 10$$

Revenue is maximized when the price is \$10.

3. $D'(p) = -0.02p - 0.03$

$$\begin{aligned}
E(p) &= \frac{-p(-0.02p - 0.03)}{-0.01p^2 - 0.03p + 600} \\
&= \frac{0.02p^2 + 0.03p}{-0.01p^2 - 0.03p + 600} \\
1 &= \frac{0.02p^2 + 0.03p}{-0.01p^2 - 0.03p + 600} \\
-0.01p^2 - 0.03p + 600 &= 0.02p^2 + 0.03p \\
0 &= 0.03p^2 + 0.06p - 600 \\
p &= \frac{-0.06 \pm \sqrt{(0.06)^2 - 4(0.03)(-600)}}{2(0.03)} \\
&= \frac{-0.06 \pm \sqrt{0.0036 + 72}}{0.06} \approx 140
\end{aligned}$$

Revenue is maximized when the price is about \$140.

CHAPTER 12 REVIEW

1. For $D(p) = 140 - 6p$, find the elasticity of demand function.

(a)

$$E(p) = \frac{140 - 6p}{6}$$

(b)

$$E(p) = \frac{6p}{140 - 6p}$$

(c)

$$E(p) = \frac{140 - 6p}{6p}$$

(d)

$$E(p) = \frac{6}{140 - 6p}$$

2. For $D(p) = 4(0.8^p)$, find the elasticity of demand function.
   (a) $E(p) = -4(\ln 0.8)p$
   (b) $E(p) = 4p$
   (c) $E(p) = -(\ln 0.8)p$
   (d) $E(p) = 0.8^p p$

3. For $D(p) = 140 - 6p$, find the elasticity of demand for $p = 9$ and determine if demand is elastic or inelastic.
   (a) $\eta \approx 0.63$, inelastic
   (b) $\eta \approx 0.63$, elastic
   (c) $\eta \approx 1.60$, inelastic
   (d) $\eta \approx 1.60$, elastic

4. For $D(p) = 500(0.8^p)$, find the elasticity of demand for $p = 7$ and determine if demand is elastic or inelastic.
   (a) $\eta \approx 1.56$, inelastic
   (b) $\eta \approx 1.56$, elastic
   (c) $\eta \approx 6.25$, inelastic
   (d) $\eta \approx 6.25$, elastic

5. Find the price for which demand is unit elastic for $D(p) = 1500 - 12p$.
   (a) \$72.00
   (b) \$87.25
   (c) \$125.00
   (d) \$62.50

6. Find the price for which demand is unit elastic for $D(p) = 200e^{-0.04p}$.
   (a) \$0.12
   (b) \$145
   (c) \$25
   (d) Demand is never unit elastic.

SOLUTIONS

1. b 2. c 3. a 4. b 5. d 6. c

CHAPTER 13

The Indefinite Integral

Until now, we found the rate of change of a given function by finding its derivative. For the rest of the book, we will work in the opposite direction. We will be given the rate of change and will construct the original function. For example, suppose we know $y' = 2x$ and want to find y. What function has a derivative of $2x$? One such function is $y = x^2$. Others are $y = x^2+1$, $y = x^2-16$, and $y = x^2+5$. The process of constructing a function from its rate of change is called *integration*.

When we see the expression $\int f(x)\,d(x)$, we want to find the function whose derivative is $f(x)$. The function is not unique. We saw earlier that if $y' = 2x$, then y could be any one of many functions—$y = x^2$, $y = x^2+1$, $y = x^2-16$, $y = x^2+5$, or $y = x^2$ plus any constant. For this reason, we say that $y = x^2 + C$ is the indefinite integral of $2x$. (We say that $2x$ is an *antiderivative* of x^2.)

$$\int 2x\,dx = x^2 + C$$

We will begin by using the differentiation formulas in reverse.

The Power Rule

The first integration formula comes from the fact that the derivative of x^n is nx^{n-1}. Rather than use the formula $\int nx^{n-1}dx = x^n + C$, we will use a more convenient formula.

$$\int x^n\,dx = \frac{1}{n+1}x^{n+1} + C \text{ except for } n = -1.$$

The power is increased by 1 and the expression is divided by the new power.

EXAMPLES

- $\int x^4\,dx$
  The new power is $4 + 1 = 5$.

$$\int x^n\,dx = \frac{1}{n+1}x^{n+1} + C$$

$$\int x^4\,dx = \frac{1}{5}x^5 + C$$

- $\int x^{10}\,dx$
  The new power is $10 + 1 = 11$.

$$\int x^{10}\,dx = \frac{1}{11}x^{11} + C$$

- $\int x^{-3}\,dx$
  The new power is $-3 + 1 = -2$.

$$\int x^{-3}\,dx = -\frac{1}{2}x^{-2} + C$$

- $\int \frac{1}{x^2}\,dx$
  Because $\dfrac{1}{x^2} = x^{-2}$, we will find $\int x^{-2}\,dx$. The new power is $-2+1 = -1$.

$$\int x^{-2}\,dx = \frac{1}{-1}x^{-1} + C = -\frac{1}{x} + C$$

When n is a fraction, $n + 1$ is a fraction, and usually $\frac{1}{n+1}$, the reciprocal of $n + 1$, is also a fraction. For example, if n is $\frac{2}{3}$, $n + 1 = \frac{2}{3} + \frac{3}{3} = \frac{5}{3}$ and $\frac{1}{n+1} = \frac{3}{5}$.

- $\int \sqrt{x}\,dx$
  As usual, we will rewrite $\sqrt{x}$ as $x^{1/2}$. This gives us $n = \frac{1}{2}$, $n+1 = \frac{3}{2}$, and $\frac{1}{n+1} = \frac{2}{3}$.

$$\int x^{1/2}\,dx = \frac{2}{3}x^{3/2} + C$$

- $\int \frac{1}{\sqrt{x}}\,dx$
  We will find $\int x^{-1/2}\,dx$. The new power is $\frac{-1}{2} + 1 = \frac{1}{2}$, and $\frac{1}{n+1}$ is 2.

$$\int x^{-1/2}\,dx = 2x^{1/2} + C \text{ or } 2\sqrt{x} + C$$

- $\int \sqrt[3]{x^4}\,dx$
  Because $\sqrt[3]{x^4} = x^{4/3}$, $n = \frac{4}{3}$, $n+1 = \frac{7}{3}$, and $\frac{1}{n+1} = \frac{3}{7}$.

$$\int x^{4/3}\,dx = \frac{3}{7}x^{7/3} + C$$

This formula can be used to integrate the simple $\int dx = \int 1 \cdot dx$ if we think of 1 as x^0. Then $n = 0$ and $n + 1 = 1$.

$$\int dx = \frac{1}{1}x^1 + C = x + C$$

We can integrate sums such as $x^3 + x$ by integrating each term separately.

$$\int (x^3 + x)\,dx = \int x^3\,dx + \int x^1\,dx = \frac{1}{4}x^4 + \frac{1}{2}x^2 + C$$

PRACTICE

1. $\int x^8\,dx$
2. $\int x^{-6}\,dx$
3. $\int x^{20}\,dx$
4. $\int \sqrt{x^5}\,dx$
5. $\int \frac{1}{x^9}\,dx$
6. $\int \frac{1}{\sqrt[3]{x}}\,dx$

7. $\int (x+1)\,dx$
8. $\int (x^4+x^3+x)\,dx$
9. $\int (x^5-x^2)\,dx$
10. $\int (\sqrt{x}-\frac{1}{x^2})\,dx$

SOLUTIONS

1. $\int x^8\,dx = \frac{1}{9}x^9 + C$
2. $\int x^{-6}\,dx = -\frac{1}{5}x^{-5} + C$
3. $\int x^{20}\,dx = \frac{1}{21}x^{21} + C$
4. $\int \sqrt{x^5}\,dx = \int x^{5/2}\,dx = \frac{2}{7}x^{7/2} + C$
5. $\int \frac{1}{x^9}\,dx = \int x^{-9}\,dx = -\frac{1}{8}x^{-8} + C$
6. $\int \frac{1}{\sqrt[3]{x}}\,dx = \int x^{-1/3}\,dx = \frac{3}{2}x^{2/3} + C$
7. $\int (x+1)\,dx = \frac{1}{2}x^2 + x + C$
8. $\int (x^4+x^3+x)\,dx = \frac{1}{5}x^5 + \frac{1}{4}x^4 + \frac{1}{2}x^2 + C$
9. $\int (x^5-x^2)\,dx = \frac{1}{6}x^6 - \frac{1}{3}x^3 + C$
10. $\int (\sqrt{x}-\frac{1}{x^2})\,dx = \int (x^{1/2}-x^{-2})\,dx = \frac{2}{3}x^{3/2} - \left(\frac{1}{-1}x^{-1}\right) + C$
    $= \frac{2}{3}x^{3/2} + x^{-1} + C$

The fact that the derivative of $a \cdot f(x)$ is $a \cdot f'(x)$ allows us to move a number either inside or outside the integral sign.

$$\int af(x)\,dx = a\int f(x)\,dx$$

We can use this fact to modify the previous formula.

$$\int ax^n\,dx = \frac{a}{n+1}x^{n+1} + C$$

EXAMPLES

- $\int 4x^5\,dx = \frac{4}{6}x^6 + C = \frac{2}{3}x^6 + C$
- $\int \frac{3}{\sqrt{x}}\,dx = \int 3x^{-1/2}\,dx = 3(2)x^{1/2} + C = 6x^{1/2} + C$
- $\int \frac{-8}{x^3}\,dx = \int -8x^{-3}\,dx = \frac{-8}{-2}x^{-2} + C = 4x^{-2} + C$
- $\int 4\,dx = 4x + C$
- $\int (17x^2 + 9)\,dx = \frac{17}{3}x^3 + 9x + C$

It is a good habit to check your answers by differentiating the right-hand side. For example, when we differentiate $\frac{17}{3}x^3 + 9x + C$, we get $17x^2 + 9$.

Sometimes we can use algebra to simplify an expression before integrating.

$$\int (x-1)(x+1)\,dx$$

We will multiply $(x-1)(x+1)$ before integrating.

$$\int (x-1)(x+1)\,dx = \int (x^2-1)\,dx = \frac{1}{3}x^3 - x + C$$

PRACTICE

1. $\int 6\sqrt{x}\,dx$
2. $\int (15x^2 + 2)\,dx$
3. $\int (6 - \frac{4}{x^2})\,dx$
4. $\int (10x + \frac{2}{\sqrt{x}} + 1)\,dx$
5. $\int \frac{x^2 - x}{x}\,dx$ (Hint: Simplify before integrating.)

SOLUTIONS

1. $\int 6\sqrt{x}\,dx = \int 6x^{1/2}\,dx = 6(\frac{2}{3})x^{3/2} + C = 4x^{3/2} + C$
2. $\int (15x^2 + 2)\,dx = \frac{15}{3}x^3 + 2x + C = 5x^3 + 2x + C$

3. $\int\left(6-\frac{4}{x^2}\right)dx = \int(6-4x^{-2})\,dx = 6x - \frac{4}{-1}x^{-1} + C$
$= 6x + 4x^{-1} + C$

4. $\int\left(10x + \frac{2}{\sqrt{x}} + 1\right)dx = \int(10x + 2x^{-1/2} + 1)\,dx$
$= \frac{10}{2}x^2 + 2(2)x^{1/2} + 1\cdot x + C$
$= 5x^2 + 4x^{1/2} + x + C$

5. $\frac{x^2-x}{x}$ simplifies to

$$\frac{x^2}{x} - \frac{x}{x} = x - 1$$

This is simpler to integrate.

$$\int \frac{x^2-x}{x}\,dx = \int (x-1)\,dx = \frac{1}{2}x^2 - x + C$$

The power rule does not automatically extend to powers of functions such as $(x^3+1)^2$. Of course, we could expand $(x^3+1)^2 = (x^3+1)(x^3+1) = x^6+2x^3+1$ and integrate term by term. However, integrating something like $\sqrt{x^3+1}$ is harder to do. We can always integrate functions of the form $f'(x)[f(x)]^n$ (for $n \neq -1$) because of the fact that $\frac{d}{dx}[f(x)^n] = n(f(x))^{n-1}f'(x)$.

$$\int f'(x)(f(x))^n\,dx = \frac{1}{n+1}(f(x))^{n+1} + C$$

EXAMPLES

- $\int(4x^3-6x)(x^4-3x^2+4)^3\,dx$

  $4x^3-6x$ is the derivative of x^4-3x^2+4, so we can use the formula.

$$\int \overbrace{(4x^3-6x)}^{f'(x)}\overbrace{(x^4-3x^2+4)^3}^{[f(x)]^n}\,dx = \overbrace{\frac{1}{4}(x^4-3x^2+4)^4}^{\frac{1}{n+1}[f(x)]^{n+1}} + C$$

- $\int 5(5x-1)^{-3}\,dx$

  $f(x) = 5x-1$, $f'(x) = 5$, $n = -3$, and $n+1 = -2$

$$\int 5(5x-1)^{-3}\,dx = \frac{1}{-2}(5x-1)^{-2} + C$$

- $\int \sqrt{x-6}\,dx$

  $f(x) = x - 6$, $f'(x) = 1$, $n = \frac{1}{2}$, $n + 1 = \frac{3}{2}$, and $\frac{1}{n+1} = \frac{2}{3}$

$$\int \sqrt{x-6}\,dx = \int (x-6)^{1/2}\,dx = \frac{2}{3}(x-6)^{3/2} + C$$

- $\int \frac{2x+2}{(x^2+2x+3)^4}\,dx = \int (2x+2)(x^2+2x+3)^{-4}\,dx$
  $= -\frac{1}{3}(x^2+2x+3)^{-3} + C$

PRACTICE

1. $\int 4x(2x^2-1)^5\,dx$
2. $\int 12x\sqrt{6x^2+8}\,dx$
3. $\int \sqrt[3]{x-5}\,dx$
4. $\int \frac{2x+10}{(x^2+10x+3)^4}\,dx$
5. $\int \frac{9x^2-2x}{\sqrt{3x^3-x^2+2}}\,dx$

SOLUTIONS

1. $\int 4x(2x^2-1)^5\,dx = \frac{1}{6}(2x^2-1)^6 + C$
2. $\int 12x\sqrt{6x^2+8}\,dx = \int 12x(6x^2+8)^{1/2}\,dx = \frac{2}{3}(6x^2+8)^{3/2} + C$
3. $\int \sqrt[3]{x-5}\,dx = \int (x-5)^{1/3}\,dx = \frac{3}{4}(x-5)^{4/3} + C$
4. $\int \frac{2x+10}{(x^2+10x+3)^4}\,dx = \int (2x+10)(x^2+10x+3)^{-4}\,dx$
   $= -\frac{1}{3}(x^2+10x+3)^{-3} + C$
5. $\int \frac{9x^2-2x}{\sqrt{3x^3-x^2+2}}\,dx = \int (9x^2-2x)(3x^3-x^2+2)^{-1/2}\,dx$
   $= 2(3x^3-x^2+2)^{1/2} + C$

Integration with Logarithms and Exponents

The derivative of $e^{f(x)}$ is $f'(x)e^{f(x)}$, making the integral of $f'(x)e^{f(x)}$ equal $e^{f(x)} + C$. The derivative of $\ln(f(x))$ is $\frac{f'(x)}{f(x)}$, making the integral of $\frac{f'(x)}{f(x)}$ equal

$\ln f(x) + C$. Technically, we should say that the integral of $\frac{f'(x)}{f(x)}$ is $\ln|f(x)| + C$ because we can only take the logarithm of positive numbers.

$$\int f'(x)e^{f(x)}\,dx = e^{f(x)} + C \quad \text{and} \quad \int \frac{f'(x)}{f(x)}\,dx = \ln|f(x)| + C$$

When we are integrating a fraction, we will determine whether or not the numerator is the derivative of the denominator. If it is, then the integral is a logarithm.

EXAMPLES

- $\int 6xe^{3x^2-8}\,dx$

  The derivative of the power is $6x$, so we are integrating a function of the form $f'(x)e^{f(x)}$, which is $e^{f(x)}$.

$$\int 6xe^{3x^2-8}\,dx = e^{3x^2-8}dx + C$$

- $\int -2xe^{10-x^2}\,dx$

  The derivative of the power is $-2x$.

$$\int -2xe^{10-x^2}\,dx = e^{10-x^2} + C$$

- $\int \frac{20}{20x-7}\,dx$

  The numerator is the derivative of the denominator, so the integral is the natural logarithm of the denominator (actually, the absolute value of the denominator).

$$\int \frac{20}{20x-7}\,dx = \ln|20x-7| + C$$

- $\int \frac{16x^3+2x+3}{4x^4+x^2+3x+2}\,dx$

  The numerator is the derivative of the denominator.

$$\int \frac{16x^3+2x+3}{4x^4+x^2+3x+2}\,dx = \ln|4x^4+x^2+3x+2| + C$$

- $\int \frac{4e^{4x-5}}{e^{4x-5}+2}\,dx$

  The numerator is the derivative of the denominator.

$$\int \frac{4e^{4x-5}}{e^{4x-5}+2}\,dx = \ln|e^{4x-5}+2| + C$$

PRACTICE

1. $\int (14x+1)e^{7x^2+x+8}\,dx$
2. $\int \frac{1}{2}x^{-1/2}e^{\sqrt{x}}\,dx$
3. $\int e^{x-4}\,dx$
4. $\int \frac{18x-4}{9x^2-4x-5}\,dx$
5. $\int \frac{6}{6x-11}\,dx$
6. $\int \frac{20x^4-18x^2+4x}{4x^5-6x^3+2x^2+8}\,dx$
7. $\int \frac{(6x+1)e^{3x^2+x+2}}{8+e^{3x^2+x+2}}\,dx$

SOLUTIONS

1. $\int (14x+1)e^{7x^2+x+8}\,dx = e^{7x^2+x+8}+C$
2. $\int \frac{1}{2}x^{-1/2}e^{\sqrt{x}}\,dx = e^{\sqrt{x}}+C$
3. $\int e^{x-4}\,dx = e^{x-4}+C$
4. $\int \frac{18x-4}{9x^2-4x-5}\,dx = \ln|9x^2-4x-5|+C$
5. $\int \frac{6}{6x-11}\,dx = \ln|6x-11|+C$
6. $\int \frac{20x^4-18x^2+4x}{4x^5-6x^3+2x^2+8}\,dx = \ln|4x^5-6x^3+2x^2+8|+C$
7. $\int \frac{(6x+1)e^{3x^2+x+2}}{8+e^{3x^2+x+2}}\,dx = \ln|8+e^{3x^2+x+2}|+C$

When we differentiate a function of the form $y = af(x)$, we multiply the derivative of $f(x)$ by a: $y' = af'(x)$. This is true with integration, which allows us to move a constant inside or outside the integral sign.

$$\int a\,f(x)\,dx = a\int f(x)\,dx$$

EXAMPLES

- $\int 5x\,dx = 5\int x\,dx$
- $\int \frac{8}{x^2}\,dx = 8\int \frac{1}{x^2}\,dx$
- $\int 10e^x\,dx = 10\int e^x\,dx$
- $\int 3(x^2-2x-6)\,dx = 3\int (x^2-2x-6)\,dx$

We will make use of this fact when we "almost" have $\int f'(x)[f(x)]^n\,dx$, $\int f'(x)e^{f(x)}\,dx$, and $\int \frac{f'(x)}{f(x)}\,dx$. For example, the integral $\int \frac{4x}{x^2+1}\,dx$ almost fits the form $\int \frac{f'(x)}{f(x)}\,dx$. If the numerator were $2x$, then the integral does fit. By writing $4x$ as $2\cdot 2x$ and moving the 2 outside of the integral sign, we can force $\int \frac{4x}{x^2+1}\,dx$ to fit the formula.

$$\int \frac{4x}{x^2+1}\,dx = \int \frac{2\cdot 2x}{x^2+1}\,dx = 2\int \frac{2x}{x^2+1}\,dx = 2\ln|x^2+1| + C$$

Only a constant can move inside or outside the integral sign; $\int x(x+1)\,dx$ is not the same as $x\int (x+1)\,dx$.

EXAMPLES

- $\int \frac{3}{2}x^2(x^3+10)^6\,dx$

  The derivative of x^3+10 is $3x^2$, but we have $\frac{3}{2}x^2$. We will rewrite $\frac{3}{2}x^2$ as $\frac{1}{2}\cdot 3x^2$ and put $\frac{1}{2}$ outside the integral sign.

  $$\int \frac{3}{2}x^2(x^3+10)^6\,dx = \int \frac{1}{2}\cdot 3x^2(x^3+10)^6\,dx = \frac{1}{2}\int 3x^2(x^3+10)^6\,dx$$

  Now this fits the form $\int f'(x)[f(x)]^n\,dx$.

  $$= \frac{1}{2}\cdot\frac{1}{7}(x^3+10)^7 + C = \frac{1}{14}(x^3+10)^7 + C$$

- $\int xe^{x^2-4}\,dx$

  The derivative of the power is $2x$, so we want to multiply x by $\frac{1}{2}\cdot 2$, which is simply 1.

  $$\int xe^{x^2-4}\,dx = \int \frac{1}{2}\cdot 2xe^{x^2-4}\,dx = \frac{1}{2}\int 2xe^{x^2-4}\,dx = \frac{1}{2}e^{x^2-4} + C$$

- $\int (5x^2 - 2x + 2)(5x^3 - 3x^2 + 6x)^8\,dx$

  The derivative of $5x^3 - 3x^2 + 6x$ is $15x^2 - 6x + 6$, which is 3 times $5x^2 - 2x + 2$.

  $$5x^2 - 2x + 2 = \frac{3}{3}(5x^2 - 2x + 2) = \frac{1}{3}[3(5x^2 - 2x + 2)] = \frac{1}{3}(15x^2 - 6x + 6)$$

  $$\begin{aligned}\int (5x^2 - 2x + 2)(5x^3 - 3x^2 + 6x)^8\,dx \\ &= \int \frac{1}{3}[3(5x^2 - 2x + 2)](5x^3 - 3x^2 + 6x)^8\,dx \\ &= \frac{1}{3}\int (15x^2 - 6x + 6)(5x^3 - 3x^2 + 6x)^8\,dx \\ &= \frac{1}{3}\cdot\frac{1}{9}(5x^3 - 3x^2 + 6x)^9 + C \\ &= \frac{1}{27}(5x^3 - 3x^2 + 6x)^9 + C\end{aligned}$$

- $\int \frac{2}{5x-3}\,dx$

  In order for us to use $\int \frac{f'(x)}{f(x)}\,dx$ we need to adjust $\frac{2}{5x-3}$. We want $\frac{5}{5x-3}$. Because we want 5 in the numerator, we will multiply 2 by $\frac{5}{5}$.

  $$2 = \frac{5}{5}\cdot 2 = \frac{2\cdot 5}{5} = \frac{2}{5}\cdot 5$$

  This gives us

  $$\frac{2}{5x-3} = \frac{2}{5}\cdot\frac{5}{5x-3}$$

  $$\int \frac{2}{5x-3}\,dx = \frac{2}{5}\int \frac{5}{5x-3}\,dx = \frac{2}{5}\ln|5x-3| + C$$

- $\int \frac{1}{2}x^2(4x^3 + 15)^{-6}\,dx$

  We need to change $\frac{1}{2}x^2$ to $12x^2$. We will multiply $\frac{1}{2}$ by $\frac{12}{12}$.

  $$\frac{1}{2} = \frac{12}{12}\cdot\frac{1}{2} = \frac{12\cdot 1}{2\cdot 12} = \frac{1\cdot 12}{24} = \frac{1}{24}\cdot 12$$

$$\int \frac{1}{2}x^2(4x^3+15)^{-6}\,dx = \frac{1}{24}\int 12x^2(4x^3+15)^{-6}\,dx$$
$$= \frac{1}{24}\cdot\frac{1}{-5}(4x^3+15)^{-5}+C$$
$$= -\frac{1}{120}(4x^3+15)^{-5}+C$$

- $\int 2x(1-x^2)^3\,dx$
  The derivative of $1-x^2$ is $-2x$, so we are off only by a factor of -1: $2x = -(-2x)$.
  $$\int 2x(1-x^2)^3\,dx = -\int -2x(1-x^2)^3\,dx = -\frac{1}{4}(1-x^2)^4+C$$

PRACTICE

1. $\int 2e^{6x+9}\,dx$
2. $\int (x+3)(x^2+6x-4)^3\,dx$
3. $\int \frac{6x^2}{7x^3+4}\,dx$
4. $\int \frac{4e^{2x}}{7-e^{2x}}\,dx$
5. $\int (x^2+1)\sqrt{x^3+3x-8}\,dx$

SOLUTIONS

1. $$\int 2e^{6x+9}\,dx = \int \frac{2}{6}\cdot 6e^{6x+9}\,dx = \frac{1}{3}\int 6e^{6x+9}\,dx = \frac{1}{3}e^{6x+9}+C$$

2. $$\int (x+3)(x^2+6x-4)^3\,dx = \int \frac{1}{2}\cdot 2(x+3)(x^2+6x-4)^3\,dx$$
$$= \frac{1}{2}\int (2x+6)(x^2+6x-4)^3\,dx$$
$$= \frac{1}{2}\cdot\frac{1}{4}(x^2+6x-4)^4+C$$
$$= \frac{1}{8}(x^2+6x-4)^4+C$$

3. $6 = \frac{21}{21} \cdot 6 = \frac{6}{21} \cdot 21 = \frac{2}{7} \cdot 21$

$$\int \frac{6x^2}{7x^3+4}\,dx = \frac{2}{7}\int \frac{21x^2}{7x^3+4}\,dx = \frac{2}{7}\ln|7x^3+4| + C$$

4. We need -2 in the numerator: $4 = (-2)(-2)$.

$$\int \frac{4e^{2x}}{7-e^{2x}}\,dx = \int \frac{(-2)(-2)e^{2x}}{7-e^{2}x}\,dx = -2\int \frac{-2e^{2x}}{7-e^{2x}}\,dx$$

$$= -2\ln|7-e^{2x}| + C$$

5. We need $3x^2+3$: $x^2+1 = \frac{3}{3}(x^2+1) = \frac{1}{3}[3(x^2+1)] = \frac{1}{3}(3x^2+3)$.

$$\int (x^2+1)\sqrt{x^3+3x-8}\,dx = \frac{1}{3}\int (3x^2+3)(x^3+3x-8)^{1/2}\,dx$$

$$= \frac{1}{3}\cdot\frac{2}{3}(x^3+3x-8)^{3/2} + C$$

$$= \frac{2}{9}(x^3+3x-8)^{3/2} + C$$

Integration by Parts

Unlike differentiation, there is no product rule that helps us to integrate a product of two or more functions. There are techniques we can use for most of the integrals found in a calculus course. *Integration by parts* is one of the most common integration techniques. At the end of the chapter there will be a brief discussion of a few other techniques.

The formula for integration by parts comes from integrating the derivative formula for a product of two functions.

$$\text{If } y = f\cdot g \text{ then } y' = f'g + fg'.$$

We will begin with the fact that $y = \int y'dx$.

$$y = \int y'\,dx$$

$$fg = \int y'\,dx \qquad \text{Replace } y \text{ with } fg.$$

$$fg = \int (f'g + fg')\,dx \qquad \text{Replace } y' \text{ with } f'g + fg'.$$

$$fg = \int f'g\,dx + \int fg'\,dx$$

$$fg - \int fg'dx = \int f'g\,dx \qquad \text{Subtract } \int fg'\,dx.$$

The formula looks like it is making a bad problem worse, but for some integrals, fg' is easier to integrate than $f'g$. When using the formula, we have to decide what f, g, f', and g' are, so it is important that you are comfortable with the integration we have done so far. We will begin by deciding what to let f, g, f', and g' represent.

EXAMPLES

Identify f, f', g, and g' so that fg' can be integrated.

- $\int 2xe^x\,dx$

  Either $f' = 2x$ and $g = e^x$ or $f' = e^x$ and $g = 2x$.

Option I	Option II
$f' = 2x$ and $g = e^x$	$f' = e^x$ and $g = 2x$
$f = x^2$ and $g' = e^x$	$f = e^x$ and $g' = 2$
$fg' = x^2e^x$	$fg' = 2e^x$

  fg' in Option II is easier to integrate.

- $\int x\sqrt{x+2}\,dx$

  Either $f' = x$ and $g = (x+2)^{1/2}$ or $f' = (x+2)^{1/2}$ and $g = x$.

Option I	Option II
$f' = x$ and $g = (x+2)^{1/2}$	$f' = (x+2)^{1/2}$ and $g = x$
$f = \frac{1}{2}x^2$ and $g' = \frac{1}{2}(x+2)^{-1/2}$	$f = \frac{2}{3}(x+2)^{3/2}$ and $g' = 1$
$fg' = \frac{1}{4}x^2(x+2)^{-1/2}$	$fg' = \frac{2}{3}(x+2)^{3/2}$

  fg' in Option II is easier to integrate.

- $\int \frac{x}{x-1}\,dx = \int x \cdot \frac{1}{x-1}\,dx$

  Either $f' = x$ and $g = \frac{1}{x-1}$ or $f' = \frac{1}{x-1}$ and $g = x$.

Option I

$f' = x$ and $g = \frac{1}{x-1} = (x-1)^{-1}$

$f = \frac{1}{2}x^2$ and $g' = -1(x-1)^{-2}$

$fg' = -\frac{1}{2}x^2(x-1)^{-2}$

Option II

$f' = \frac{1}{x-1}$ and $g = x$

$f = \ln(x-1)$ and $g' = 1$

$fg' = \ln(x-1)$

fg' in Option II looks like it might be easier to integrate.

PRACTICE

Identify f, f', g, and g' so that fg' can be integrated.

1. $\int xe^{x-2}\,dx$
2. $\int 4x(1-x)^5\,dx$
3. $\int (2x+1)(x-3)^{10}\,dx$
4. $\int x^3e^{2x^2-1}\,dx$

SOLUTIONS

1.

Option I

$f' = x$ and $g = e^{x-2}$

$f = \frac{1}{2}x^2$ and $g' = e^{x-2}$

$fg' = \frac{1}{2}x^2e^{x-2}$

Option II

$f' = e^{x-2}$ and $g = x$

$f = e^{x-2}$ and $g' = 1$

$fg' = e^{x-2}$

fg' in Option II is easier to integrate.

2.

Option I

$f' = 4x$ and $g = (1-x)^5$

$f = 2x^2$ and $g' = -5(1-x)^4$

$fg' = -10x^2(1-x)^4$

Option II

$f' = (1-x)^5$ and $g = 4x$

$f = -\frac{1}{6}(1-x)^6$ and $g' = 4$

$fg' = -\frac{2}{3}(1-x)^6$

fg' in Option II is easier to integrate.

3.

Option I

$f' = 2x+1$ and $g = (x-3)^{10}$

$f = x^2 + x$ and $g' = 10(x-3)^9$

$fg' = 10(x^2+x)(x-3)^9$

Option II

$f' = (x-3)^{10}$ and $g = 2x+1$

$f = \frac{1}{11}(x-3)^{11}$ and $g' = 2$

$fg' = \frac{2}{11}(x-3)^{11}$

fg' in Option II is easier to integrate.

4. The derivative of $2x^2 - 1$ is $4x$, so we want x in the product with e^{2x^2-1}. We can view x^3 as $x \cdot x^2$, giving us $x^3 e^{2x^2-1} = x^2 \cdot xe^{2x^2-1}$.

Option I

$$f' = xe^{2x^2-1} \text{ and } g = x^2$$
$$f = \tfrac{1}{4}e^{2x^2-1} \text{ and } g' = 2x$$
$$fg' = \tfrac{1}{2}xe^{2x^2-1}$$

Option II

$$f' = x^3 \text{ and } g = e^{2x^2-1}$$
$$f = \tfrac{1}{4}x^4 \text{ and } g' = 4xe^{2x^2-1}$$
$$fg' = x^5 e^{2x^2-1}$$

Option III

$$f' = x^2 \text{ and } g = xe^{2x^2-1}$$
$$f = \tfrac{1}{3}x^3 \text{ and } g' = e^{2x^2-1} + 4x^2e^{2x^2-1}$$
$$fg' = \tfrac{1}{3}x^3(e^{2x^2-1} + 4x^2e^{2x^2-1})$$

fg' in Option I is easier to integrate.

We are ready to integrate by parts.

EXAMPLES

- $\int xe^x\,dx$ We will let $f' = e^x$ and $g = x$, which gives us $f = e^x$ and $g' = 1$.

$$\int f'(x)g(x)\,dx = f(x)g(x) - \int f(x)g'(x)\,dx$$

$$\overbrace{\int xe^x\,dx}^{\int f'g\,dx} = \overbrace{xe^x}^{fg} - \overbrace{\int 1 \cdot e^x\,dx}^{\int fg'\,dx}$$

$\int e^x\,dx$ is easy to integrate: $\int e^x\,dx = e^x + C$.

$$\int xe^x\,dx = xe^x - \int 1 \cdot e^x\,dx = xe^x - e^x + C$$

We will differentiate $y = xe^x - e^x + C$ to make sure that $y' = xe^x$.

$$y' = 1 \cdot e^x + xe^x - e^x + 0 = xe^x \checkmark$$

- $\int x(2x+7)^8\,dx$
  We will let $f' = (2x+7)^8$ and $g = x$. This gives us $f = \frac{1}{18}(2x+7)^9$ and $g' = 1$.

$$\begin{aligned}\int x(2x+7)^8\,dx &= \frac{1}{18}(2x+7)^9 \cdot x - \int \frac{1}{18}(2x+7)^9 \cdot 1\,dx \\ &= \frac{1}{18}x(2x+7)^9 - \frac{1}{18}\cdot\frac{1}{2}\int 2(2x+7)^9\,dx \\ &= \frac{1}{18}x(2x+7)^9 - \frac{1}{36}\cdot\frac{1}{10}(2x+7)^{10} + C \\ &= \frac{1}{18}x(2x+7)^9 - \frac{1}{360}(2x+7)^{10} + C\end{aligned}$$

- $\int x \ln x\,dx$
  We will let $f' = x$ and $g = \ln x$. This gives us $f = \frac{1}{2}x^2$ and $g' = \frac{1}{x}$.

$$\begin{aligned}\int x\ln x\,dx &= \frac{1}{2}x^2 \cdot \ln x - \int \frac{1}{2}x^2 \cdot \frac{1}{x}\,dx \\ &= \frac{1}{2}x^2 \ln x - \frac{1}{2}\int x\,dx \\ &= \frac{1}{2}x^2 \ln x - \frac{1}{2}\cdot\frac{1}{2}x^2 + C = \frac{1}{2}x^2 \ln x - \frac{1}{4}x^2 + C\end{aligned}$$

- $\int \ln x\,dx$
  Integration by parts works on products, so we have to think of $\ln x$ as the product of two functions 1 and $\ln x$. We will let $f' = 1$ and $g = \ln x$. This gives us $f = x$ and $g' = \frac{1}{x}$.

$$\int \ln x\,dx = x\ln x - \int x \cdot \frac{1}{x}\,dx = x\ln x - \int 1\,dx = x\ln x - x + C$$

PRACTICE

1. $\int 3xe^{4x+9}\,dx$
2. $\int 6x(2x+5)^4\,dx$
3. $\int x^3 e^{x^2}\,dx$
4. $\int \frac{x}{(x-1)^4}\,dx$
5. $\int (3x+7)\sqrt{x+2}\,dx$

6. $\int x \ln 3x \, dx$
7. $\int \ln \sqrt{x} \, dx$ (Hint: use a logarithm property before integrating.)

SOLUTIONS

1. We will let $f' = e^{4x+9}$ and $g = 3x$, giving us $f = \frac{1}{4}e^{4x+9}$ and $g' = 3$.

$$\begin{aligned}\int 3xe^{4x+9}\,dx &= \left(\frac{1}{4}e^{4x+9}\right)(3x) - \int \left(\frac{1}{4}e^{4x+9}\right)(3)\,dx \\ &= \frac{3}{4}xe^{4x+9} - \frac{3}{4}\int e^{4x+9}\,dx \\ &= \frac{3}{4}xe^{4x+9} - \frac{3}{4}\cdot\frac{1}{4}e^{4x+9} + C \\ &= \frac{3}{4}xe^{4x+9} - \frac{3}{16}e^{4x+9} + C\end{aligned}$$

2. We will let $f' = (2x+5)^4$ and $g = 6x$, giving us $f = \frac{1}{10}(2x+5)^5$ and $g' = 6$.

$$\begin{aligned}\int 6x(2x+5)^4\,dx &= \frac{1}{10}(2x+5)^5(6x) - \int \frac{1}{10}(2x+5)^5(6)\,dx \\ &= \frac{3}{5}x(2x+5)^5 - \frac{3}{5}\int (2x+5)^5\,dx \\ &= \frac{3}{5}x(2x+5)^5 - \frac{3}{5}\cdot\frac{1}{2}\int 2(2x+5)^5\,dx \\ &= \frac{3}{5}x(2x+5)^5 - \frac{3}{10}\cdot\frac{1}{6}(2x+5)^6 + C \\ &= \frac{3}{5}x(2x+5)^5 - \frac{1}{20}(2x+5)^6 + C\end{aligned}$$

3. We will let $f' = xe^{x^2}$ and $g = x^2$, giving us $f = \frac{1}{2}e^{x^2}$ and $g' = 2x$.

$$\begin{aligned}\int x^3e^{x^2}\,dx &= \frac{1}{2}e^{x^2}(x^2) - \frac{1}{2}\int 2xe^{x^2}\,dx \\ &= \frac{1}{2}x^2e^{x^2} - \frac{1}{2}e^{x^2} + C\end{aligned}$$

4. We will let $f' = (x-1)^{-4}$ and $g = x$, giving us $f = -\frac{1}{3}(x-1)^{-3}$ and $g' = 1$.

$$\begin{aligned}
\int \frac{x}{(x-1)^4}\,dx &= -\frac{1}{3}(x-1)^{-3}(x) - \int -\frac{1}{3}(x-1)^{-3}(1)\,dx \\
&= -\frac{1}{3}x\frac{1}{(x-1)^3} + \frac{1}{3}\int (x-1)^{-3}\,dx \\
&= -\frac{1}{3}x\frac{1}{(x-1)^3} + \frac{1}{3}\cdot -\frac{1}{2}(x-1)^{-2} + C \\
&= -\frac{1}{3}x\frac{1}{(x-1)^3} - \frac{1}{6}\frac{1}{(x-1)^2} + C \\
&= -\frac{x}{3(x-1)^3} - \frac{1}{6(x-1)^2} + C
\end{aligned}$$

5. We will let $f' = (x+2)^{1/2}$ and $g = 3x+7$, giving us $f = \frac{2}{3}(x+2)^{3/2}$ and $g' = 3$.

$$\begin{aligned}
\int (3x+7)\sqrt{x+2}\,dx &= \frac{2}{3}(x+2)^{3/2}(3x+7) - \int \frac{2}{3}(x+2)^{3/2}(3)\,dx \\
&= \frac{2}{3}(x+2)^{3/2}(3x+7) - 2\int (x+2)^{3/2}\,dx \\
&= \frac{2}{3}(x+2)^{3/2}(3x+7) - 2\cdot\frac{2}{5}(x+2)^{5/2} + C \\
&= \frac{2}{3}(x+2)^{3/2}(3x+7) - \frac{4}{5}(x+2)^{5/2} + C
\end{aligned}$$

6. We will let $f' = x$ and $g = \ln 3x$, giving us $f = \frac{1}{2}x^2$ and $g' = \frac{3}{3x} = \frac{1}{x}$.

$$\begin{aligned}
\int x\ln 3x\,dx &= \frac{1}{2}x^2\ln 3x - \int \frac{1}{2}x^2\left(\frac{1}{x}\right)\,dx \\
&= \frac{1}{2}x^2\ln 3x - \frac{1}{2}\int x\,dx
\end{aligned}$$

$$\begin{aligned}
&= \frac{1}{2}x^2\ln 3x - \frac{1}{2}\cdot\frac{1}{2}x^2 + C \\
&= \frac{1}{2}x^2\ln 3x - \frac{1}{4}x^2 + C
\end{aligned}$$

7. We can rewrite $\ln\sqrt{x}$ as $\ln x^{1/2} = \frac{1}{2}\ln x$. From an earlier example, we found $\int \ln x\,dx = x\ln x - x + C$.

$$\int \ln\sqrt{x}\,dx = \frac{1}{2}\int \ln x\,dx$$
$$= \frac{1}{2}(x\ln x - x) + C$$

Sometimes an integral can be found after using integration by parts more than once.

EXAMPLE

- $\int x^2e^x\,dx$
  We will let $f' = e^x$ and $g = x^2$, giving us $f = e^x$ and $g' = 2x$.

$$\int x^2e^x\,dx = x^2e^x - 2\int xe^x\,dx$$

For $\int xe^x\,dx$, we will let $f' = e^x$ and $g = x$, giving us $f = e^x$ and $g' = 1$.

$$\int xe^x\,dx = xe^x - \int e^x(1)\,dx = xe^x - e^x + C$$

Now we can finish.

$$\int x^2e^x\,dx = x^2e^x - 2\int xe^x\,dx = x^2e^x - 2(xe^x - e^x) + C$$
$$= x^2e^x - 2xe^x + 2e^x + C$$

Miscellaneous Techniques

There are several other integration techniques. We will briefly discuss three of them.

A simple substitution can change a difficult integral into an easy one. For example, $\int x\sqrt{1+x}\,dx$ is not easy to integrate. An expression such as $\sqrt{1+\text{variable}}$ is harder to integrate than $\sqrt{\text{variable}}$. We can get around this making the substitution $u = x + 1$, which gives us $u - 1 = x$ and $du = dx$. Instead of integrating

$x\sqrt{1+x}$, we will integrate

$$(u-1)\sqrt{u} = u\sqrt{u} - \sqrt{u} = u \cdot u^{1/2} - u^{1/2} = u^1 \cdot u^{1/2} - u^{1/2}$$
$$= u^{1+1/2} - u^{1/2} = u^{3/2} - u^{1/2}$$

$u^{3/2} - u^{1/2}$ is much easier to integrate.

$$\int x\sqrt{1+x}\,dx = \int (u^{3/2} - u^{1/2})\,du = \frac{2}{5}u^{5/2} - \frac{2}{3}u^{3/2} + C$$

Because $u = 1 + x$, we will replace u with $1 + x$.

$$\frac{2}{5}u^{5/2} - \frac{2}{3}u^{3/2} + C = \frac{2}{5}(1+x)^{5/2} - \frac{2}{3}(1+x)^{3/2} + C$$

EXAMPLE

- $\int \ln(2x+1)\,dx$

  We will let $u = 2x + 1$, which gives us $dx = \frac{1}{2}du$ (which comes from the fact that $du = 2dx$).

$$\int \ln(2x+1)\,dx = \int \ln u \left(\frac{1}{2}du\right)$$
$$= \frac{1}{2}\int \ln u\,du$$
$$= \frac{1}{2}(u \ln u - u) + C \quad \text{(From earlier, we know}$$
$$\int \ln x\,dx = x\ln x - x + C.\text{)}$$
$$= \frac{1}{2}[(2x+1)\ln(2x+1) - (2x+1)] + C$$

(Replace u with $2x + 1$.)

A Shortcut for Integration by Parts

When integrating a product of two functions where one is a polynomial and the other can be integrated several times (without too much trouble), integration by parts can be used multiple times. There is a shortcut that eliminates some of the tedious steps. Create a table that has two columns. The polynomial and its derivatives go in the first column. The other function and its integrals go in the other column. For example, say we want to integrate $\int x^4 e^x \, dx$. The polynomial function is x^4, and the other function is e^x, which integrates as many times as we want without any trouble. The derivatives of x^4 are $4x^3$, $12x^2$, $24x$, 24, and 0. All of the integrals of e^x are e^x. A table is constructed (Table 13.1). The signs in the solution alternate from plus to minus, so we will record these symbols in the table (Table 13.2). In order to make it clear which expressions are multiplied and which are added, we will label each entry (Table 13.3). The integral is found by computing the following.

$$(1) \cdot B + (2) \cdot C + (3) \cdot D + (4) \cdot E + (5) \cdot F$$

$$\int x^4 e^x \, dx = x^4 e^x - 4x^3 e^x + 12x^2 e^x - 24x e^x + 24e^x + C$$

Table 13.1

Polynomial and its derivatives	Other function and its integrals
x^4	e^x
$4x^3$	e^x
$12x^2$	e^x
$24x$	e^x
24	e^x
0	e^x

Table 13.2

Polynomial and its derivatives	Other function and its integrals
$+x^4$	e^x
$-4x^3$	e^x
$+12x^2$	e^x
$-24x$	e^x
$+24$	e^x
-0	e^x

Table 13.3

Polynomial and its derivatives	Other function and its integrals
$+x^4$ (1)	e^x (A)
$-4x^3$ (2)	e^x (B)
$+12x^2$ (3)	e^x (C)
$-24x$ (4)	e^x (D)
$+24$ (5)	e^x (E)
-0 (6)	e^x (F)

EXAMPLE

- $\int x^3(x+1)^4\,dx$

  The derivatives of x^3 are $3x^2$, $6x$, 6, and 0. The first three integrals of $(x+1)^4$ are $\frac{1}{5}(x+1)^5$, $\frac{1}{30}(x+1)^6$, $\frac{1}{210}(x+1)^7$ and $\frac{1}{1680}(x+1)^8$ (Table 13.4). The integral can be found by computing

$$(1)\cdot B+(2)\cdot C+(3)\cdot D+(4)\cdot E$$

$$\int x^3(x+1)^4\,dx=\frac{1}{5}x^3(x+1)^5-3x^2\cdot\frac{1}{30}(x+1)^6+6x\cdot\frac{1}{210}(x+1)^7$$
$$-6\cdot\frac{1}{1680}(x+1)^8+C$$

Table 13.4

Polynomial and its derivatives	Other function and its integrals
$+x^3$ (1)	$(x+1)^4$ (A)
$-3x^2$ (2)	$\frac{1}{5}(x+1)^5$ (B)
$+6x$ (3)	$\frac{1}{30}(x+1)^6$ (C)
-6 (4)	$\frac{1}{210}(x+1)^7$ (D)
$+0$ (5)	$\frac{1}{1680}(x+1)^8$ (E)

Tables of Integrals

There are books with hundreds of integral formulas. Using a table can make finding an integral very easy. However, we might have to use algebra on our expression to make it look like one of the formulas. Below is a small table of three integrals.

Integral Formulas

A $$\int \frac{1}{a^2 - x^2}\,dx = \frac{1}{2a}\ln\left|\frac{x+a}{x-a}\right| + C$$

B $$\int x^2\sqrt{a^2+x^2}\,dx = \frac{1}{8}x(a^2+2x^2)\sqrt{a^2+x^2} - \frac{1}{8}a^4\ln\left|x+\sqrt{a^2+x^2}\right| + C$$

C $$\int \frac{x^2}{\sqrt{a^2+x^2}}\,dx = -\frac{1}{2}a^2\ln\left|x+\sqrt{a^2+x^2}\right| + \frac{1}{2}x\sqrt{a^2+x^2} + C$$

EXAMPLES

- $$\int \frac{x^2}{\sqrt{x^2+9}}\,dx$$

  The integral fits Formula C, where $a^2 = 9$. All we need to do is to replace a^2 with 9 in $-\frac{1}{2}a^2\ln\left|x+\sqrt{a^2+x^2}\right| + \frac{1}{2}x\sqrt{a^2+x^2} + C$.

$$\int \frac{x^2}{\sqrt{x^2+9}}\,dx = -\frac{1}{2}\cdot 9\ln\left|x+\sqrt{9+x^2}\right| + \frac{1}{2}x\sqrt{9+x^2} + C$$
$$= -\frac{9}{2}\ln\left|x+\sqrt{9+x^2}\right| + \frac{1}{2}x\sqrt{9+x^2} + C$$

- $\int x^2\sqrt{1+x^2}\,dx$ The integral fits Formula B with $a^2 = 1$.

$$\int x^2\sqrt{1+x^2}\,dx = \frac{1}{8}x(1+2x^2)\sqrt{1+x^2} - \frac{1}{8}\cdot 1^4\cdot\ln\left|x+\sqrt{1+x^2}\right| + C$$

- $$\int \frac{3}{25-x^2}\,dx$$

When we move the 3 outside the integral sign, the integral fits Formula A, with $a^2 = 25$.

$$\int \frac{3}{25 - x^2}\,dx = 3\int \frac{1}{5^2 - x^2}\,dx = 3\left(\frac{1}{10}\ln\left|\frac{x+5}{x-5}\right|\right) + C$$

- $\displaystyle\int \frac{1}{1 - 4x^2}\,dx$

The integral almost fits Formula A. We will use algebra to rewrite $1 - 4x^2$ in the form $a^2 - x^2$.

$$\begin{aligned}
1 - 4x^2 &= \frac{4}{4} - 4x^2 && \text{Replace 1 with } \frac{4}{4}. \\
&= 4\cdot\frac{1}{4} - 4x^2 \\
&= 4\left(\frac{1}{4} - x^2\right) && \text{Factor 4.} \\
&= 4\left[\left(\frac{1}{2}\right)^2 - x^2\right] && \frac{1}{4} \text{ needs to be a square.}
\end{aligned}$$

Now we have $a = \frac{1}{2}$ and

$$\frac{1}{1 - 4x^2} = \frac{1}{4[(\frac{1}{2})^2 - x^2]}$$

We will move the 4 out of the denominator and then outside the integral sign.

$$\frac{1}{4[(\frac{1}{2})^2 - x^2]} = \frac{1}{4}\cdot\frac{1}{(\frac{1}{2})^2 - x^2}$$

$$\begin{aligned}
\int \frac{1}{1 - 4x^2}\,dx &= \int \frac{1}{4}\frac{1}{(\frac{1}{2})^2 - x^2}\,dx \\
&= \frac{1}{4}\int \frac{1}{(\frac{1}{2})^2 - x^2}\,dx
\end{aligned}$$

$$= \frac{1}{4} \cdot \frac{1}{2 \cdot \frac{1}{2}} \ln \left| \frac{x + 1/2}{x - 1/2} \right| + C$$

$$= \frac{1}{4} \ln \left| \frac{x + 1/2}{x - 1/2} \right| + C$$

CHAPTER 13 REVIEW

1. $\int 4x^7\,dx =$
   (a) $\frac{1}{8}x^8 + C$
   (b) $28x^6 + C$
   (c) $4x^8 + C$
   (d) $\frac{1}{2}x^8 + C$
2. $\int (x - 3)\,dx =$
   (a) $\frac{1}{2}x^2 - 3x + C$
   (b) $x^2 - 3x + C$
   (c) $\frac{1}{2}x^2 - 3 + C$
   (d) $\frac{1}{2}x^2 + C$
3. $\int \sqrt[4]{x^3}\,dx =$
   (a) $\frac{4}{3}x^{7/4} + C$
   (b) $\frac{3}{4}x^{-1/4} + C$
   (c) $\frac{4}{7}x^{7/4} + C$
   (d) $\frac{7}{4}x^{7/4} + C$
4. $\int 4e^{3x+2}\,dx =$
   (a) $12e^{3x+2} + C$
   (b) $\frac{4}{3}e^{3x+2} + C$
   (c) $3e^{3x+2} + C$
   (d) $\frac{3}{4}e^{3x+2} + C$

5. $\int xe^{2x}\,dx =$ (Hint: use integration by parts.)
   (a) $\frac{1}{2}x^2e^{2x} - \frac{1}{2}e^{2x} + C$
   (b) $\frac{1}{2}x^2e^{2x} + \frac{1}{2}xe^{2x} - \frac{1}{2}e^{2x} + C$
   (c) $\frac{1}{2}x^2e^{2x} + C$
   (d) $\frac{1}{2}xe^{2x} - \frac{1}{4}e^{2x} + C$
6. $\int (3x^4 - x^2 + 3x + 5)\,dx =$
   (a) $12x^3 - 2x + 3 + C$
   (b) $\frac{3}{4}x^5 - \frac{1}{3}x^3 + 3x^2 + 5x + C$
   (c) $\frac{3}{5}x^5 - \frac{1}{3}x^3 + \frac{3}{2}x^2 + 5x + C$
   (d) $\frac{3}{5}x^5 - \frac{1}{3}x^3 + \frac{3}{2}x + 5 + C$
7. $\displaystyle\int \frac{4x}{5 + 2x^2}\,dx =$
   (a) $\ln|5 + 2x^2| + C$
   (b) $\ln|\frac{1}{5+2x^2}| + C$
   (c) $\ln|5 + 4x + 2x^2| + C$
   (d) Cannot be integrated
8. $\int x^2e^{x^3}\,dx =$
   (a) $\frac{1}{3}x^3e^{x^3} + C$
   (b) $2xe^{x^3} + C$
   (c) $\frac{1}{3}e^{x^3} + C$
   (d) $\frac{1}{3}x^2e^{1/4x^4} + C$
9. $\displaystyle\int \frac{3}{(2x + 1)^2}\,dx =$
   (a) $\frac{3}{2}\ln|2x + 1| + C$
   (b) $\frac{-3}{4x+2} + C$
   (c) $\frac{-3}{2x+1} + C$
   (d) $\frac{-12}{(2x+1)^3} + C$

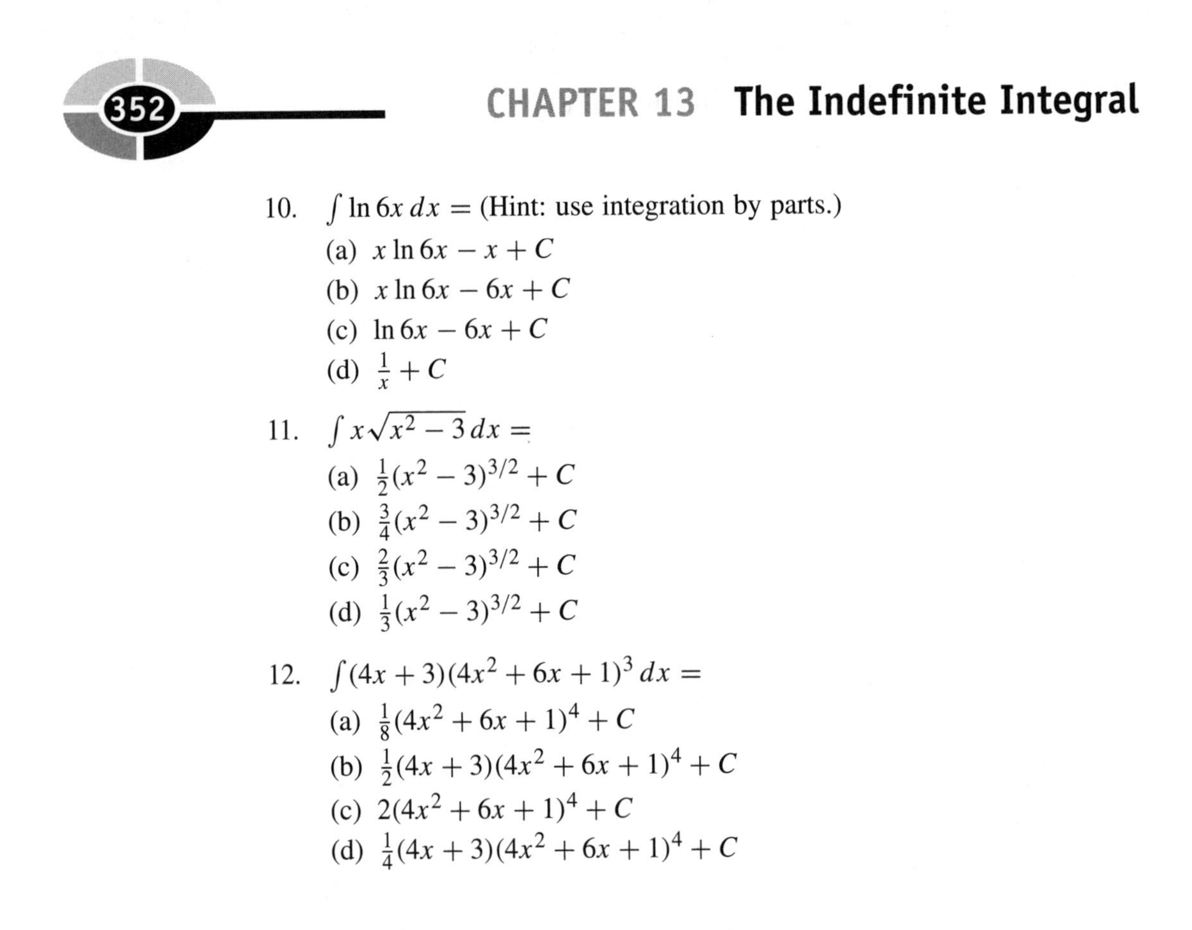

10. $\int \ln 6x \, dx =$ (Hint: use integration by parts.)
   (a) $x \ln 6x - x + C$
   (b) $x \ln 6x - 6x + C$
   (c) $\ln 6x - 6x + C$
   (d) $\frac{1}{x} + C$

11. $\int x\sqrt{x^2 - 3}\, dx =$
   (a) $\frac{1}{2}(x^2 - 3)^{3/2} + C$
   (b) $\frac{3}{4}(x^2 - 3)^{3/2} + C$
   (c) $\frac{2}{3}(x^2 - 3)^{3/2} + C$
   (d) $\frac{1}{3}(x^2 - 3)^{3/2} + C$

12. $\int (4x + 3)(4x^2 + 6x + 1)^3 \, dx =$
   (a) $\frac{1}{8}(4x^2 + 6x + 1)^4 + C$
   (b) $\frac{1}{2}(4x + 3)(4x^2 + 6x + 1)^4 + C$
   (c) $2(4x^2 + 6x + 1)^4 + C$
   (d) $\frac{1}{4}(4x + 3)(4x^2 + 6x + 1)^4 + C$

SOLUTIONS

1. d	2. a	3. c	4. b	5. d	6. c
7. a	8. c	9. b	10. a	11. d	12. a

CHAPTER 14

The Definite Integral and the Area Under the Curve

The indefinite integral is an algebraic expression. The definite integral is a number. The notation "$\int_a^b f(x)\,dx$" means the difference of the integral when it is evaluated at $x = a$ and $x = b$. The numbers a and b are called the *limits of integration*. Suppose $F(x)$ is an antiderivative of $f(x)$ (in other words, $F'(x) = f(x)$). Then $\int_a^b f(x)\,dx = F(b) - F(a)$. This is called the *Fundamental Theorem of Calculus.*

In the example below, we will be integrating $F(x) = 2x$. An antiderivative of $2x$ is x^2.

$$\begin{aligned}\int_1^2 2x\,dx &= F(2) - F(1)\\ &= 2^2 - 1^2\\ &= 3\end{aligned}$$

We could change $F(x)$ by adding or subtracting a constant, but this constant would not change the definite integral. For example, if we say that $F(x) = x^2 + 10$, then

$$\begin{aligned}\int_1^2 2x\,dx &= F(2) - F(1)\\ &= 2^2 + 10 - (1^2 + 10)\\ &= 14 - 11\\ &= 3.\end{aligned}$$

For this reason, we do not need to worry about "$+C$" when finding the definite integral.

Instead of writing $F(b) - F(a)$, we will use notation that allows us not to refer to $F(x)$ by name. The notation

$$F(x)\,\Big|_a^b$$

means $F(b) - F(a)$.

EXAMPLES

- $\int_2^5 3x^2\,dx$

  An antiderivative of $3x^2$ is x^3.

$$\begin{aligned}\int_2^5 3x^2\,dx &= x^3\,\Big|_2^5\\ &= 5^3 - 2^3 = 117\end{aligned}$$

- $\int_{-1}^{2} x\,dx$

$$\int_{-1}^{2} x\,dx = \frac{1}{2}x^2 \Big|_{-1}^{2}$$

$$= \frac{1}{2}(2^2) - \left(\frac{1}{2}(-1)^2\right)$$

$$= 2 - \left(\frac{1}{2}\right) = \frac{3}{2}$$

- $\int_{0}^{4}(2x+3)\,dx$

$$\int_{0}^{4}(2x+3)\,dx = \left(x^2+3x\right)\Big|_{0}^{4}$$

$$= 4^2 + 3(4) - (0^2 + 3(0))$$

$$= 28 - 0 = 28$$

- $\int_{-2}^{3} 2e^{2x+1}\,dx$

$$\int_{-2}^{3} 2e^{2x+1}\,dx = e^{2x+1}\Big|_{-2}^{3}$$

$$= e^{2(3)+1} - e^{2(-2)+1}$$

$$e^7 - e^{-3} \approx 1096.5834$$

- $\displaystyle\int_{2}^{5} \frac{4x}{(2x^2+1)^2}\,dx$

$$\int_{2}^{5} 4x(2x^2+1)^{-2}\,dx = -1(2x^2+1)^{-1}\Big|_{2}^{5}$$

$$= \frac{-1}{2x^2+1}\Big|_{2}^{5}$$

$$= \frac{-1}{2(5^2)+1} - \left(\frac{-1}{2(2^2)+1}\right)$$

$$= \frac{-1}{51} + \frac{1}{9} = \frac{-1}{51} \cdot \frac{3}{3} + \frac{1}{9} \cdot \frac{17}{17}$$

$$= \frac{-3+17}{153} = \frac{14}{153}$$

PRACTICE

1. $\int_{-1}^{2} 6x\,dx$
2. $\int_{1}^{3} (10x+7)\,dx$
3. $\int_{0}^{1} \frac{1}{x+1}$ (Give your answer accurate to three decimal places.)
4. $\int_{1}^{4} \sqrt{x}\,dx$
5. $\int_{1}^{2} (4x^3-6)\,dx$
6. $\int_{3}^{8} \sqrt{x+1}\,dx$
7. $\int_{-6}^{-2} (5x-2)\,dx$
8. $\int_{0}^{3} 2e^{4x}\,dx$ (Give your answer accurate to three decimal places.)
9. $\int_{2}^{3} (x^2+8x-1)\,dx$

SOLUTIONS

1.

$$\int_{-1}^{2} 6x\,dx = 3x^2 \Big|_{-1}^{2}$$

$$3(2)^2 - [3(-1)^2] = 12 - 3 = 9$$

2.

$$\int_{1}^{3} (10x+7)\,dx = (5x^2+7x)\Big|_{1}^{3}$$

$$= 5(3)^2 + 7(3) - [5(1)^2 + 7(1)]$$

$$= 66 - (12) = 54$$

3.

$$\int_0^1 \frac{1}{x+1} = \ln|x+1| \Big|_0^1$$

$$= \ln(1+1) - \ln(1+0) = \ln 2 - \ln 1 \approx 0.693$$

4.

$$\int_1^4 \sqrt{x}\, dx = \int_1^4 x^{1/2}\, dx$$

$$= \frac{2}{3}x^{3/2} \Big|_1^4 = \frac{2}{3}\sqrt{x^3} \Big|_1^4$$

$$= \frac{2}{3}\sqrt{4^3} - \frac{2}{3}\sqrt{1^3}$$

$$= \frac{16}{3} - \frac{2}{3} = \frac{14}{3}$$

5.

$$\int_1^2 (4x^3 - 6)\, dx = (x^4 - 6x) \Big|_1^2$$

$$= 2^4 - 6(2) - [1^4 - 6(1)]$$

$$= 4 - (-5) = 9$$

6.

$$\int_3^8 \sqrt{x+1}\, dx = \int_3^8 (x+1)^{1/2}\, dx$$

$$= \frac{2}{3}(x+1)^{3/2} \Big|_3^8 = \frac{2}{3}\sqrt{(x+1)^3} \Big|_3^8$$

$$= \frac{2}{3}\sqrt{(8+1)^3} - \frac{2}{3}\sqrt{(3+1)^3} = \frac{2}{3}(27) - \frac{2}{3}(8)$$

$$= \frac{54}{3} - \frac{16}{3} = \frac{38}{3}$$

7.

$$\int_{-6}^{-2} (5x - 2)\,dx = \left(\frac{5}{2}x^2 - 2x\right)\Big|_{-6}^{-2}$$

$$\frac{5}{2}(-2)^2 - 2(-2) - \left[\frac{5}{2}(-6)^2 - 2(-6)\right]$$

$$= 14 - 102 = -88$$

8.

$$\int_0^3 2e^{4x}\,dx = \frac{1}{2}e^{4x}\Big|_0^3$$

$$= \frac{1}{2}e^{4(3)} - \frac{1}{2}e^{4(0)} = \frac{1}{2}e^{12} - \frac{1}{2}e^0$$

$$\approx 81,376.896$$

9.

$$\int_2^3 (x^2 + 8x - 1)\,dx = \left(\frac{1}{3}x^3 + 4x^2 - x\right)\Big|_2^3$$

$$= \frac{1}{3}(3)^3 + 4(3)^2 - 3 - \left[\frac{1}{3}(2)^3 + 4(2)^2 - 2\right]$$

$$= 42 - \frac{50}{3} = \frac{126}{3} - \frac{50}{3} = \frac{76}{3}$$

Area Under the Curve

Finding the area of a figure is a common problem in mathematics. When a figure is made from standard shapes, we can use geometry formulas to compute its area. But if we want to find the area of shapes with curved edges, we probably need to use calculus. The area of a shape with one or more curved edges can always be approximated by the area of rectangles. The area of the six rectangles in Figure 14.1 closely approximates the shaded area under the curve. The approximation is better when the rectangles are more narrow (see Figure 14.2).

In the same way that the slopes of secant lines approximate the slope of a tangent line, the area under a curve can be approximated by the area of rectangles. The area under the curve is the limit of the total area of the rectangles, as the width of the rectangles shrinks to zero.

The curve in Figures 14.1 and 14.2 is the graph of $f(x) = \frac{1}{x}$. The areas of the rectangles in Figures 14.1 and 14.2 approximate the area under the curve

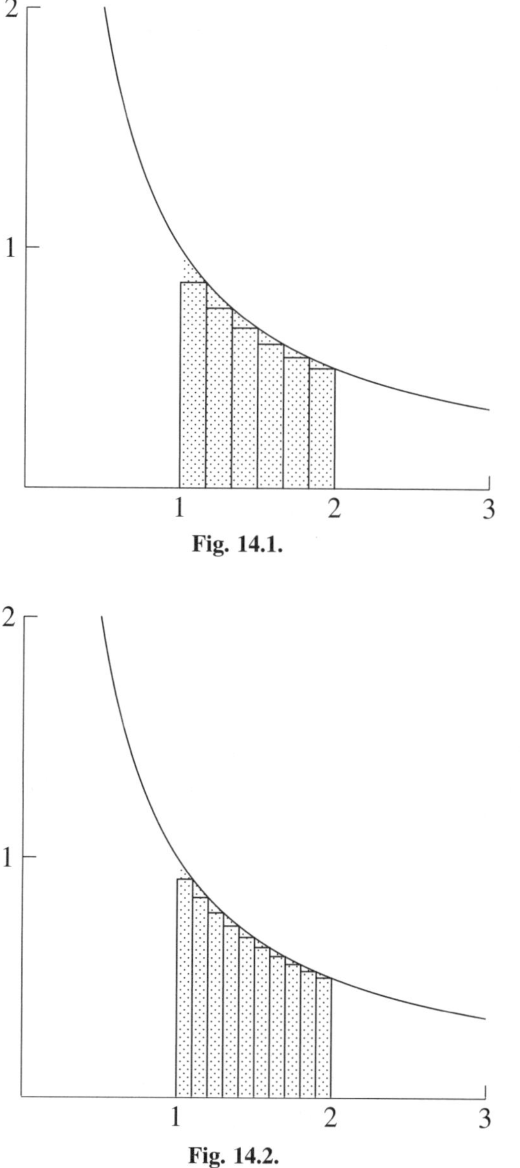

Fig. 14.1.

Fig. 14.2.

between $x = 1$ and $x = 2$. As we shall see later, the area is exactly $\ln 2$, which is approximately 0.6931. From the rectangles in Figure 14.1, the approximation is 0.6532. From the rectangles in Figure 14.2, the approximation is 0.6688.

Suppose we want to find the area under the curve of $f(x)$ between $x = a$ and $x = b$. We can subdivide the interval $[a, b]$ into n equal subintervals to create the base of each rectangle. The width of each rectangle is $\frac{1}{n}$. The graph in Figure 14.3

shows how the interval [1, 2] is divided into ten equal subintervals. The width of each rectangle in this example is $\frac{1}{10}$.

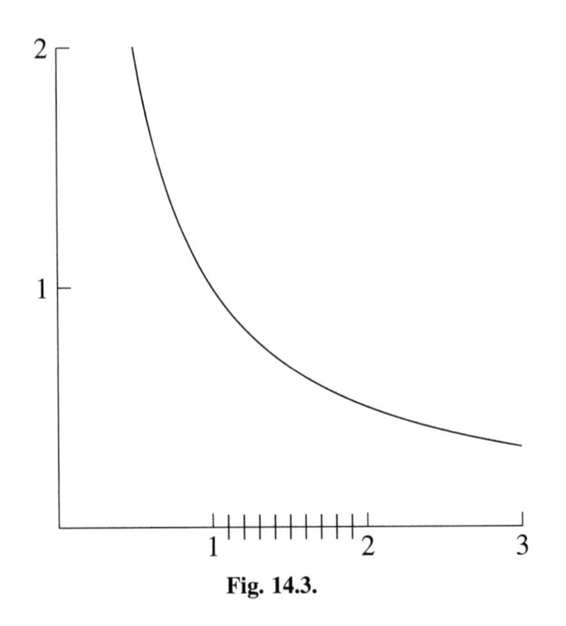

Fig. 14.3.

From the subdivisions, we can construct the rectangles shown in Figure 14.2. The height of each rectangle is the y-value of a point on the curve (Figure 14.4).

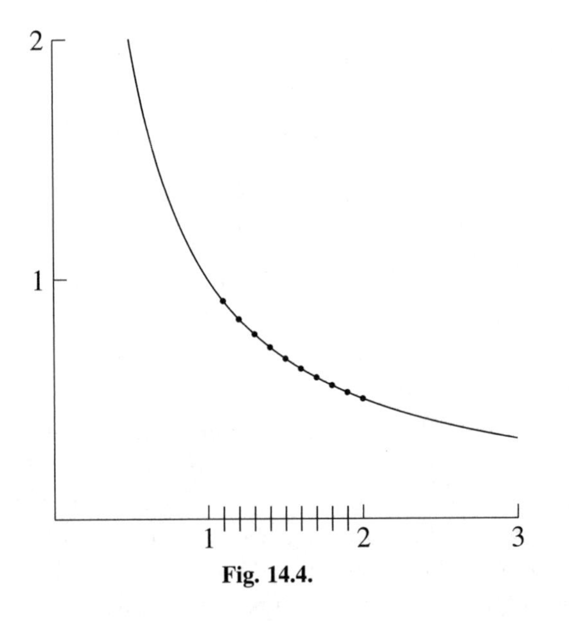

Fig. 14.4.

When the interval $[a, b]$ is subdivided into n equal subintervals, the subintervals are $[a, x_1]$, $[x_1, x_2]$, ... $[x_i, x_{i+1}]$, ... $[x_{n-1}, b]$. The width of the ith interval is $x_{i+1} - x_i$, which is equal to $\frac{1}{n}$. The height of each rectangle is either $f(x_i)$ or $f(x_{i+1})$, it really does not matter which. One choice usually overestimates the area, and the other usually underestimates it (Figure 14.5).

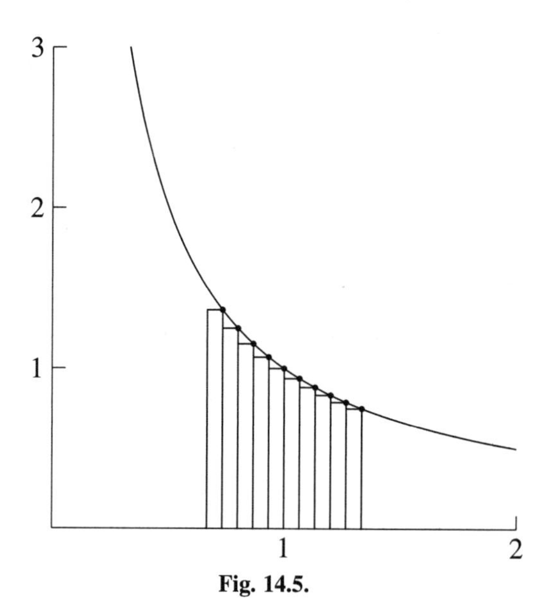

Fig. 14.5.

The area of a typical rectangle is

$$\text{“width} \times \text{height”} = \overbrace{(x_{i+1} - x_i)}^{\text{Width}} \cdot \overbrace{f(x_i)}^{\text{Height}}$$

The sum of these n areas is

$$\sum_{i=1}^{n} (x_{i+1} - x_i) \cdot f(x_i)$$

The exact area between the curve and the x-axis (from $x = a$ to $x = b$) is the limit of this sum as n gets large without bound.

$$\text{Area} = \lim_{n \to \infty} \sum_{i=1}^{n} (x_{i+1} - x_i) \cdot f(x_i)$$

It can be shown with some advanced calculus techniques that this limit is the definite integral $\int_a^b f(x)\,dx$.

$$\text{Area} = \lim_{n\to\infty} \sum_{i=1}^{n} (x_{i+1} - x_i) \cdot f(x_i) = \int_a^b f(x)\,dx$$

The quantity "$x_{i+1} - x_i$" becomes "dx" in the limit, and the summation symbol "Σ" becomes the integral symbol "$\int$."

EXAMPLES

Find the indicated area.

- The graph of $y = \sqrt{x}$ is shown in Figure 14.6.

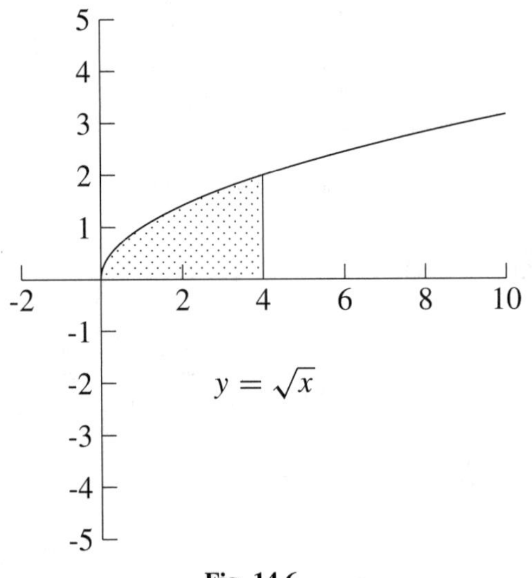

Fig. 14.6.

We want to find the area between the curve of $y = \sqrt{x}$ and the x-axis, from $x = 0$ to $x = 4$. We will find the definite integral of $\sqrt{x}$, with $a = 0$ and $b = 4$.

$$\int_0^4 x^{1/2}\,dx = \frac{2}{3}x^{3/2}\Big|_0^4$$

$$= \frac{2}{3}\sqrt{x^3}\Big|_0^4$$

$$= \frac{2}{3}\sqrt{4^3} - \frac{2}{3}\sqrt{0^3}$$

$$= \frac{2}{3}(8) - \frac{2}{3}(0) = \frac{16}{3}$$

- The graph of $y = -x^4 + 3x^2 + 4$ is shown in Figure 14.7.

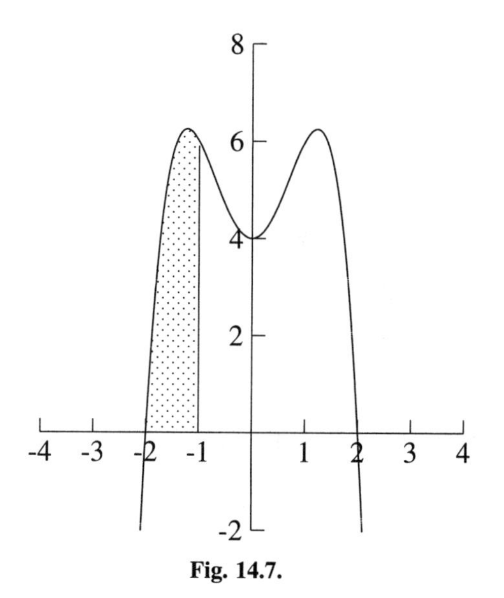

Fig. 14.7.

The area is the definite integral of $-x^4 + 3x^2 + 4$ from $a = -2$ to $b = -1$.

$$\int_{-2}^{-1} (-x^4 + 3x^2 + 4)\,dx = \left(-\frac{1}{5}x^5 + x^3 + 4x\right)\Big|_{-2}^{-1}$$

$$= -\frac{1}{5}(-1)^5 + (-1)^3 + 4(-1)$$

$$- \left(-\frac{1}{5}(-2)^5 + (-2)^3 + 4(-2)\right)$$

$$= -\frac{24}{5} - \left(-\frac{48}{5}\right) = \frac{24}{5}$$

- The graph of $y = e^x$ is shown in Figure 14.8.

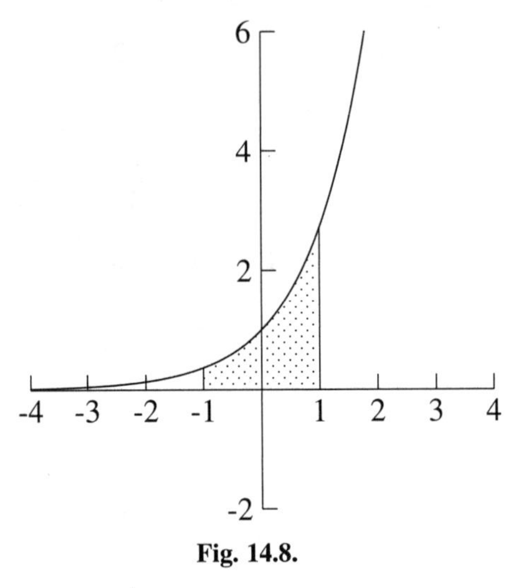

Fig. 14.8.

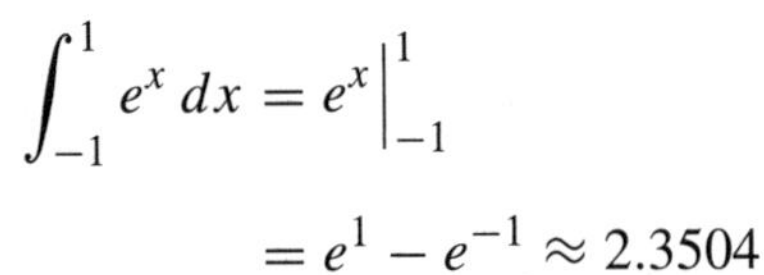

$$\int_{-1}^{1} e^x \, dx = e^x \Big|_{-1}^{1}$$

$$= e^1 - e^{-1} \approx 2.3504$$

When the area is below the x-axis, the definite integral is a negative number.

- The graph of $y = x^2 - 5x + 4$ is shown in Figure 14.9.

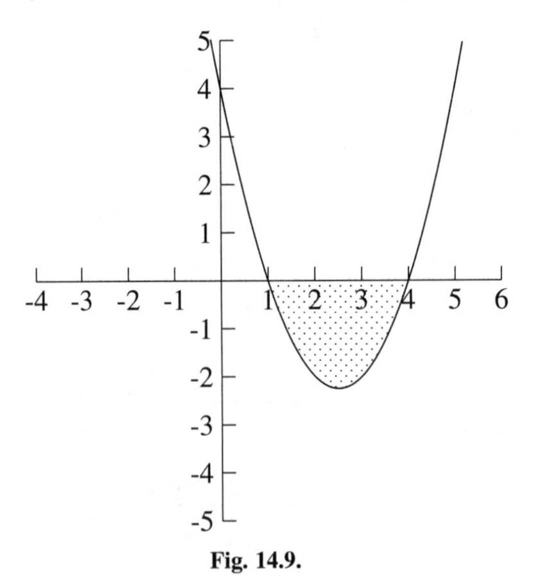

Fig. 14.9.

The shaded area lies between $a = 1$ and $b = 4$.

$$\int_1^4 (x^2 - 5x + 4)\,dx = \left(\frac{1}{3}x^3 - \frac{5}{2}x^2 + 4x\right)\Big|_1^4$$

$$= \frac{1}{3}(4)^3 - \frac{5}{2}(4)^2 + 4(4)$$

$$- \left(\frac{1}{3}(1)^3 - \frac{5}{2}(1)^2 + 4(1)\right)$$

$$= -\frac{8}{3} - \left(\frac{11}{6}\right) = -\frac{9}{2}$$

PRACTICE

Find the indicated area.

1. The graph of $y = x^2 + 2$ is given in Figure 14.10.

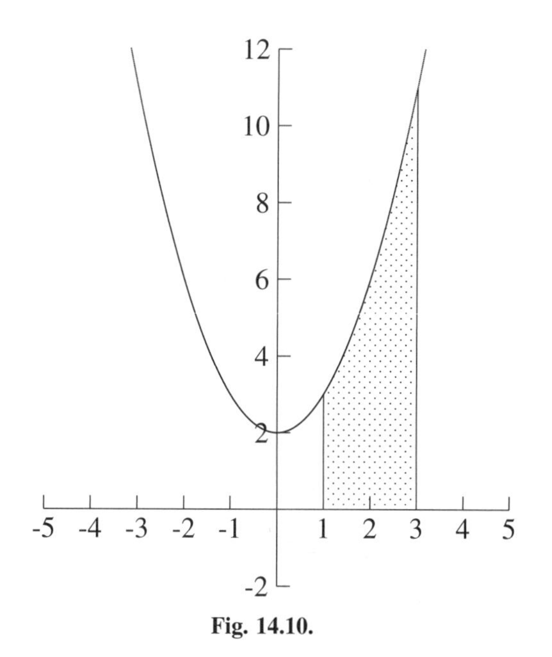

Fig. 14.10.

2. The graph of $y = \frac{1}{x^2}$ is given in Figure 14.11.

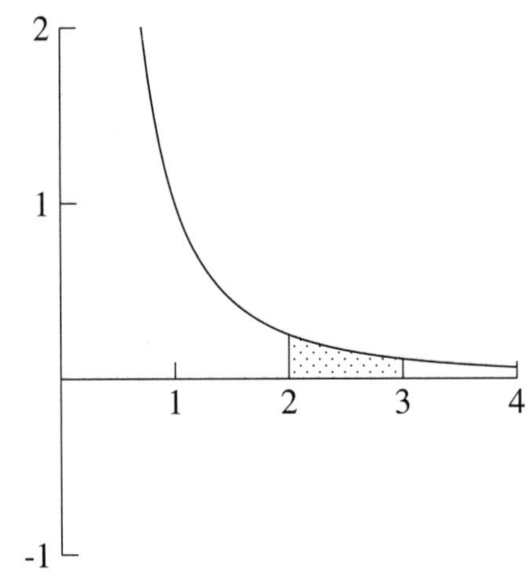

Fig. 14.11.

3. The graph of $y = 2x^3 + 3x^2 - 12x - 10$ is given in Figure 14.12. (Use $a = -3$ and $b = -1$.)

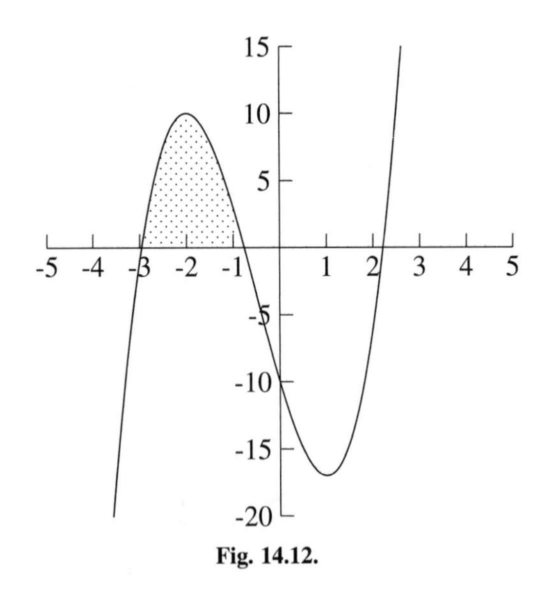

Fig. 14.12.

4. The graph of $y = \ln x$ is given in Figure 14.13. Give your answer accurate to three decimal places. (Hint: use integration by parts.)

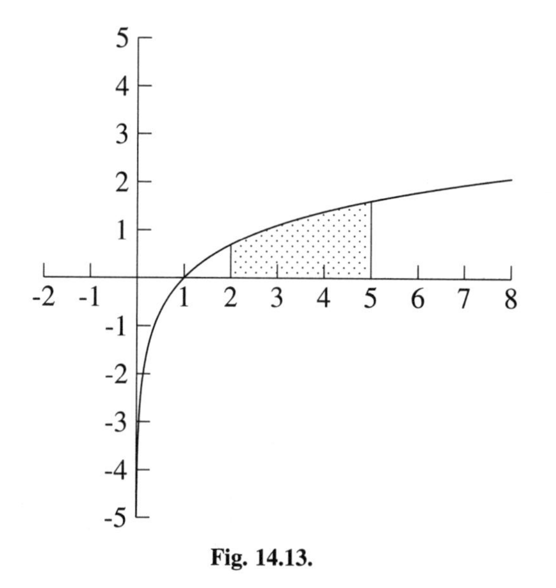

Fig. 14.13.

5. The graph of $y = \frac{2x}{x^2+1}$ is given in Figure 14.14. Give your answer accurate to three decimal places.

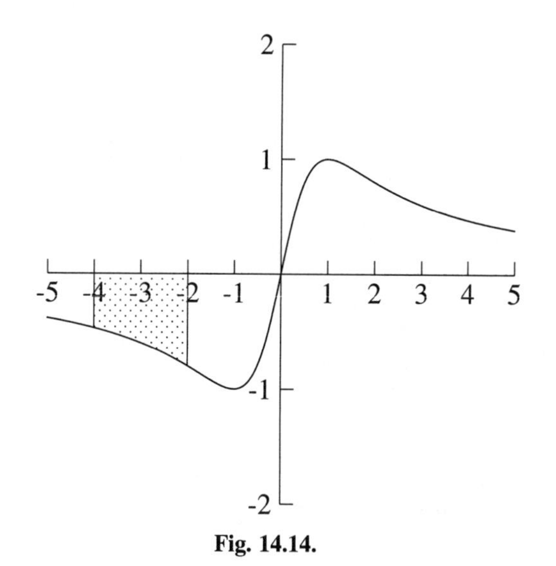

Fig. 14.14.

SOLUTIONS

1.

$$\int_1^3 (x^2 + 2)\,dx = \left(\frac{1}{3}x^3 + 2x\right)\Big|_1^3$$
$$= \left(\frac{1}{3}(3)^3 + 2(3)\right) - \left(\frac{1}{3}(1)^3 + 2(1)\right)$$
$$= 15 - \frac{7}{3} = \frac{38}{3}$$

2.

$$\int_2^3 \frac{1}{x^2}\,dx = \int_2^3 x^{-2}\,dx = -x^{-1}\Big|_2^3$$
$$= -\frac{1}{x}\Big|_2^3 = -\frac{1}{3} - \left(-\frac{1}{2}\right)$$
$$= \frac{1}{6}$$

3.

$$\int_{-3}^{-1} (2x^3 + 3x^2 - 12x - 10)\,dx = \left(\frac{1}{2}x^4 + x^3 - 6x^2 - 10x\right)\Big|_{-3}^{-1}$$
$$= \frac{1}{2}(-1)^4 + (-1)^3 - 6(-1)^2 - 10(-1)$$
$$- \left(\frac{1}{2}(-3)^4 + (-3)^3 - 6(-3)^2 - 10(-3)\right)$$
$$= \frac{7}{2} - \left(-\frac{21}{2}\right) = 14$$

4.

$$\int_2^5 \ln x\,dx = (x\ln x - x)\Big|_2^5$$
$$= 5\ln 5 - 5 - (2\ln 2 - 2) \approx 3.661$$

5.

$$\int_{-4}^{-2} \frac{2x}{x^2+1}\,dx = \ln(x^2+1)\Big|_{-4}^{-2}$$
$$= \ln((-2)^2+1) - \ln((-4)^2+1)$$
$$= \ln 5 - \ln 17 \approx -1.224$$

The negative sign indicates that the area is below the x-axis.

When some of the area between the curve and the x-axis is above the x-axis and some is below, the definite integral subtracts the area below the axis from the area above the axis. If more area is above the x-axis than below it, the definite integral is positive. If more area is below the x-axis than above it, the definite integral is negative.

EXAMPLE

- The graph of $y = -x^3 + 2x^2 + 8x$ is given in Figure 14.15.

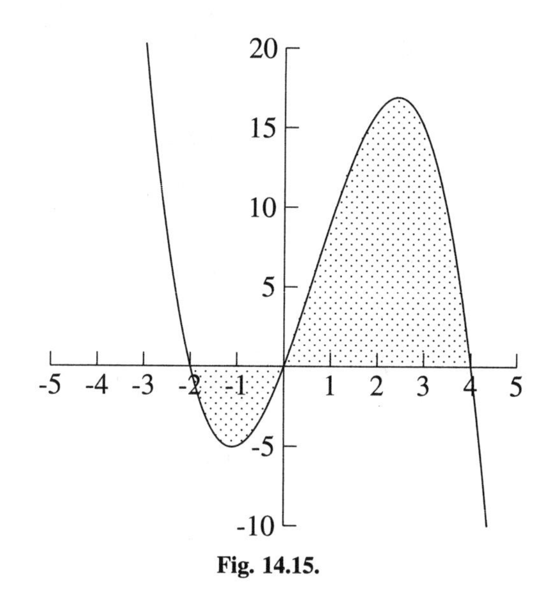

Fig. 14.15.

The shaded area above the x-axis is more than the shaded area below, so the definite integral will be positive.

$$\int_{-2}^{4} (-x^3 + 2x^2 + 8x)\,dx = \left(-\frac{1}{4}x^4 + \frac{2}{3}x^3 + 4x^2\right)\Big|_{-2}^{4}$$

$$= -\frac{1}{4}(4)^4 + \frac{2}{3}(4)^3 + 4(4)^2$$

$$- \left(-\frac{1}{4}(-2)^4 + \frac{2}{3}(-2)^3 + 4(-2)^2\right)$$

$$= \frac{128}{3} - \left(\frac{20}{3}\right) = 36$$

If we compute each area separately and then add them, we have the same answer.

$$\int_{-2}^{0} (-x^3 + 2x^2 + 8x)\,dx + \int_{0}^{4} (-x^3 + 2x^2 + 8x)\,dx$$

$$= \left(-\frac{1}{4}x^4 + \frac{2}{3}x^3 + 4x^2\right)\Big|_{-2}^{0} + \left(-\frac{1}{4}x^4 + \frac{2}{3}x^3 + 4x^2\right)\Big|_{0}^{4}$$

$$= -\frac{1}{4}(0)^4 + \frac{2}{3}(0)^3 + 4(0)^2$$

$$- \left(-\frac{1}{4}(-2)^4 + \frac{2}{3}(-2)^3 + 4(-2)^2\right)$$

$$+ \left(-\frac{1}{4}\right)(4)^4 + \frac{2}{3}(4)^3 + 4(4)^2$$

$$- \left(-\frac{1}{4}(0)^4 + \frac{2}{3}(0)^3 + 4(0)^2\right)$$

$$= 0 - \left(\frac{20}{3}\right) + \frac{128}{3} - (0) = 36$$

PRACTICE

1. The graph of $y = x^4 - 5x^2 + 4$ is given in Figure 14.16. Find the shaded area.

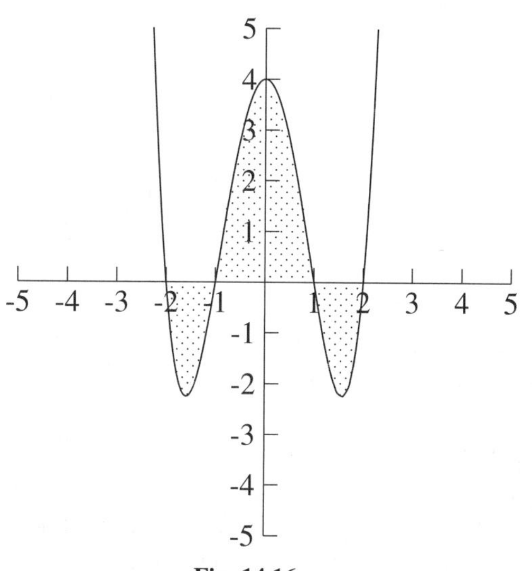

Fig. 14.16.

SOLUTION

1.

$$\int_{-2}^{2} (x^4 - 5x^2 + 4)\, dx = \left(\frac{1}{5}x^5 - \frac{5}{3}x^3 + 4x \right) \Big|_{-2}^{2}$$

$$= \frac{1}{5}(2)^5 - \frac{5}{3}(2)^3 + 4(2)$$

$$- \left(\frac{1}{5}(-2)^5 - \frac{5}{3}(-2)^3 + 4(-2) \right)$$

$$= \frac{16}{15} - \left(-\frac{16}{15} \right) = \frac{32}{15}$$

If we split the area into three separate integrals, we have

$$\int_{-2}^{-1} (x^4 - 5x^2 + 4)\, dx + \int_{-1}^{1} (x^4 - 5x^2 + 4)\, dx + \int_{1}^{2} (x^4 - 5x^2 + 4)\, dx$$

$$= \left(\frac{1}{5}x^5 - \frac{5}{3}x^3 + 4x\right)\Big|_{-2}^{-1}$$

$$+ \left(\frac{1}{5}x^5 - \frac{5}{3}x^3 + 4x\right)\Big|_{-1}^{1}$$

$$+ \left(\frac{1}{5}x^5 - \frac{5}{3}x^3 + 4x\right)\Big|_{1}^{2}$$

$$= -\frac{22}{15} + \frac{76}{15} + \left(-\frac{22}{15}\right) = \frac{32}{15}$$

We can find the area between two curves by subtracting the area of one curve from the area of the other curve. Suppose the area lies above the x-axis. The area between the two curves is the area under the top curve with the area under the bottom curve deleted (see Figures 14.17–14.19). The area between a line, $y = -x + 7$, and curve, $y = x^2 - 6x + 11$, is shaded in Figure 14.17. The total area under to line is shaded in Figure 14.18. The area under the curve is deleted in Figure 14.19, leaving the area between the line and curve.

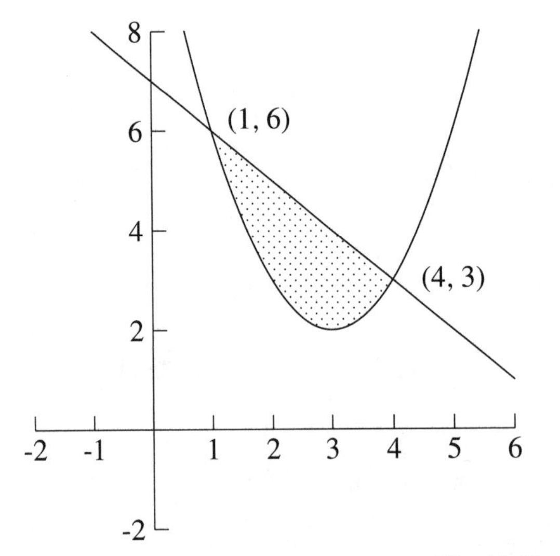

Fig. 14.17.

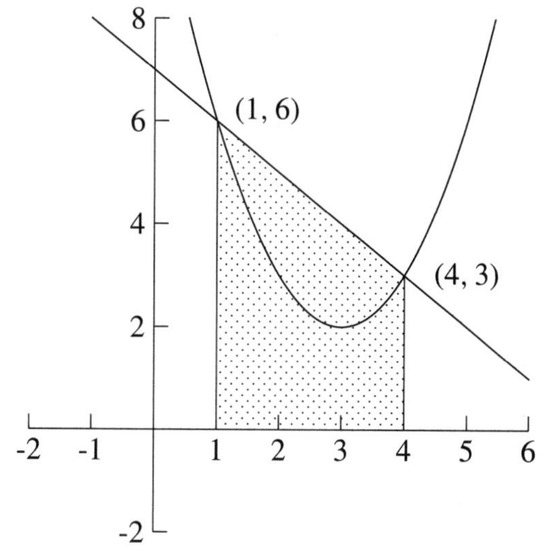

Fig. 14.18.

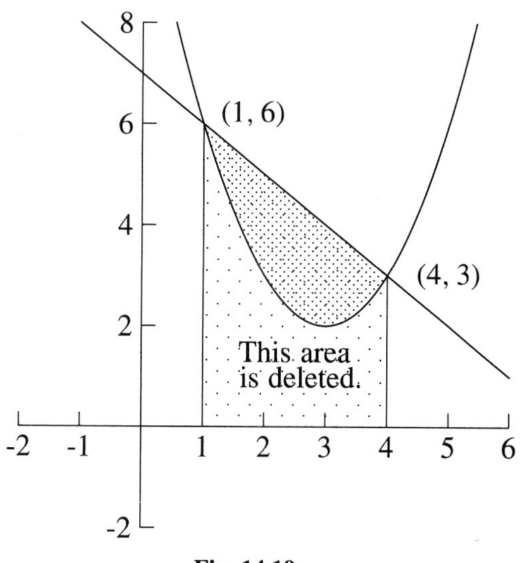

Fig. 14.19.

The area between these curves is computed by subtracting the definite integral of the bottom curve from the definite integral of the top curve.

$$\text{Area} = \int_1^4 (-x + 7)\,dx - \int_1^4 (x^2 - 6x + 11)\,dx$$

We can combine and simplify these integrals.

$$\begin{aligned}
\text{Area} &= \int_1^4 (-x+7)\,dx - \int_1^4 (x^2-6x+11)\,dx \\
&= \int_1^4 [(-x+7) - [(x^2-6x+11)]\,dx = \int_1^4 (-x^2+5x-4)\,dx \\
&= \left(-\frac{1}{3}x^3 + \frac{5}{2}x^2 - 4x\right)\Big|_1^4 \\
&= -\frac{1}{3}(4)^3 + \frac{5}{2}(4)^2 - 4(4) - \left(-\frac{1}{3}(1)^3 + \frac{5}{2}(1)^2 - 4(1)\right) \\
&= \frac{8}{3} - \left(-\frac{11}{6}\right) = \frac{9}{2}
\end{aligned}$$

No matter where the area between two curves lies, above the x-axis, below the x-axis, or both above and below, the area between the curves is always computed by subtracting the definite integral of the bottom curve from the definite integral of the top curve.

$$\text{Area} = \int_a^b (\text{Top curve} - \text{Bottom curve})\,dx$$

EXAMPLES

Find the shaded area.

- The curves in Figure 14.20 are $y = -x^2+4x-3$ (top) and $y = x^2-4x+3$ (bottom).

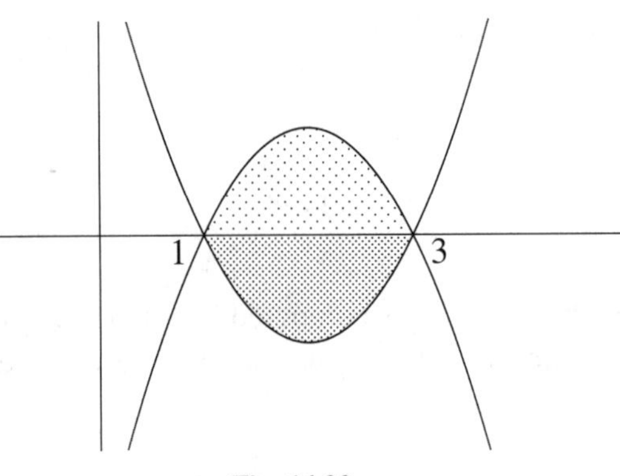

Fig. 14.20.

$$\begin{aligned}
\text{Area} &= \int_a^b (\text{Top curve} - \text{Bottom curve})\,dx \\
&= \int_1^3 [(-x^2+4x-3)-(x^2-4x+3)]\,dx \\
&= \int_1^3 (-2x^2+8x-6)\,dx \\
&= \left(-\frac{2}{3}x^3+4x^2-6x\right)\Big|_1^3 \\
&= -\frac{2}{3}(3)^3+4(3)^2-6(3)-\left(-\frac{2}{3}(1)^3+4(1)^2-6(1)\right) \\
&= 0-\left(-\frac{8}{3}\right)=\frac{8}{3}
\end{aligned}$$

The area above the x-axis in Figure 14.20 appears to be equal to the area below it. This would mean that the total shaded area is also twice the area above the x-axis. Let us see if this is true.

$$\begin{aligned}
\int_1^3 [(-x^2+4x-3)\,dx &= \left(-\frac{1}{3}x^3+2x^2-3x\right)\Big|_1^3 \\
&= -\frac{1}{3}(3)^3+2(3)^2-3(3) \\
&\quad -\left(-\frac{1}{3}(1)^3+2(1)^2-3(1)\right) \\
&= 0-\left(-\frac{4}{3}\right)=\frac{4}{3}
\end{aligned}$$

$\frac{4}{3}$ is half of $\frac{8}{3}$. The area under the x-axis is $-\frac{4}{3}$. The integral $\int_1^3[(-x^2+4x-3)-(x^2-4x+3)]\,dx$ is computing $\frac{4}{3}-(-\frac{4}{3})=\frac{8}{3}$.

- The top curve in Figure 14.21 is $y=\sqrt{x}$, and the bottom curve is $y=x^3$. We want to find the area between these curves between $a=0$ and $b=1$.

$$\begin{aligned}
\int_0^1 (x^{1/2}-x^3)\,dx &= \left(\frac{2}{3}x^{3/2}-\frac{1}{4}x^4\right)\Big|_0^1 \\
&= \left(\frac{2}{3}\sqrt{x^3}-\frac{1}{4}x^4\right)\Big|_0^1
\end{aligned}$$

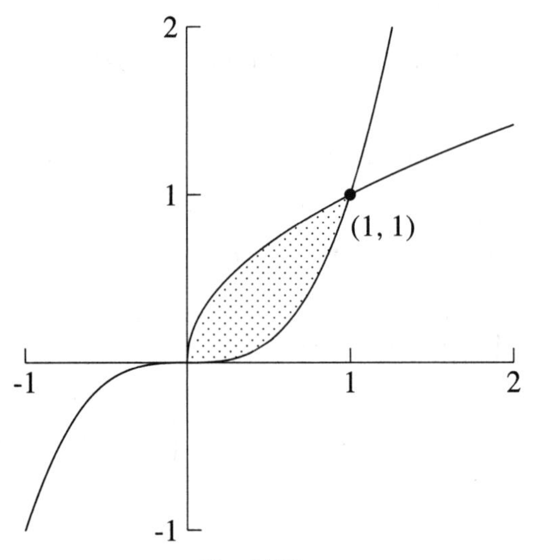

Fig. 14.21.

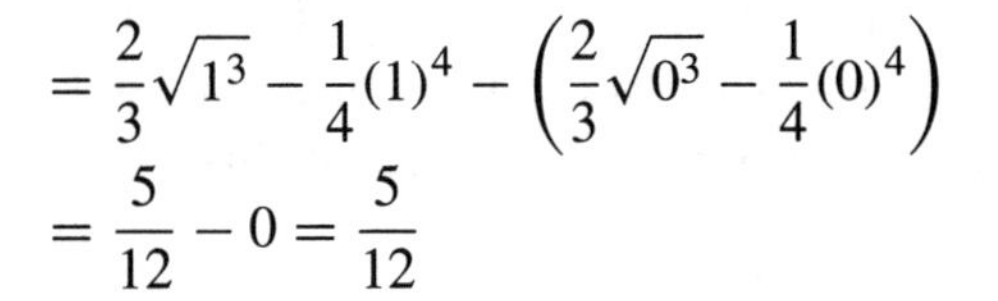

$$= \frac{2}{3}\sqrt{1^3} - \frac{1}{4}(1)^4 - \left(\frac{2}{3}\sqrt{0^3} - \frac{1}{4}(0)^4\right)$$
$$= \frac{5}{12} - 0 = \frac{5}{12}$$

PRACTICE

Find the shaded areas.

1. The top curve in Figure 14.22 is $y = -x^2 - 3x$, and the bottom curve is $y = x$.

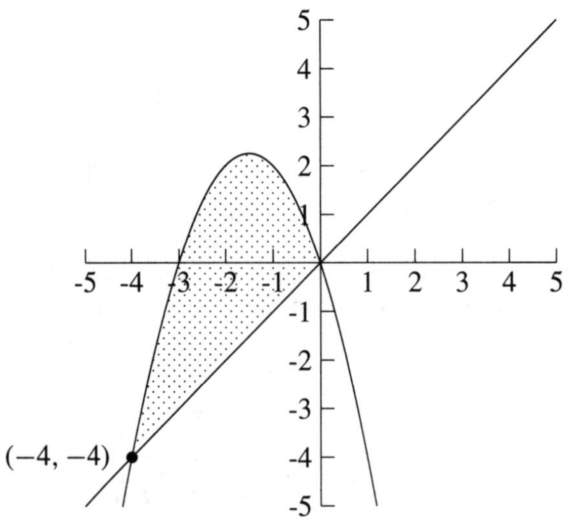

Fig. 14.22.

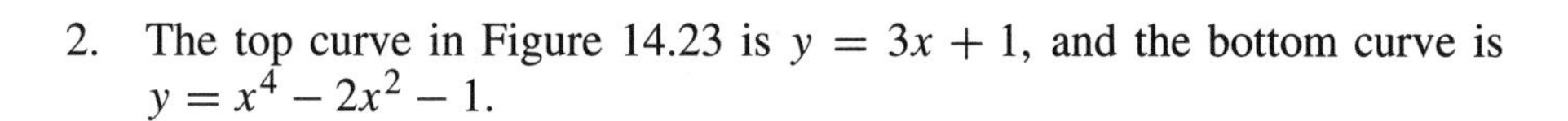

2. The top curve in Figure 14.23 is $y = 3x + 1$, and the bottom curve is $y = x^4 - 2x^2 - 1$.

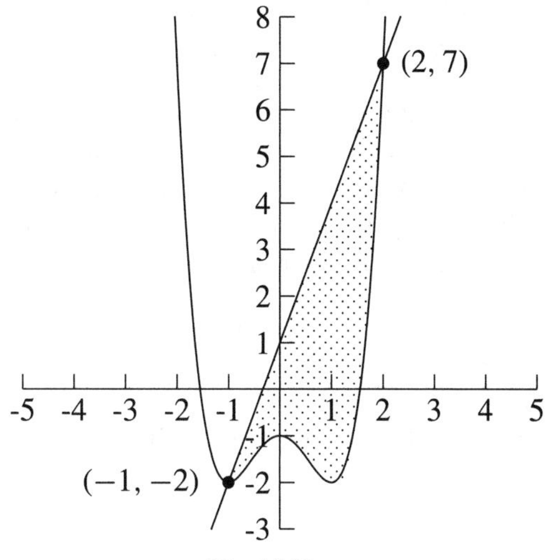

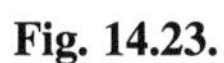

Fig. 14.23.

SOLUTIONS

1.
$$\int_{-4}^{0} (-x^2 - 3x - x)\, dx = \int_{-4}^{0} (-x^2 - 4x)\, dx$$
$$= \left(-\frac{1}{3}x^3 - 2x^2\right) \Big|_{-4}^{0}$$
$$= -\frac{1}{3}(0)^3 - 2(0)^2 - \left(-\frac{1}{3}(-4)^3 - 2(-4)^2\right)$$
$$= 0 - \left(-\frac{32}{3}\right) = \frac{32}{3}$$

2.
$$\int_{-1}^{2} [(3x + 1) - (x^4 - 2x^2 - 1)]\, dx = \int_{-1}^{2} (-x^4 + 2x^2 + 3x + 2)\, dx$$
$$= \left(-\frac{1}{5}x^5 + \frac{2}{3}x^3 + \frac{3}{2}x^2 + 2x\right) \Big|_{-1}^{2}$$
$$= -\frac{1}{5}(2)^5 + \frac{2}{3}(2)^3 + \frac{3}{2}(2)^2 + 2(2)$$

$$-\left(-\frac{1}{5}(-1)^5+\frac{2}{3}(-1)^3+\frac{3}{2}(-1)^2+2(-1)\right)$$

$$=\frac{134}{15}-\left(-\frac{29}{30}\right)=\frac{297}{30}$$

When we cannot use the graph to find $x = a$ and $x = b$, we can find them algebraically. When we set two functions equal to each other and solve for x, we get the x-value or values where the graphs intersect (cross each other). For example, if we want the x-value where the lines $y = 2x$ and $y = x + 2$ intersect, we set $2x$ and $x + 1$ equal to each other and solve for x.

$$2x = x + 1$$

$$x = 1$$

The lines $y = 2x$ and $y = x + 1$ intersect at $x = 1$.

EXAMPLE

- Find the shaded area in Figure 14.24.

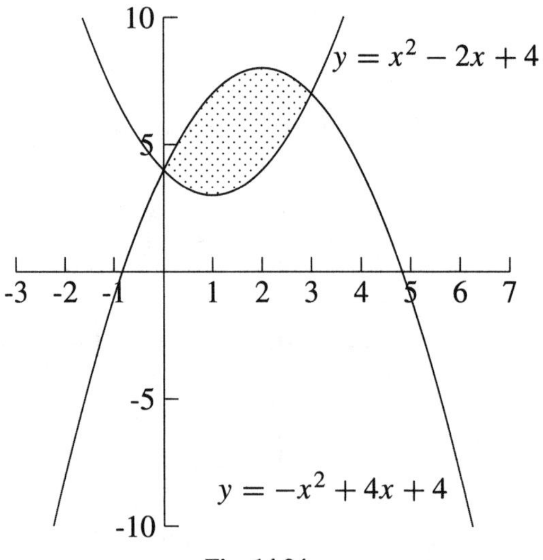

Fig. 14.24.

We will find $x = a$ and $x = b$ by solving $x^2 - 2x + 4 = -x^2 + 4x + 4$.

$$x^2 - 2x + 4 = -x^2 + 4x + 4$$

$$2x^2 - 6x = 0$$

$$2x(x-3)=0$$

$$2x=0 \qquad x-3=0$$

$$x=0 \qquad x=3$$

Now we know that $a=0$ and $b=3$.

$$\int_0^3 [(-x^2+4x+4)-(x^2-2x+4)]\,dx = \int_0^3 (-2x^2+6x)\,dx$$

$$= \left(-\frac{2}{3}x^3+3x^2\right)\Big|_0^3$$

$$= -\frac{2}{3}(3)^3+3(3)^2$$

$$-\left(-\frac{2}{3}(0)^3+3(0)^2\right)$$

$$= 9-0=9$$

PRACTICE

1. Find the shaded area for the graph in Figure 14.25. The curves are the graphs of $y=x^2+x-10$ and $y=-x^2+5x+6$.

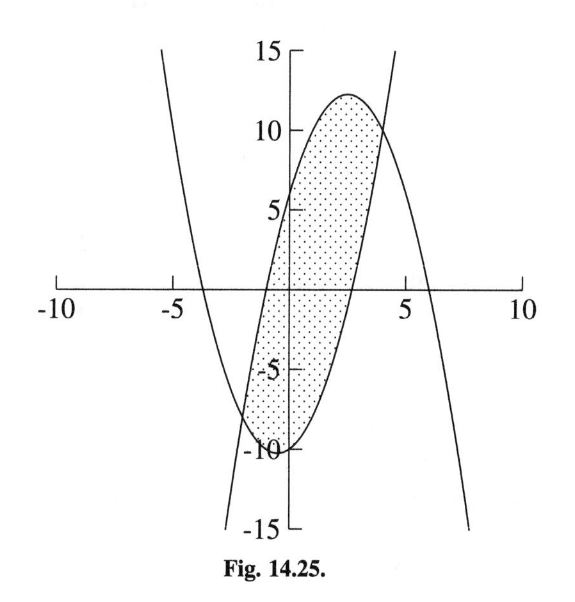

Fig. 14.25.

SOLUTION

1. We will find $x = a$ and $x = b$ by solving $x^2 + x - 10 = -x^2 + 5x + 6$.

$$x^2 + x - 10 = -x^2 + 5x + 6$$

$$2x^2 - 4x - 16 = 0 \quad \text{We will divide both sides of the equation by 2.}$$

$$x^2 - 2x - 8 = 0$$

$$(x + 2)(x - 4) = 0$$

$$x + 2 = 0 \qquad x - 4 = 0$$

$$x = -2 \qquad x = 4$$

The graphs intersect at $a = -2$ and $b = 4$. From the graph, we see that $y = -x^2 + 5x + 6$ is the top curve.

$$\int_{-2}^{4} [(-x^2 + 5x + 6) - (x^2 + x - 10)]\, dx$$

$$= \int_{-2}^{4} (-2x^2 + 4x + 16)\, dx$$

$$= \left(-\frac{2}{3}x^3 + 2x^2 + 16x\right) \Big|_{-2}^{4}$$

$$= -\frac{2}{3}(4)^3 + 2(4)^2 + 16(4)$$

$$- \left(-\frac{2}{3}(-2)^3 + 2(-2)^2 + 16(-2)\right)$$

$$= \frac{160}{3} - \left(-\frac{56}{3}\right) = 72$$

Sometimes the area between two curves occurs in more than one region.Usually, one curve is on the top in one region and the other curve is on top in the other. When this happens, we need to evaluate more than one definite integral.

EXAMPLES

Find the shaded area.

- The line in Figure 14.26 is $y = x - 3$, and the curve is $y = -x^3 + 5x - 3$. From $a = -2$ to $b = 0$, the line $y = x - 3$ is on top. From $a = 0$ to $b = 2$, the curve $y = -x^3 + 5x - 3$ is on top.

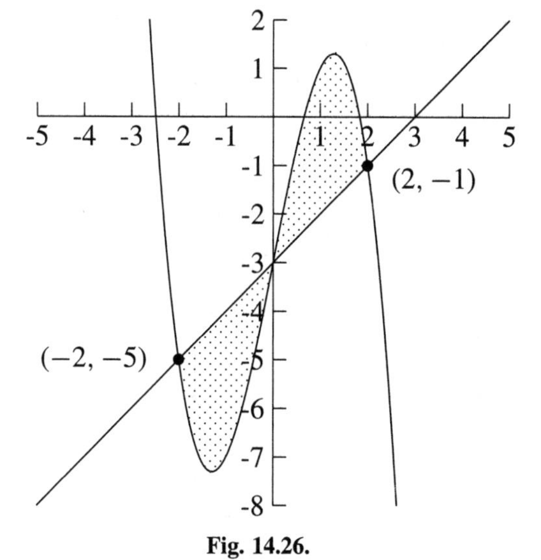

Fig. 14.26.

$$\int_{-2}^{0} [(x-3) - (-x^3 + 5x - 3)]\,dx + \int_{0}^{2} [(-x^3 + 5x - 3) - (x - 3)]\,dx$$

$$= \int_{-2}^{0} (x^3 - 4x)\,dx + \int_{0}^{2} (-x^3 + 4x)\,dx$$

$$= \left(\frac{1}{4}x^4 - 2x^2\right)\Big|_{-2}^{0} + \left(\left(-\frac{1}{4}\right)x^4 + 2x^2\right)\Big|_{0}^{2}$$

$$= \frac{1}{4}(0)^4 - 2(0)^2 - \left(\frac{1}{4}(-2)^4 - 2(-2)^2\right) + \left(-\frac{1}{4}\right)(2)^4 + 2(2)^2$$

$$- \left(-\frac{1}{4}(0)^4 + 2(0)^2\right)$$

$$= 0 - (-4) + 4 - 0 = 8$$

- The graphs in Figure 14.27 intersect at three points. We will find these points by solving $x^3 + 2x^2 - 24x = 2x^2 + x$.

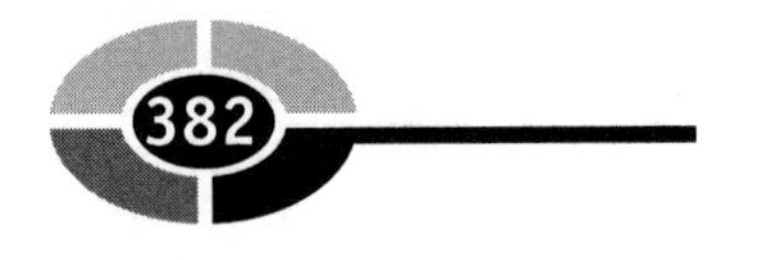

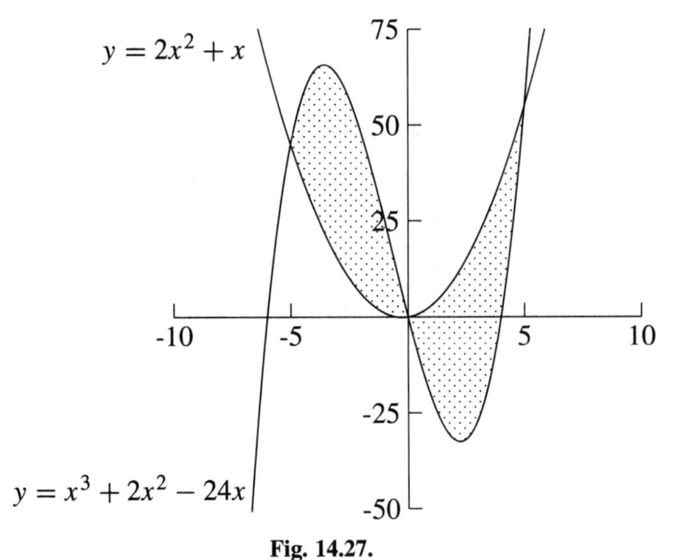

Fig. 14.27.

$$x^3 + 2x^2 - 24x = 2x^2 + x$$

$$x^3 - 25x = 0$$

$$x(x^2 - 25) = 0$$

$$x(x + 5)(x - 5) = 0$$

$$x = 0 \qquad x + 5 = 0 \qquad x - 5 = 0$$

$$x = -5 \qquad x = 5$$

The curve $y = x^3 + 2x^2 - 24x$ is on top between $a = -5$ and $b = 0$. The curve $y = 2x^2 + x$ is on top between $a = 0$ and $b = 5$.

$$\int_{-5}^{0} [(x^3 + 2x^2 - 24x) - (2x^2 + x)]\,dx$$

$$+ \int_{0}^{5} [(2x^2 + x) - (x^3 + 2x^2 - 24x)]\,dx$$

$$= \int_{-5}^{0} (x^3 - 25x)\,dx + \int_{0}^{5} (-x^3 + 25x)\,dx$$

$$= \left(\frac{1}{4}x^4 - \frac{25}{2}x^2\right)\Big|_{-5}^{0} + \left(\left(-\frac{1}{4}\right)x^4 + \frac{25}{2}x^2\right)\Big|_{0}^{5}$$

$$= \frac{1}{4}(0)^4 - \frac{25}{2}(0)^2 - \left(\frac{1}{4}(-5)^4 - \frac{25}{2}(-5)^2\right)$$

$$+ \left(-\frac{1}{4}\right)(5)^4 + \frac{25}{2}(5)^2 - \left(\left(-\frac{1}{4}\right)(0)^4 + \frac{25}{2}(0)^2\right)$$

$$= 0 - \left(-\frac{625}{4}\right) + \frac{625}{4} - 0 = \frac{625}{2}$$

PRACTICE

Find the shaded area.

1. The line in Figure 14.28 is $y = 3x$, and the curve is $y = -x^3 + 8x^2 - 12x$.

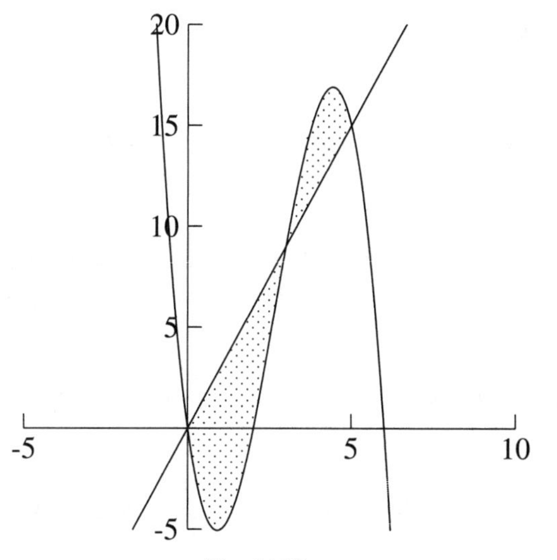

Fig. 14.28.

2. Refer to Figure 14.29.

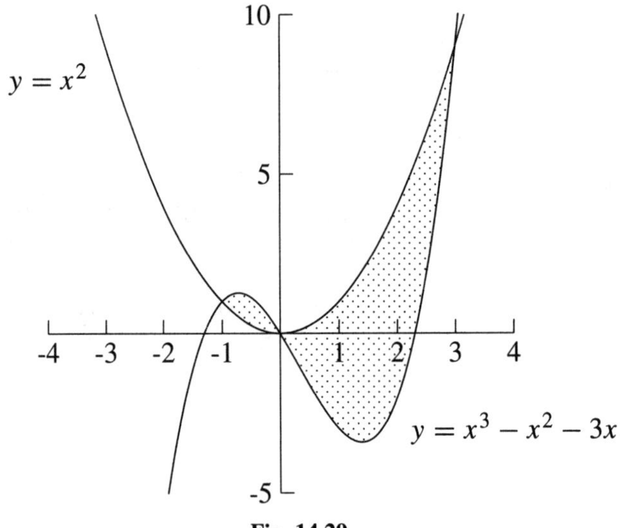

Fig. 14.29.

SOLUTIONS

1. We will find where the graphs intersect by solving $3x = -x^3+8x^2-12x$.

$$3x = -x^3 + 8x^2 - 12x$$

$$x^3 - 8x^2 + 15x = 0$$

$$x(x^2 - 8x + 15) = 0$$

$$x(x - 3)(x - 5) = 0$$

$$x = 0 \qquad x - 3 = 0 \qquad x - 5 = 0$$

$$x = 3 \qquad x = 5$$

The curves intersect at $x = 0$, $x = 3$, and $x = 5$. The line $y = 3x$ is on top between $a = 0$ and $b = 3$. The curve $y = -x^3 + 8x^2 - 12x$ is on top between $a = 3$ and $b = 5$.

$$\int_0^3 [3x - (-x^3 + 8x^2 - 12x)]\,dx + \int_3^5 (-x^3 + 8x^2 - 12x - 3x)\,dx$$

$$= \int_0^3 (x^3 - 8x^2 + 15x)\,dx + \int_3^5 (-x^3 + 8x^2 - 15x)\,dx$$

$$= \left(\frac{1}{4}x^4 - \frac{8}{3}x^3 + \frac{15}{2}x^2\right)\Big|_0^3 + \left(\left(-\frac{1}{4}\right)x^4 + \frac{8}{3}x^3 - \frac{15}{2}x^2\right)\Big|_3^5$$

$$= \frac{1}{4}(3)^4 - \frac{8}{3}(3)^3 + \frac{15}{2}(3)^2 - \left(\frac{1}{4}(0)^4 - \frac{8}{3}(0)^3 + \frac{15}{2}(0)^2\right)$$

$$+ \left(-\frac{1}{4}\right)(5)^4 + \frac{8}{3}(5)^3 - \frac{15}{2}(5)^2$$

$$- \left(\left(-\frac{1}{4}\right)(3)^4 + \frac{8}{3}(3)^3 - \frac{15}{2}(3)^2\right)$$

$$= \frac{63}{4} - 0 + \left(-\frac{125}{12}\right) - \left(-\frac{63}{4}\right) = \frac{189}{12} - \frac{125}{12} + \frac{189}{12} = \frac{253}{12}$$

2. We will find where the graphs intersect by solving $x^3 - x^2 - 3x = x^2$.

$$x^3 - x^2 - 3x = x^2$$

$$x^3 - 2x^2 - 3x = 0$$

$$x(x^2 - 2x - 3) = 0$$

$$x(x+1)(x-3) = 0$$

$$x = 0 \qquad x + 1 = 0 \qquad x - 3 = 0$$

$$x = -1 \qquad x = 3$$

From $a = -1$ to $b = 0$, the curve $y = x^3 - x^2 - 3x$ is on top. From $a = 0$ to $b = 3$, the curve $y = x^2$ is on top.

$$\int_{-1}^{0} (x^3 - x^2 - 3x - x^2)\,dx + \int_0^3 [x^2 - (x^3 - x^2 - 3x)]\,dx$$

$$= \int_{-1}^{0} (x^3 - 2x^2 - 3x)\,dx + \int_0^3 (-x^3 + 2x^2 + 3x)\,dx$$

$$= \left(\frac{1}{4}x^4 - \frac{2}{3}x^3 - \frac{3}{2}x^2\right)\Big|_{-1}^{0} + \left(\left(-\frac{1}{4}\right)x^4 + \frac{2}{3}x^3 + \frac{3}{2}x^2\right)\Big|_0^3$$

$$= \frac{1}{4}(0)^4 - \frac{2}{3}(0)^3 - \frac{3}{2}(0)^2 - \left(\frac{1}{4}(-1)^4 - \frac{2}{3}(-1)^3 - \frac{3}{2}(-1)^2\right)$$

$$+ \left(-\frac{1}{4}\right)(3)^4 + \frac{2}{3}(3)^3 + \frac{3}{2}(3)^2$$

$$- \left(\left(-\frac{1}{4}\right)(0)^4 + \frac{2}{3}(0)^3 + \frac{3}{2}(0)^2\right)$$

$$= 0 - \left(-\frac{7}{12}\right) + \frac{45}{4} - 0 = \frac{71}{6}$$

CHAPTER 14 REVIEW

1. $\int_{-1}^{3}(4x^3 - 6x^2 + 1)\,dx =$

   (a) 64
   (b) 46
   (c) 32
   (d) 28

2. $\int_2^4 \dfrac{3x^2+1}{x^3+x-2}\,dx =$

   (a) $\ln 4/33 \approx -2.1102$
   (b) $-7/36$
   (c) $\ln 33/4 \approx 2.1102$
   (d) Does not exist

3. What is the shaded area in Figure 14.30?

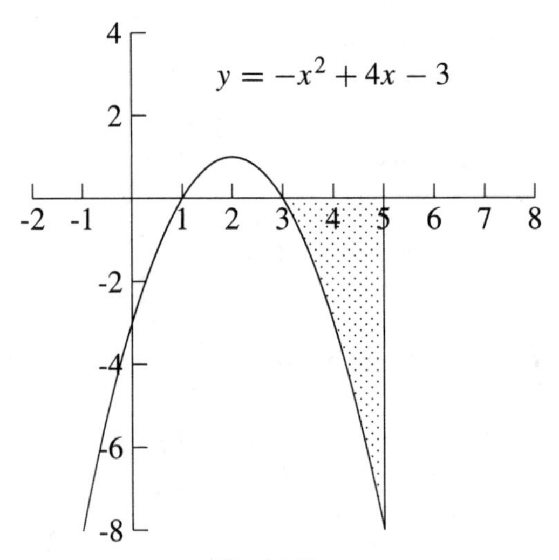

Fig. 14.30.

   (a) $6\frac{2}{3}$
   (b) $-6\frac{2}{3}$
   (c) -8
   (d) 8

4. What is the shaded area in Figure 14.31?

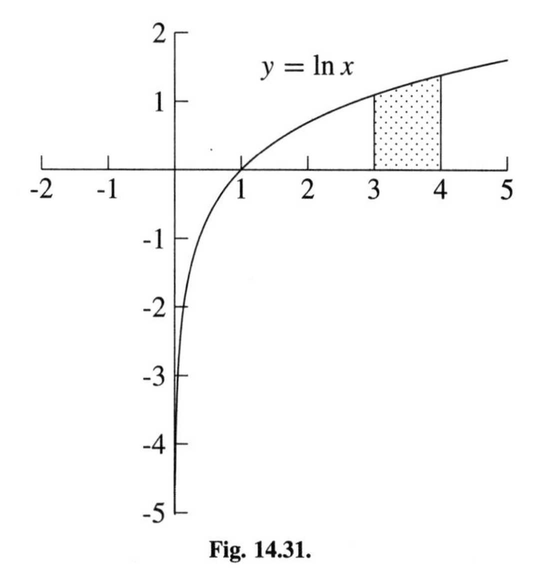

Fig. 14.31.

(a) $4 \ln 4 - 3 \ln 3 - 1 \approx 1.2493$
(b) $\ln 4 - \ln 3 - 1 \approx -0.7123$
(c) $\ln 4 - \ln 3 \approx 0.2877$
(d) Does not exist

5. What is the shaded area in Figure 14.32? (The line is $y = -x$ and the curve is $y = -3\sqrt{x}$.)

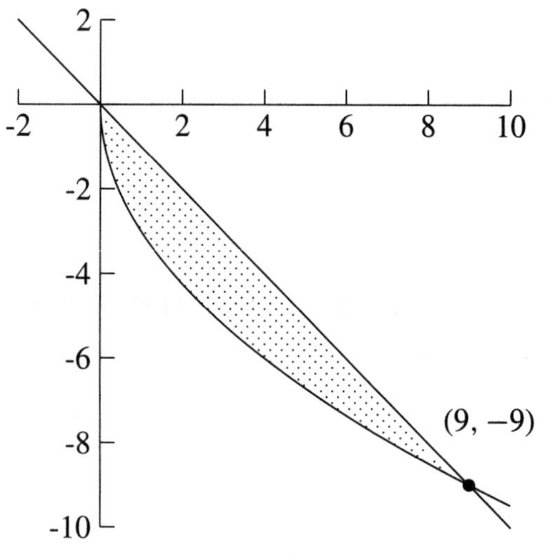

Fig. 14.32.

(a) 13.5
(b) −13.5
(c) −18
(d) 0

6. What is the shaded area in Figure 14.33? (The line is $y = -x + \frac{5}{2}$ and the curve is $y = \frac{1}{x}$.)

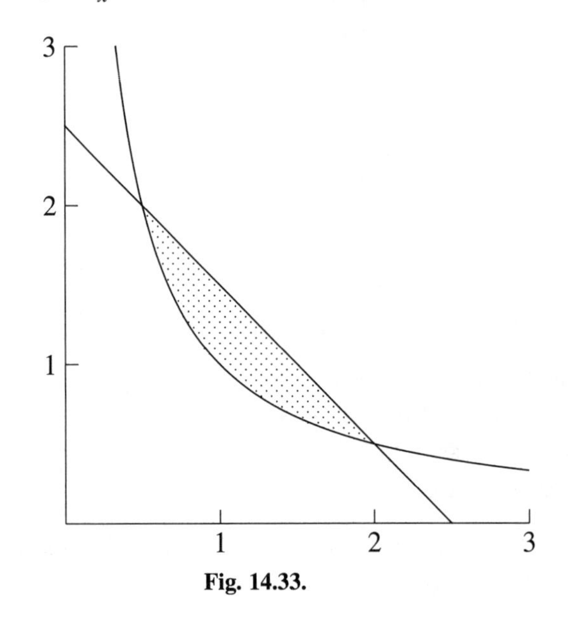

Fig. 14.33.

(a) About 0.4887
(b) About 2.9887
(c) About 0.3069
(d) Does not exist

7. What is the shaded area in Figure 14.34? (The curves are $y = \frac{3}{2}x^2 - 3x - \frac{9}{2}$ and $y = x^3 + x^2 - 12x$, and they intersect at $x = -3$, $x = \frac{1}{2}$, and $x = 3$.

(a) About 18
(b) About 42.7396
(c) About 39.4115
(d) About −10.1398

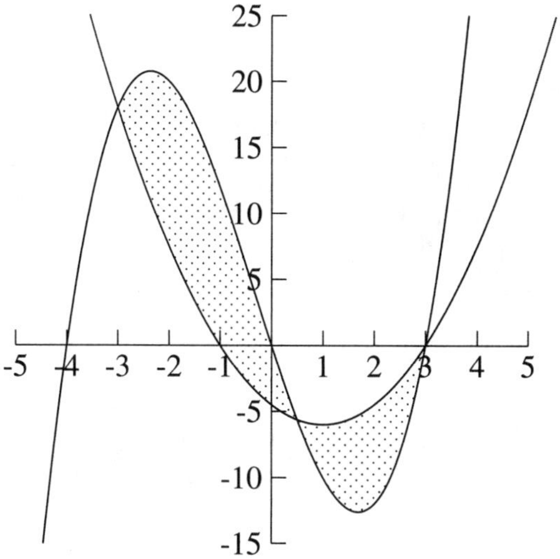

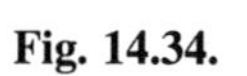

Fig. 14.34.

SOLUTIONS

1. d 2. c 3. b 4. a 5. a 6. a 7. b

15

CHAPTER

Applications of the Integral

There are many applications of the integral in science, engineering, business, and economics. This chapter introduces four of them. Many applications use the power of the definite integral to instantly sum many numbers. In the first application, we will construct a function based on information we have on the rate of change of the function. As we know from Chapter 13, finding $f(x)$ from $f'(x)$ gives us a family of functions, all of which differ by a constant. If we have one functional value, we can find the specific function. In the problems that follow, we will be given the derivative and one functional value. We will integrate the derivative and use the functional value in the indefinite integral to find C.

EXAMPLES

- The marginal cost function for a product is $C'(x) = 0.08x + 1$, where x is the number produced in a month. It costs $6800 to produce 200 units.

We will begin with the indefinite integral of $0.08x + 1$.

$$C(x) = \int (0.08x + 1)\,dx = 0.04x^2 + x + C$$

The cost function is $C(x) = 0.04x^2 + x + C$. We will use the fact that when $x = 200$, $C(x) = 6800$ to find the constant.

$$6800 = 0.04(200)^2 + 200 + C$$

$$5000 = C$$

The cost function is $C(x) = 0.04x^2 + x + 5000$.

- The velocity of an object is $v(t) = 14t + 3$ feet per second after t seconds. Assume that at the beginning, $t = 0$, the object has traveled 0 feet. Find the position function.
  Velocity is the derivative of the position function, sometimes denoted $s(t)$.

$$s(t) = \int (14t + 3)\,dt = 7t^2 + 3t + C$$

At $t = 0$, the distance traveled is 0, so we will substitute 0 for t as well as for $s(t)$ to find C.

$$0 = 7(0)^2 + 3(0) + C$$

$$0 = C$$

The distance function is $s(t) = 7t^2 + 3t$.

- The slope of the tangent line for a function is found by computing $y' = 12x^3 - 6x + 5$, and the point $(1, -3)$ is on the curve. What is the function?

$$y = \int (12x^3 - 6x + 5)\,dx = 3x^4 - 3x^2 + 5x + C$$

The point $(1, -3)$ is on the curve, which means that when $x = 1$, $y = -3$.

$$-3 = 3(1)^4 - 3(1)^2 + 5(1) + C$$

$$-8 = C$$

The function is $y = 3x^4 - 3x^2 + 5x - 8$.

PRACTICE

1. The velocity of an object after t seconds is $v(t) = 4t + 10$ feet per second. Assume that the object traveled 0 feet at 0 seconds. What is the position function?
2. The value of an investment t months after purchase is changing during the first year at the rate of $V'(t) = -0.978t^2 + 12.786t - 26.871$. The initial investment is \$750. What is the investment value function?
3. The marginal revenue for selling x units of a product is $R'(x) = 2x - 4$. The revenue for selling 10 units is \$70. What is the revenue function?
4. The marginal revenue t weeks after a product is introduced is

$$R'(t) = \frac{-1000}{(t+1)^2}$$

   After 9 weeks, the revenue is \$300. What is the revenue function?
5. The rate at which a culture of bacteria is growing is $N'(t) = 300e^{0.20t}$ bacteria per hour after t hours. There were 1500 bacteria in the culture initially. What is the function that gives the number of bacteria present after t hours?
6. The slope of the tangent line for a function is found by computing $y' = 5e^{5x+10}$, and the point $(-2, 4)$ is on the curve. What is the function?

SOLUTIONS

1.

$$s(t) = \int (4t + 10)\,dt = 2t^2 + 10t + C$$

$s(t) = 2t^2 + 10t + C$ Now let $t = 0$ and $s(t) = 0$.

$$0 = 2(0)^2 + 10(0) + C$$

$$0 = C$$

$$s(t) = 2t^2 + 10t$$

2. "The initial investment is \$750" means that at $t = 0$, $V(t)$ is 750.

$$V(t) = \int (-0.978t^2 + 12.786t - 26.871)\,dt$$

$$= -0.326t^3 + 6.393t^2 - 26.871t + C$$

$$V(t) = -0.326t^3 + 6.393t^2 - 26.871t + C$$

Now let $t = 0$ and $V(t) = 750$.

$$750 = -0.326(0)^3 + 6.393(0)^2 - 26.871(0) + C$$

$$750 = C$$

$$V(t) = -0.326t^3 + 6.393t^2 - 26.871t + 750$$

3.

$$R(x) = \int (2x - 4)\,dx = x^2 - 4x + C$$

$R(x) = x^2 - 4x + C$ Now let $x = 10$ and $R(x) = 70$.

$$70 = (10)^2 - 4(10) + C$$

$$10 = C$$

$$R(x) = x^2 - 4x + 10$$

4.

$$R(t) = \int \frac{-1000}{(t+1)^2}\,dt = \int -1000(t+1)^{-2}\,dt$$

$$= \frac{-1000}{-1}(t+1)^{-1} + C = \frac{1000}{t+1} + C$$

$R(t) = \frac{1000}{t+1} + C$ Now let $t = 9$ and $R(t) = 300$.

$$300 = \frac{1000}{9+1} + C$$

$$200 = C$$

$$R(t) = \frac{1000}{t+1} + 200$$

5.

$$N(t) = \int 300e^{0.20t}\,dt = \frac{300}{0.20}e^{0.20t} + C$$

$N(t) = 1500e^{0.20t} + C$ Now let $t = 0$ and $N(t) = 1500$.

$$1500 = 1500e^{0.20(0)} + C = 1500(1) + C$$

$$0 = C$$

$$N(t) = 1500e^{0.20t}$$

6.

$$y = \int 5e^{5x+10}\,dx = e^{5x+10} + C$$

$y = e^{5x+10} + C$ Now let $x = -2$ and $y = 4$.

$$4 = e^{5(-2)+10} + C = e^0 + C = 1 + C$$

$$3 = C$$

$$y = e^{5x+10} + 3$$

Continuous Money Flow

When money is regularly deposited into an investment over a period of time, we can find an investment's *accumulated value* with formulas from algebra. However, the formulas can be a little awkward. It is easier to approximate the accumulated value by finding the area under a curve.

For example, if we want to deposit \$600 per year into an account paying 5% annual interest, compounded annually, for five years, the accumulated value is about \$3315. The graph in Figure 15.1 shows what each deposit is worth after five years, when \$600 is deposited at the end of the year. The first \$600 grows to \$729.30 in four years; the second \$600 grows to \$694.58 in three years; the third grows to \$661.50 in two years; and the fourth grows to \$630. The \$600 deposit in the fifth year does not earn interest until the sixth year.

The value of the investments after five years is the total area of the rectangles, which is about 3315. If \$600 is deposited continuously throughout the year and earns interest continuously, the accumulated value is the area under the curve of the function $V(t) = 600e^{0.05t}$ from $x = 0$ to $x = 5$ (see Figure 15.2).

$$\text{Area} = \int_0^5 600e^{0.05t}\,dt$$

$$= \frac{600}{0.05}e^{0.05t}\Big|_0^5$$

$$= 12{,}000(e^{0.05(5)} - e^{0.05(0)})$$

$$= 12{,}000(e^{0.25} - 1) \approx 3408.31$$

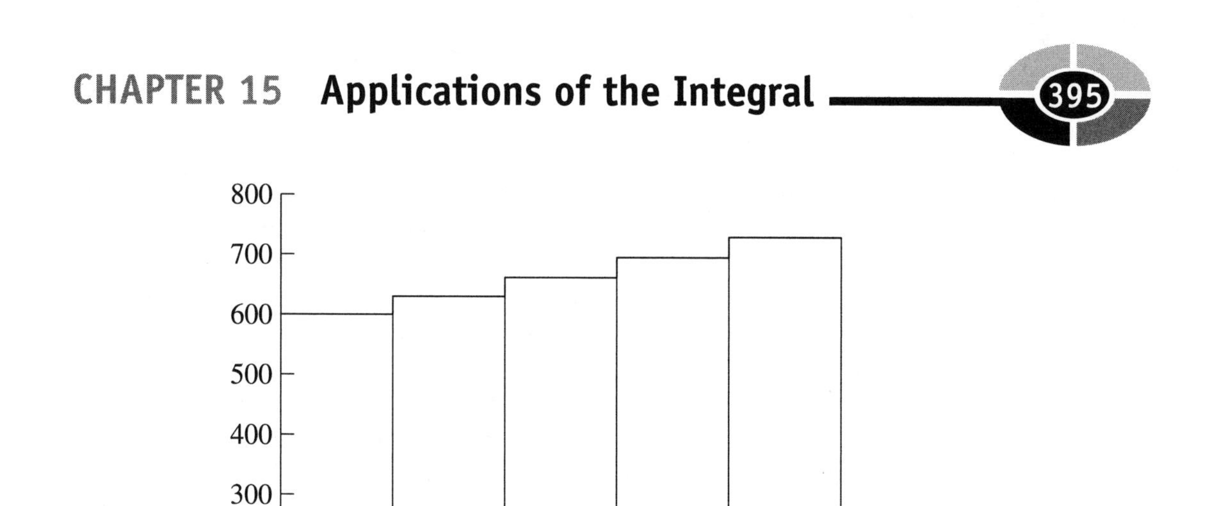

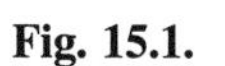

Fig. 15.1.

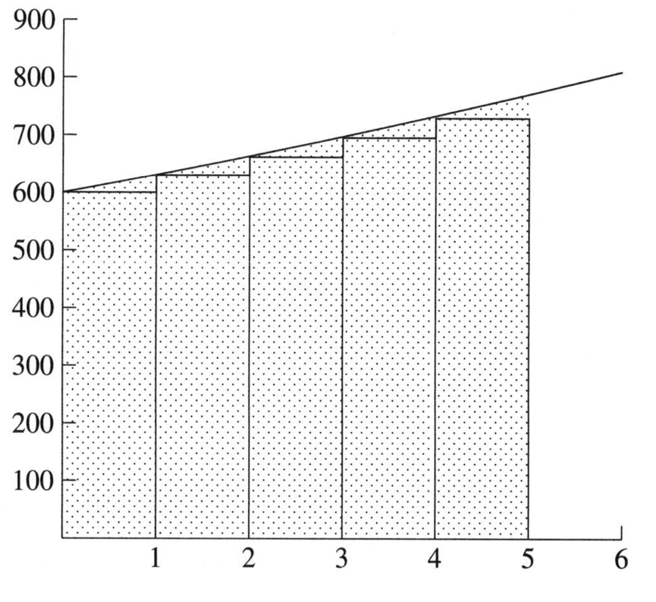

Fig. 15.2.

This approximation overestimates the accumulated value somewhat. If \$600 is deposited a little each day through the year (about \$1.64 per day), then the accumulated value is about \$3400, which is much closer to the approximation above. In fact, the accumulated value of daily deposits (whose interest is compounded daily) is closely approximated by continuous deposits (whose interest is compounded continuously).

For the problems in this section, we will assume that money is *continuously flowing* into an account that earns interest continuously. The accumulated value of the account is the definite integral $\int_0^T Pe^{rt}\,dt$, where T is the length of time, in years, P is the amount that is regularly deposited, and r is the annual interest rate.

$$\text{Area} = \int_0^T Pe^{rt}\,dt = \frac{P}{r}e^{rt}\Big|_0^T = \frac{P}{r}[e^{rT} - e^{r(0)}] = \frac{P}{r}[e^{rT} - 1]$$

EXAMPLES

- Find the accumulated value of \$912.50 that flows continuously into an account that pays 8% annual interest, compounded continuously, for 20 years. (\$912.50 per year is \$2.50 per day.)

  We will use the integral above with $P = 912.50$, $r = 0.08$, and $T = 20$.

$$\int_0^{20} 912.50e^{0.08t}\,dt = \frac{912.50}{0.08}e^{0.08t}\Big|_0^{20}$$

$$= 11{,}406.25\left(e^{0.08(20)} - e^{0.08(0)}\right)$$

$$= 11{,}406.25\left(e^{1.6} - 1\right) \approx 45{,}089.28$$

  The accumulated value is \$45,089.28.

- What is the future value of \$3000 deposited continuously throughout each year into an account that pays 6% annual interest, compounded continuously, for 10 years? For 30 years?

  The *future value* is another term for *accumulated value*. The future value after 10 years is the definite integral $\int_0^{10} 3000e^{0.06t}\,dt$.

$$\int_0^{10} 3000e^{0.06t}\,dt = \frac{3000}{0.06}e^{0.06t}\Big|_0^{10}$$

$$= 50{,}000\left(e^{0.06(10)} - e^{0.06(0)}\right)$$

$$= 50{,}000\left(e^{0.6} - 1\right) \approx 41{,}105.94$$

The future value is \$41,105.94.
The future value after 30 years is the definite integral $\int_0^{30} 3000e^{0.06t}\,dt$.

$$\int_0^{30} 3000e^{0.06t}\,dt = \frac{3000}{0.06}e^{0.06t}\Big|_0^{30}$$

$$= 50{,}000\left(e^{0.06(30)} - e^{0.06(0)}\right)$$

$$= 50{,}000\left(e^{1.8} - 1\right) \approx 252{,}482.37$$

The future value is \$252,482.37.

- A homeowner expects to receive \$8000 per year in gas royalties for a gas well on the property. If the money flows continuously into an account paying 6.25% annual interest, compounded continuously, for 15 years, what is the accumulated value?

$$\int_0^{15} 8000e^{0.0625t}\,dt = \frac{8000}{0.0625}e^{0.0625t}\Big|_0^{15}$$

$$= 128{,}000\left(e^{0.0625(15)} - e^{0.0625(0)}\right)$$

$$= 128{,}000\left(e^{0.9375} - 1\right) \approx 198{,}859.45$$

The accumulated value is \$198,859.45.

PRACTICE

1. Find the accumulated value of \$5000 per year that flows continuously into an account paying 8% annual interest, compounded continuously, for 10 years.
2. Find the accumulated value of \$30,000 per year that flows continuously into an account that pays 12% annual interest, compounded continuously, for 8 years.
3. An inventor expects to receive \$40,000 annual royalties for a product sold to a manufacturer. The money will continuously flow into an account paying 12% annual interest, compounded continuously for 10 years. What is the accumulated value of the royalty payments?

SOLUTIONS

1.

$$\int_0^{10} 5000e^{0.08t}\,dt = \frac{5000}{0.08}e^{0.08t}\Big|_0^{10}$$
$$= 62{,}500\left(e^{0.08(10)} - e^{0.08(0)}\right)$$
$$= 62{,}500\left(e^{0.8} - 1\right) \approx 76{,}596.31$$

The accumulated value is \$76,596.31.

2.

$$\int_0^{8} 30{,}000e^{0.12t}\,dt = \frac{30{,}000}{0.12}e^{0.12t}\Big|_0^{8}$$
$$= 250{,}000\left(e^{0.12(8)} - e^{0.12(0)}\right)$$
$$= 250{,}000\left(e^{0.96} - 1\right) \approx 402{,}924.12$$

The accumulated value is \$402,924.12.

3.

$$\int_0^{10} 40{,}000e^{0.12t}\,dt = \frac{40{,}000}{0.12}e^{0.12t}\Big|_0^{10}$$
$$= 333{,}333.33\left(e^{0.12(10)} - e^{0.12(0)}\right)$$
$$= 333{,}333.33\left(e^{1.2} - 1\right) \approx 773{,}372.31$$

The accumulated value is \$773,372.31.

If we know how much money we need for a future date, we can use the integral $\int_0^T Pe^{rt}\,dt$ to find how much money we must invest over time to reach our goal. We will set the integral equal to the amount of money we need. And then will solve the equation for P.

EXAMPLE

- A grandmother wants to give her newborn grandson a gift of \$50,000 on his 20th birthday. She will let the money continuously flow into an account that pays $7\frac{1}{2}\%$ annual interest, compounded continuously. How much should be deposited each year?

We know that $r = 0.075$, and $T = 20$, but we do not know P. We will solve the equation $50{,}000 = \int_0^{20} Pe^{0.075t}\,dt$ for P.

$$50{,}000 = \int_0^{20} Pe^{0.075t}\,dt$$

$$50{,}000 = \frac{P}{0.075}e^{0.075}\Big|_0^{20}$$

$$50{,}000 = \frac{P}{0.075}\left(e^{0.075(20)} - e^{0.075(0)}\right)$$

$$50{,}000 = \frac{P}{0.075}\left(e^{1.5} - 1\right)$$

$$3750 = P\left(e^{1.5} - 1\right) \quad \text{Multiply both sides by 0.075.}$$

$$3750 \approx P(3.48168907)$$

$$\frac{3750}{3.48168907} \approx P$$

$$1077.06 \approx P$$

The grandmother should continuously invest \$1077.06 each year for 20 years so that it grows to \$50,000 in 20 years.

PRACTICE

1. The parents of a five-year-old child want to start a college fund so that their child will have \$150,000 at age 18. The money will continuously flow into an account which earns 8% annual interest, compounded continuously. How much should be deposited each year?

SOLUTION

1.

$$150{,}000 = \int_0^{13} Pe^{0.08t}\,dt$$

$$150{,}000 = \frac{P}{0.08} e^{0.08t} \Big|_0^{13}$$

$$150{,}000 = \frac{P}{0.08} \left(e^{0.08(13)} - e^{0.08(0)}\right)$$

$$150{,}000 = \frac{P}{0.08} \left(e^{1.04} - 1\right)$$

$$12{,}000 = P \left(e^{1.04} - 1\right) \quad \text{Multiply both sides by 0.08.}$$

$$12{,}000 \approx P(1.829217014)$$

$$\frac{12{,}000}{1.829217014} \approx P$$

$$6560.18 \approx P$$

The parents should continuously save \$6560.18 per year for 13 years so that it grows to \$150,000.

Many lottery winners choose to take the *cash value* of their winnings rather than annual payments that can last 20, even 25, years. The cash value is the amount of money that the state needs to have on hand in order to make the payments over 20 or 25 years. During the years, the amount of money is declining because of the payments but is growing from interest earned. With the right amount of money, the account will last long enough to make all the payments.

Suppose one such jackpot is worth \$1.2 million, and the state makes 20 annual payments of \$60,000. How much should be invested now, if it can earn 5% annual interest? For 20 annual payments, the state would need \$747,733. If the payments were made monthly (and interest compounded monthly), the state would need \$757,627. And if the payments were made daily (and interest compounded daily), the state would need \$758,514. If the payments were made continuously (and interest compounded continuously), the amount the state would need can be found with the definite integral $\int_0^{20} 60{,}000e^{-0.05t}\,dt$. This formula computes the area under the curve $y = 60{,}000e^{-0.05t}$ from $t = 0$ to $t = 20$. This is the *accumulated present value* of \$60,000 continuously flowing from an account earning 5% annual interest, compounded continuously. As we will see, this amount is very close to the amount needed for daily payments.

$$\int_0^{20} 60{,}000e^{-0.05t}\,dt = \frac{60{,}000}{-0.05} e^{-0.05t} \Big|_0^{20}$$

$$= \frac{60{,}000}{-0.05}\left(e^{-0.05(20)} - e^{-0.05(0)}\right)$$

$$= \frac{60{,}000}{-0.05}\left(e^{-1} - 1\right) \approx 758{,}545$$

Another application for the integral $\int_0^T Pe^{-rt}\,dt$ is to compute how much future payments made into an account are worth now.

EXAMPLES

- What is the accumulated present value of \$6000 that will continuously flow into an account each year for a total of 15 years if the account pays $7\frac{1}{2}\%$ annual interest, compounded continuously?

$$\int_0^{15} 6000e^{-0.075t}\,dt = \frac{6000}{-0.075}e^{-0.075t}\Big|_0^{15}$$

$$= -80{,}000\left(e^{-0.075(15)} - e^{-0.075(0)}\right)$$

$$= -80{,}000\left(e^{-1.125} - 1\right) \approx 54{,}027.80$$

The accumulated present value is \$54,027.80.

- A homeowner expects to receive \$8000 each year for 15 years from gas royalties. It will continuously flow into an account that pays 6.25% annual interest, compounded continuously. An investor approaches the homeowner and wants to purchase the payments. How much are the payments worth today?

We want the present value of the payments.

$$\int_0^{15} 8000e^{-0.0625t}\,dt = \frac{8000}{-0.0625}e^{-0.0625t}\Big|_0^{15}$$

$$= -128{,}000\left(e^{-0.0625(15)} - e^{-0.0625(0)}\right)$$

$$= -128{,}000\left(e^{-0.9375} - 1\right) \approx 77{,}874.48$$

Today, the payments are worth \$77,874.48. (\$77,874.48 would need to be on account, earning 6.25% annual interest, compounded continuously, in order to make \$8000 annual payments for 15 years.)

PRACTICE

1. What is the accumulated present value of \$3000 annual payments that continuously flow into an account paying 6% annual interest, compounded continuously, for 25 years?
2. What is the accumulated present value of an investment with a continuous money flow of \$12,000 per year into an account that pays 10% annual interest, compounded continuously, for ten years?
3. A woman won a \$10 million lottery. She can either take the cash value or 25 annual payments of \$400,000. Assuming 4% annual interest, use the accumulated present value to estimate the cash value of her jackpot. (The actual cash value is \$6,248,831.98.)
4. The manager of a small company has promised to pay one of its retirees an income of \$1800 per month. It is assumed that the retiree will live 20 years after retiring. How much should be deposited now into an account earning $7\frac{1}{2}$% annual interest, compounded continuously? (Assume that the money is paid continuously throughout each month.)

SOLUTIONS

1.

$$\begin{aligned}
\int_0^{25} 3000e^{-0.06t}\,dt &= \frac{3000}{-0.06}e^{-0.06t}\Big|_0^{25} \\
&= -50{,}000\left(e^{-0.06(25)} - e^{-0.06(0)}\right) \\
&= -50{,}000\left(e^{-1.5} - 1\right) \approx 38{,}843.49
\end{aligned}$$

The accumulated present value is \$38,843.49.

2.

$$\begin{aligned}
\int_0^{10} 12{,}000e^{-0.10t}\,dt &= \frac{12{,}000}{-0.10}e^{-0.10t}\Big|_0^{10} \\
&= -120{,}000\left(e^{-0.10(10)} - e^{-0.10(0)}\right) \\
&= -120{,}000\left(e^{-1} - 1\right) \approx 75{,}854.47
\end{aligned}$$

The accumulated present value is \$75,854.47.

3.

$$\int_0^{25} 400{,}000e^{-0.04t}\,dt = \frac{400{,}000}{-0.04}e^{-0.04t}\Big|_0^{25}$$

$$= -10{,}000{,}000\left(e^{-0.04(25)} - e^{-0.04(0)}\right)$$

$$= -10{,}000{,}000\left(e^{-1} - 1\right) \approx 6{,}321{,}205.59$$

The cash value is estimated at $6,321,205.59.

4. When $1800 is paid each month, then the annual payments amount to $21,600.

$$\int_0^{20} 21{,}600e^{-0.075t}\,dt = \frac{21{,}600}{-0.075}e^{-0.075t}\Big|_0^{20}$$

$$= -288{,}000\left(e^{-0.075(20)} - e^{-0.075(0)}\right)$$

$$= -288{,}000\left(e^{-1.5} - 1\right) \approx 223{,}738.51$$

$223,738.51 should be deposited into the account.

The amount of money that continuously flows into (or out from) an account does not need to be the same throughout the year. The amount could vary. If the annual flow is given by the function $f(t)$, then the accumulated amount is $\int_0^T f(t)e^{rt}\,dt$. If the flow is out of the account, then the accumulated present value is $\int_0^T f(t)e^{-rt}\,dt$.

EXAMPLE

- A business is earning a continuous profit of $100,000 per year and is growing at the rate of $8000 per year. This makes the profit function $f(t) = 100{,}000 + 8000t$, after t years. The profit continuously flows into an account that earns 5% annual interest, compounded continuously. The owner is considering selling the business. The owner wants $1 million plus the accumulated present value of the profit for six years. What is the selling price?

  The accumulated present value for six years is the integral $\int_0^6 (100{,}000 + 8000t)e^{-0.05t}\,dt$. We will use integration by parts with $f'(t) = e^{0.05t}$

(so $f(t) = -\frac{1}{0.05}e^{-0.05t} = -20e^{-0.05t}$), and $g(t) = 100{,}000 + 8000t$ (so $g'(t) = 8000$).

$$\begin{aligned}
\int_0^6 (100{,}000 + 8000t)e^{-0.05t}\,dt &= -20e^{-0.05t}(100{,}000 + 8000t)\Big|_0^6 \\
&\quad - \int_0^6 -20e^{-0.05t}(8000)\,dt \\
&= -20e^{-0.05t}(100{,}000 + 8000t)\Big|_0^6 \\
&\quad + \int_0^6 160{,}000e^{-0.05t}\,dt \\
&= -20e^{-0.05t}(100{,}000 + 8000t)\Big|_0^6 \\
&\quad - 3{,}200{,}000e^{-0.05t}\Big|_0^6 \\
&\approx 636{,}560
\end{aligned}$$

The selling price is \$1 million + \$0.637 million = \$1.637 million.

Consumers' Surplus and Suppliers' Surplus

Suppose that there are eight people in a store wanting to buy a can of chili and that the price of the chili is not marked. Each person pays as much as he or she is willing to pay. One person pays \$6. Another pays \$5.50. The third pays \$5; the fourth, \$4.50; the fifth, \$4.00; the sixth, \$3; the seventh, \$2; and the eighth, \$1. The store collects \$6 + 5.50 + 5 + 4.50 + 4.00 + 3.00 + 2.00 + 1.00 = \$31. If the store had marked a price of \$1, then all eight would have paid \$1, and the store would have only collected \$8, a reduction of \$23. In other words, sales worth \$31 to the consumers would have been purchased for \$8. The difference is called the *consumers' surplus*. It is the difference between what consumers are willing to pay for a product or service and what they actually do pay. Let us look at this situation graphically. The graph in Figure 15.3 shows how much each person is willing to pay for the chili. Each rectangle represents one customer and what he or she is willing to pay for the chili. The total area of the rectangles is 31. In Figure 15.4, the shaded area represents what each customer does pay if the price of \$1 is marked on the can. The unshaded area represents the consumers' surplus.

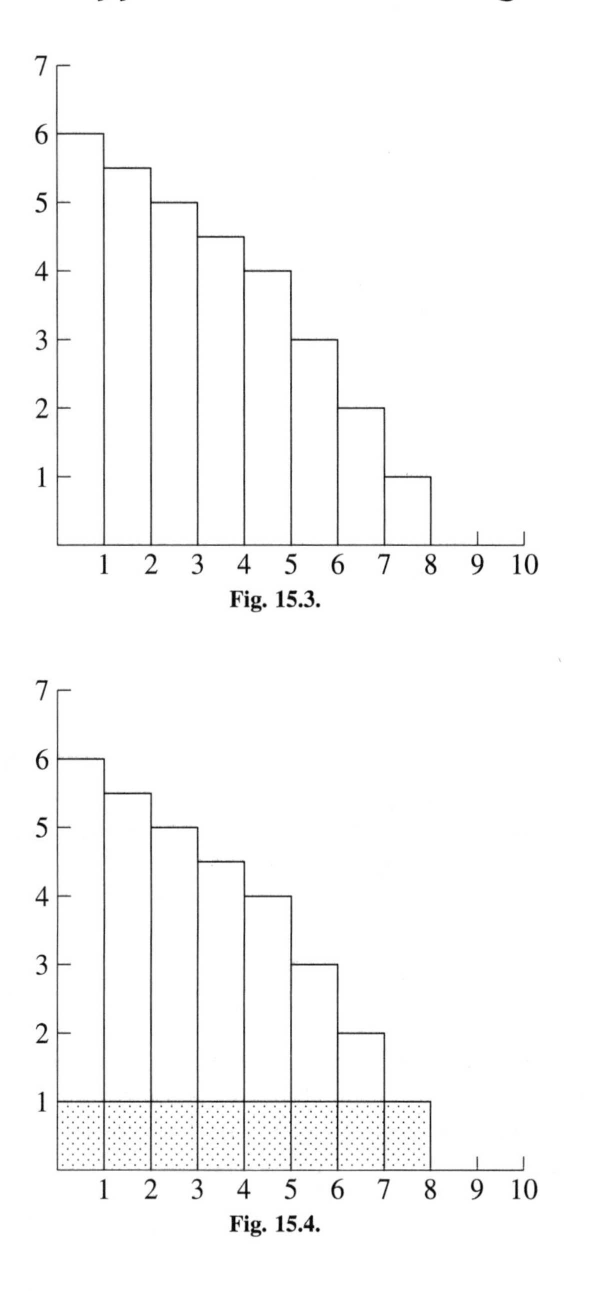

Fig. 15.3.

Fig. 15.4.

The function $y = -0.0596x^2 - 0.16667x + 6.14285$ approximates the heights of the rectangles, and the area under the curve (about 33.6) approximates the area of the rectangles (Figure 15.5). When we subtract the amount consumers would have spent on chili, $1 \times 8 = 8$ from the integral $\int_0^8 (-0.0596x^2 - 0.16667x + 6.14285)\,dx$, we get an approximation for the consumers' surplus.

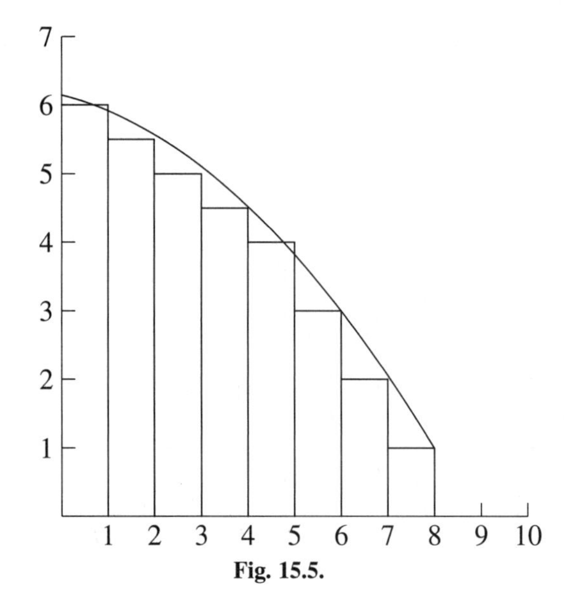

Fig. 15.5.

The function $y = -0.0596x^2 - 0.16667x + 6.14285$ is the demand function for this particular group of consumers. Economists use the demand curve for large groups of consumers to compute the consumers' surplus for a product. It is computed the same way—the area under the curve minus the revenue. The revenue is computed as $p \cdot q$, the price times the quantity sold.

$$\text{Consumers' surplus} = \int_0^q (\text{Demand})\, dx - pq$$

In the problems below, we will be given a demand function, $D(x)$, and a quantity. We will use the demand function and the given quantity to compute the price that consumers are willing to pay for q units. Once we have the demand function, quantity, and price, we can find the consumers' surplus.

EXAMPLES

Find the consumers' surplus.

- $D(x) = -\frac{1}{20}x + 25$; $q = 150$.
  The price at $q = 150$ is $D(150) = -\frac{1}{20}(150) + 25 = 17.50$. The revenue collected when 150 units are demanded at \$17.50 is (\$17.50)(150) = \$2625.

We will have the consumers' surplus when we subtract 2625 from $\int_0^{150}(-\frac{1}{20}x + 25)\,dx$.

$$\int_0^{150}\left(-\frac{1}{20}x+25\right)dx - 2625 = \left(-\frac{1}{40}x^2+25x\right)\Big|_0^{150} - 2625$$

$$= -\frac{1}{40}(150)^2 + 25(150)$$

$$-\left(-\frac{1}{40}(0)^2+25(0)\right) - 2625$$

$$= 3187.5 - 0 - 2625 = 562.50$$

Consumers paid \$2625 for a product or service that was worth \$3187.50, which produced a consumers' surplus of \$562.50.

- $D(x) = 9.54e^{-0.229x}$; $q = 9$.
  The price when 9 units are demanded is $D(9) = 9.54e^{-0.229(9)} \approx 1.21$. The revenue is (\$9)(1.21) = 10.89.

$$\int_0^9 9.54e^{-0.229x}\,dx - 10.89 = \frac{9.54}{-0.229}e^{-0.229x}\Big|_0^9 - 10.89$$

$$= \frac{9.54}{-0.229}\left(e^{-0.229(9)} - e^{-0.229(0)}\right)$$

$$- 10.89 \approx 25.47$$

The consumers' surplus is \$25.47.

PRACTICE

Find the consumers' surplus for the given demand function and quantity.

1. $D(x) = -0.01x^2 - x + 20$; $q = 15$
2.

$$D(x) = \frac{20x+40}{x^2+4x+5};\; q = 12$$

SOLUTIONS

1. When 15 units are demanded, the price is $D(15) = -0.01(15)^2 - 15 + 20 = 2.75$, and the revenue is (\$2.75)(15) = \$41.25.

$$\int_0^{15} (-0.01x^2 - x + 20)\,dx - 41.25$$

$$= \left(\frac{-0.01}{3}x^3 - \frac{1}{2}x^2 + 20x\right)\Bigg|_0^{15} - 41.25$$

$$= \frac{-0.01}{3}(15)^3 - \frac{1}{2}(15)^2 + 20(15)$$

$$- \left(\frac{-0.01}{3}(0)^3 - \frac{1}{2}(0)^2 + 20(0)\right) - 41.25$$

$$= 176.25 - 0 - 41.25 = 135$$

The consumers' surplus is \$135.

2. The price for $q = 12$ is $D(12) = \frac{20(12)+40}{(12)^2+4(12)+5} \approx 1.42$. The revenue for selling 12 units is (\$1.42)(12) = \$17.04.

$$\int_0^{12} \frac{20x + 40}{x^2 + 4x + 5}\,dx - 17.04 = \int_0^{12} \frac{10(2x + 4)}{x^2 + 4x + 5}\,dx - 17.04$$

$$= 10\int_0^{12} \frac{(2x + 4)}{x^2 + 4x + 5}\,dx - 17.04$$

$$= 10[\ln(x^2 + 4x + 5)]\Big|_0^{12} - 17.04$$

$$= 10[\ln((12)^2 + 4(12) + 5) - \ln((0)^2$$

$$+ 4(0) + 5)] - 17.04$$

$$\approx 10(5.283 - 1.609) - 17.04 \approx 19.7$$

The consumers' surplus is \$19.70.

The suppliers' surplus measures the difference between the amount of money a supplier is willing to accept at a given price for a product and the amount the supplier actually does receive. Suppose a supplier is willing to sell 100 units at \$5 each but is able to sell 100 units at \$8 each. This gives the supplier a surplus of (\$8)(100) − (\$5)(100) = \$300. If the price is p for selling q units, then the revenue is pq and the suppliers' surplus is found by computing $pq - \int_0^q (\text{Supply})\,dx$.

In the problems below, we will be asked to find the suppliers' surplus for a given supply function and quantity. The supply function, $S(x)$, gives the price a supplier is willing to sell x units. We will use the given quantity to find the price using the supply function.

EXAMPLE

- Find the suppliers' surplus for $S(x) = 0.05x^2 + x + 10$ and $q = 20$.
  The price for $q = 20$ units is $S(20) = 0.05(20)^2 + 20 + 10 = 50$, and the revenue is (\$50)(20) = \$1000.

$$\begin{aligned}
\text{Suppliers' surplus} &= 1000 - \int_0^{20} (0.05x^2 + x + 10)\,dx \\
&= 1000 - \left[\frac{0.05}{3}x^3 + \frac{1}{2}x^2 + 10x\right]\Bigg|_0^{20} \\
&= 1000 - \left[\frac{0.05}{3}(20)^3 + \frac{1}{2}(20)^2 + 10(20)\right. \\
&\quad \left. - \left(\frac{0.05}{3}(0)^3 + \frac{1}{2}(0)^2 + 10(0)\right)\right] \\
&= 1000 - \left(\frac{1600}{3} - 0\right) \approx 466.67
\end{aligned}$$

The suppliers' surplus is \$466.67.

The graph in Figure 15.6 shows what the formula $pq - \int_0^q (\text{Supply})\,dx$ is computing. The area of the rectangle is the revenue, $20 \times 50 = 1000$. The shaded area is the suppliers' surplus, the difference between the area under the line $y = 50$ and the curve $y = 0.05x^2 + x + 10$.

The market for a product or service is in equilibrium when supply equals demand: both suppliers and consumers agree on a price and quantity. We can find the equilibrium by setting the supply function equal to the demand function and solving for x, the equilibrium quantity. We can find the equilibrium price by putting x into either the supply or demand function (we would get the same price from both functions). After we have the equilibrium point, we can find the consumers' surplus and the suppliers' surplus.

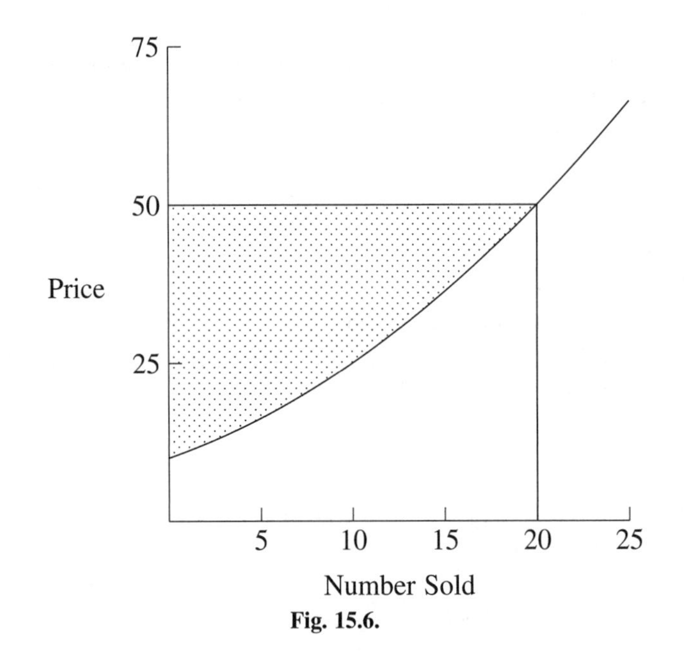

Fig. 15.6.

EXAMPLE

- Find the equilibrium point, consumers' surplus, and suppliers' surplus for $D(x) = -x + 107.5$ and $S(x) = 0.02x^2 + 2x + 20$.

  We will find the equilibrium point by solving the equation $S(x) = D(x)$ for x.

$$0.02x^2 + 2x + 20 = -x + 107.5$$

$$0.02x^2 + 3x - 87.50 = 0$$

$$x = \frac{-3 \pm \sqrt{3^2 - 4(0.02)(-87.50)}}{2(0.02)} = \frac{-3 \pm \sqrt{16}}{0.04} = 25$$

($x = -175$ is not a solution.) The equilibrium quantity is $q = 25$. We can find the equilibrium price with $q = 25$ in either $D(x)$ or $S(x)$.

$$D(25) = -25 + 107.50 = 82.50$$

$$S(25) = 0.02(25)^2 + 2(25) + 20 = 82.50$$

The equilibrium price is \$82.50, and the revenue is (\$82.50)(25) = \$2062.50.

$$\text{Consumers' surplus} = \int_0^{25} (-x + 107.5)\, dx - 2062.50$$

$$= \left(-\frac{1}{2}x^2 + 107.50x\right)\Big|_0^{25} - 2062.50$$

$$= -\frac{1}{2}(25)^2 + 107.50(25) - \left(-\frac{1}{2}(0)^2 + 107.50(0)\right)$$

$$- 2062.50$$

$$= 2375 - 0 - 2062.50 = 312.50$$

The consumers' surplus is \$312.50.

$$\text{Suppliers' surplus} = 2062.50 - \int_0^{25} (0.02x^2 + 2x + 20)\,dx$$

$$= 2062.50 - \left(\frac{0.02}{3}x^3 + x^2 + 20x\right)\Big|_0^{25}$$

$$= 2062.50 - \left(\frac{0.02}{3}(25)^3 + (25)^2 + 20(25)\right.$$

$$\left. - \left(\frac{0.02}{3}(0)^3 + (0)^2 + 20(0)\right)\right)$$

$$\approx 2062.50 - (1229.17 - 0) \approx 833.33$$

The suppliers' surplus is \$833.33.

PRACTICE

Find the equilibrium point, consumers' surplus, and suppliers' surplus for the given supply and demand functions.

1. $D(x) = -\frac{1}{2}x + 100$ and $S(x) = 2x + 50$
2. $D(x) = -0.2x + 219$ and $S(x) = 0.01x^2 + x + 30$

SOLUTIONS

1. We will first find the equilibrium quantity and price.

$$2x + 50 = -\frac{1}{2}x + 100$$

$$\frac{5}{2}x = 50$$

$$x = \frac{2}{5} \cdot 50 = 20 \quad \text{The equilibrium quantity is 20.}$$

$$S(20) = 2(20) + 50 = 90 \quad \text{The equilibrium price is \$90.}$$

$$D(20) = -\frac{1}{2}(20) + 100 = 90$$

The revenue is (\$90)(20) = \$1800.

$$\begin{aligned}
\text{Consumers' surplus} &= \int_0^{20} \left(-\frac{1}{2}x + 100\right) dx - 1800 \\
&= \left(-\frac{1}{4}x^2 + 100x\right) \Big|_0^{20} - 1800 \\
&= -\frac{1}{4}(20)^2 + 100(20) \\
&\quad - \left(-\frac{1}{4}(0)^2 + 100(0)\right) - 1800 \\
&= 1900 - 0 - 1800 = 100
\end{aligned}$$

$$\begin{aligned}
\text{Suppliers' surplus} &= 1800 - \int_0^{20} (2x + 50)\, dx \\
&= 1800 - \left(x^2 + 50x\right) \Big|_0^{20} \\
&= 1800 - \left[(20)^2 + 50(20) - \left((0)^2 + 50(0)\right)\right] \\
&= 1800 - 1400 = 400
\end{aligned}$$

The consumers' surplus is \$100, and the suppliers' surplus is \$400.

2.

$$0.01x^2 + x + 30 = -0.2x + 219$$

$$0.01x^2 + 1.2x - 189 = 0$$

$$x = \frac{-1.2 \pm \sqrt{1.2^2 - 4(0.01)(-189)}}{2(0.01)} = \frac{-1.2 \pm 3}{0.02} = 90$$

($x = -210$ is not a solution). The equilibrium quantity is 90. The equilibrium price is $D(90) = -0.2(90) + 219 = 201$, and the revenue is (\$201)(90) = \$18,090.

$$\text{Consumers' surplus} = \int_0^{90} (-0.2x + 219)\,dx - 18{,}090$$

$$= (-0.1x^2 + 219x)\Big|_0^{90} - 18{,}090$$

$$= -0.1(90)^2 + 219(90)$$

$$- \left(-0.1(0)^2 + 219(0)\right) - 18{,}090$$

$$= 18{,}900 - 0 - 18{,}090 = 810$$

$$\text{Suppliers' surplus} = 18{,}090 - \int_0^{90} (0.01x^2 + x + 30)\,dx$$

$$= 18{,}090 - \left(\frac{0.01}{3}x^3 + \frac{1}{2}x^2 + 30x\right)\Big|_0^{90}$$

$$= 18{,}090 - \left[\frac{0.01}{3}(90)^3 + \frac{1}{2}(90)^2 + 30(90)\right.$$

$$\left. - \left(\frac{0.01}{3}(0)^3 + \frac{1}{2}(0)^2 + 30(0)\right)\right]$$

$$= 18{,}090 - (9180 - 0) = 8910$$

The consumers' surplus is \$810, and the suppliers' surplus is \$8910.

The Average Value of a Function

We can find the average value of a function on an interval from $x = a$ to $x = b$ using the definite integral. We might want to find the average temperature during a 24-hour period, the average revenue over the course of a year, or the average balance of a checking account.

The average value of a function $f(x)$ over the interval $[a, b]$ is

$$\frac{1}{b-a}\int_a^b f(x)\,dx.$$

For example, the average value of the linear function $f(x) = x - 1$ on the interval $[2, 4]$ is the y-value of the midpoint, which is 2 (see Figure 15.7). This agrees with the formula.

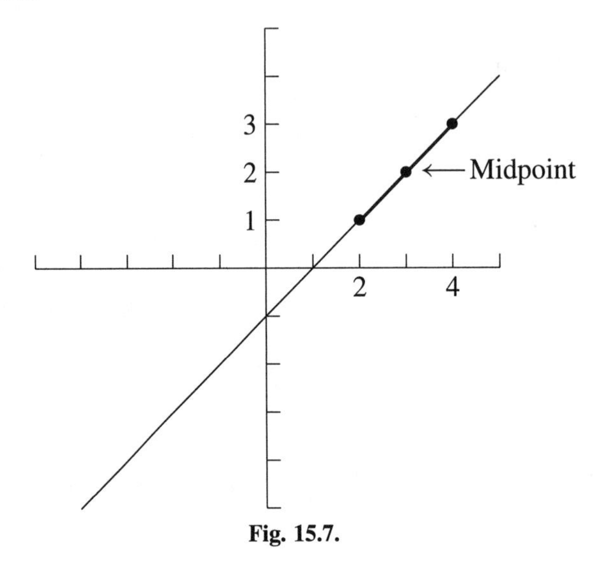

Fig. 15.7.

$$\begin{aligned}
\text{Average} &= \frac{1}{4-2}\int_2^4 (x-1)\,dx \\
&= \frac{1}{2}\left(\frac{1}{2}x^2 - x\right)\Big|_2^4 \\
&= \frac{1}{2}\left[\frac{1}{2}(4)^2 - (4) - \left(\frac{1}{2}(2)^2 - (2)\right)\right] \\
&= \frac{1}{2}(4-0) = 2
\end{aligned}$$

EXAMPLES

- Find the average functional value of $f(x) = x^2 - x - 2$ on the interval $[1, 3]$.

$$\begin{aligned}
\frac{1}{3-1}\int_1^3 (x^2 - x - 2)\,dx &= \frac{1}{2}\left(\frac{1}{3}x^3 - \frac{1}{2}x^2 - 2x\right)\Big|_1^3 \\
&= \frac{1}{2}\left[\frac{1}{3}(3)^3 - \frac{1}{2}(3)^2 - 2(3) - \left(\frac{1}{3}(1)^3 - \frac{1}{2}(1)^2 - 2(1)\right)\right]
\end{aligned}$$

$$= \frac{1}{2}\left[-\frac{3}{2} - \left(-\frac{13}{6}\right)\right] = \frac{1}{3}$$

The average of all y-values between $x = 1$ and $x = 3$ is $\frac{1}{3}$.

- The temperature during a 24-hour period in a certain city can be approximated by the function $T(t) = 0.0005t^4 - 0.0346t^3 + 0.7014t^2 - 5.0101t + 38.844$, t hours after midnight. What is the average temperature for the entire day? What is the average temperature between noon and 6:00 pm?

$$\frac{1}{24-0}\int_0^{24} (0.0005t^4 - 0.0346t^3 + 0.7014t^2 - 5.0101t + 38.844)\,dt$$

$$= \frac{1}{24}\left(0.0001t^5 - 0.00865t^4 + 0.2338t^3 - 2.50505t^2 + 38.844t\right)\Big|_0^{24}$$

$$= \frac{1}{24}(647.7984 - 0) \approx 27$$

The average temperature during the entire 24-hour period is 27 degrees. Noon is 12 hours after midnight, and 6:00 pm is 18 hours after midnight.

$$\frac{1}{18-12}\int_{12}^{18} (0.0005t^4 - 0.0346t^3 + 0.7014t^2 - 5.0101t + 38.844)\,dt$$

$$= \frac{1}{6}\left(0.0001t^5 - 0.00865t^4 + 0.2338t^3 - 2.50505t^2 + 38.844t\right)\Big|_{12}^{18}$$

$$= \frac{1}{6}(531.9918 - 354.924) = \frac{1}{6}(177.0678) \approx 29.5$$

The average temperature between noon and 6:00 pm is 29.5 degrees.

- The velocity of a particle after t seconds is $v(t) = 4t + 10$ feet per second. Find the average velocity for the first 15 seconds and between 10 and 20 seconds.

$$\frac{1}{15-0}\int_0^{15} (4t + 10)\,dt = \frac{1}{15}(2t^2 + 10t)\Big|_0^{15}$$

$$= \frac{1}{15}(600 - 0) = 40$$

The average velocity between 0 and 15 seconds is 40 feet per second.

$$\frac{1}{20-10}\int_{10}^{20} (4t + 10)\,dt = \frac{1}{10}(2t^2 + 10t)\Big|_{10}^{20}$$

$$= \frac{1}{10}(1000 - 300) = 70$$

The average velocity between 10 and 20 seconds is 70 feet per second.

- The revenue for a product during its first year is approximated by $R(t) = \frac{1000}{t+1} + 200$, t weeks after its introduction. What is the average weekly revenue during its first year?

$$\frac{1}{52-0}\int_0^{52}\left(\frac{1000}{t+1} + 200\right)\,dt$$

$$= \frac{1}{52}\left(1000\int_0^{52}\frac{1}{t+1}\,dt + \int_0^{52} 200\,dt\right)$$

$$= \frac{1}{52}(1000\ln(t+1) + 200t)\Big|_0^{52}$$

$$= \frac{1}{52}(1000\ln 53 - 1000\ln 1 + 200(52) - 200(0))$$

$$\approx \frac{1}{52}(14,370.29191) \approx 276$$

The average weekly revenue is \$276.

PRACTICE

1. Find the average value of the function $f(x) = 6x^2 + 8x - 7$ on the interval $[-1, 3]$.
2. Find the average value of the function $f(x) = 4x^3 + 18x^2 + x$ on the interval $[-2, 3]$.
3. The temperature during one day can be approximated by the function $T(t) = 0.00124t^4 - 0.07512t^3 + 1.4115t^2 - 7.8916t + 75.1013$, t hours after midnight. What is the average temperature over the 24-hour period? Between 9 am and noon?
4. For the first year, the value of an investment t months after purchase can be approximated by the function $V(t) = -0.326t^3 + 6.393t^2 - 26.871t + 750$. What is the average value during the first year?
5. The value of a piece of office equipment over the first eight years of its life can be approximated by the function $V(t) = 10,000e^{-0.105t}$. What is the average value of the equipment over the eight years?

SOLUTIONS

1.

$$\frac{1}{3-(-1)}\int_{-1}^{3}(6x^2+8x-7)\,dx = \frac{1}{4}(2x^3+4x^2-7x)\Big|_{-1}^{3}$$

$$= \frac{1}{4}(69-9) = 15$$

The average functional value on the interval $[-1, 3]$ is 15.

2.

$$\frac{1}{3-(-2)}\int_{-2}^{3}(4x^3+18x^2+x)\,dx = \frac{1}{5}\left(x^4+6x^3+\frac{1}{2}x^2\right)\Big|_{-2}^{3}$$

$$= \frac{1}{5}\left(\frac{495}{2}-(-30)\right) = \frac{111}{2}$$

The average functional value on the interval $[-2, 3]$ is $\frac{111}{2}$.

3.

$$\frac{1}{24-0}\int_{0}^{24}(0.00124t^4 - 0.07512t^3 + 1.4115t^2$$

$$- 7.8916t + 75.1013)\,dt$$

$$= \frac{1}{24}(0.000248t^5 - 0.01878t^4$$

$$+ 0.4705t^3 - 3.9458t^2 + 75.1013t)\Big|_{0}^{24}$$

$$= \frac{1}{24}(1777.8199 - 0) \approx 74.1$$

$$\frac{1}{12-9}\int_{9}^{12}(0.00124t^4 - 0.07512t^3 + 1.4115t^2$$

$$- 7.8916t + 75.1013)\,dt$$

$$= \frac{1}{3}(0.000248t^5 - 0.01878t^4$$

$$+ 0.4705t^3 - 3.9458t^2 + 75.1013t)\Big|_{9}^{12}$$

$$= \frac{1}{3}(227.60768) \approx 75.9$$

The average temperature during the entire day is 74.1 degrees, and the average temperature between 9 am and noon is 75.9 degrees.

4.

$$\frac{1}{12-0}\int_0^{12} (-0.326t^3 + 6.393t^2 - 26.871t + 750)\,dt$$

$$= \frac{1}{12}(-0.0815t^4 + 2.131t^3$$

$$- 13.4355t^2 + 750t)\Big|_0^{12}$$

$$= \frac{1}{12}(9057.672 - 0) \approx 754.81$$

The average value of the investment for the first 12 months is \$754.81.

5.

$$\frac{1}{8-0}\int_0^8 (10{,}000e^{-0.105t})\,dt = \frac{1}{8}\left(\frac{10{,}000}{-0.105}e^{-0.105t}\right)\Big|_0^8$$

$$\approx \frac{1}{8}\Big(-95{,}238e^{-0.105(8)}$$

$$- \left(-95{,}238e^{-0.105(0)}\right)\Big)$$

$$\approx \frac{1}{8}(-41{,}115 - (-95{,}238)) \approx 6765$$

The average value of the equipment for the first eight years is \$6765.

CHAPTER 15 REVIEW

1. The slope of the tangent line for a function is found by computing $y' = 8x^3 - 9x^2 + 6x - 4$, and $(2, 7)$ is a point on the graph. What is the function?
   (a) $y = 2x^4 - 3x^3 + 3x^2 - 4x - 5$
   (b) $y = 24x^2 - 18x - 53$
   (c) $y = 8x^3 - 9x^2 + 6x - 33$
   (d) $y = \frac{8}{3}x^3 - \frac{9}{2}x^2 + 6x - \frac{25}{3}$

2. The cost for producing x units of a product is $C(x) = 0.04x^2 + x + 5000$. Find the average cost for producing the first 150 units.
   (a) \$875
   (b) \$6350
   (c) \$6050
   (d) \$5375

3. An insurance agent collects \$7200 per year for selling a particular policy. If the money is flowing continuously into an account that pays 5% annual interest, compounded continuously, what is the accumulated value after eight years?
   (a) \$11,934.60
   (b) \$5901.90
   (c) \$70,822.76
   (d) \$10,741.14

4. The demand for a service sold by the hour is $D(x) = -0.8x + 150$ for x hours. Find the consumers' surplus if $x = 100$ hours are sold.
   (a) \$11,000
   (b) \$70
   (c) \$4000
   (d) There is no consumers' surplus

5. A grandfather wants to give his newborn granddaughter a \$25,000 gift on her 21st birthday. How much money per year should continuously flow into an account that pays 6% annual interest, compounded continuously?
   (a) \$1500
   (b) \$594
   (c) \$338
   (d) \$14,208

6. Find the equilibrium point for a product whose demand function is $D(x) = -0.005x^2 + 2x + 30{,}000$ and whose supply function is $S(x) = 0.02x^2 - x + 5090$.
   (a) The equilibrium quantity is $q = 1060$, and the equilibrium price is \$26,502.
   (b) The equilibrium quantity is $q = 1000$, and the equilibrium price is \$27,000.

(c) The equilibrium quantity is $q = 840$, and the equilibrium price is \$18,362.
(d) The equilibrium point does not exist.

7. The monthly balance for a bank account during one year can be approximated by $B(x) = 7.1615x^4 - 260.1273x^3 + 3129.3402x^2 - 14{,}214.9741x + 25{,}916.6667$ (where $x = 1$ is January). For the eleven months from January to December, what is the average monthly balance?
(a) \$8902
(b) \$8160
(c) \$6511
(d) \$7103

8. A woman won a \$15 million lottery jackpot. She is considering taking the cash value instead of 20 annual payments of \$750,000. Assuming that the money is flowing continuously from an account paying 4% annual interest, compounded continuously, approximate the cash value.
(a) \$12,413,205
(b) \$9,978,096
(c) \$10,325,082
(d) \$13,011,250

9. Find the suppliers' surplus for $S(x) = 100e^{0.02x}$ and $q = 150$.
(a) \$301,290
(b) \$95,428
(c) \$205,855
(d) There is no suppliers' surplus

10. The marginal revenue for selling x units of a product is $R'(x) = 100 \ln x$. The revenue for selling 50 units is \$14,760. What is the revenue function?
(a) $R(x) = 100x \ln x - x - 4750$
(b) $R(x) = 100x \ln x - 100x + 200$
(c) $R(x) = 100 \ln x - x + 14,419$
(d) $R(x) = 100x \ln x - 4800$

SOLUTIONS

1. a	2. d	3. c	4. c	5. b
6. a	7. d	8. c	9. c	10. b

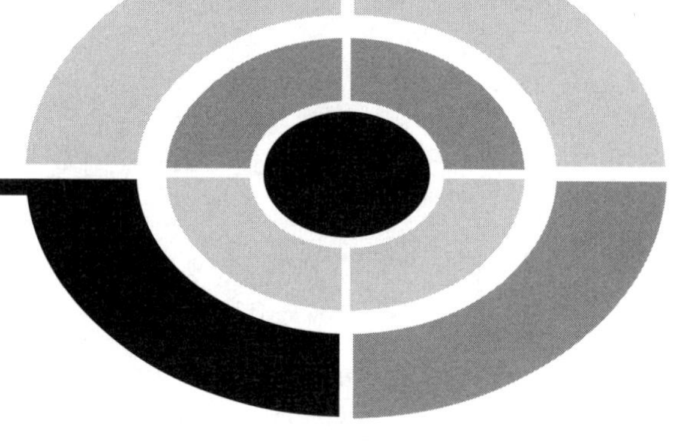

Final Exam

1. Find $\lim_{x\to 3}(4x^2 + 1)$.

   (a) 13
   (b) 169
   (c) 0
   (d) 37

2. Find y' if $y = \sqrt{x} + \frac{1}{\sqrt{x}}$.

   (a)
   $$y' = \frac{1}{2\sqrt{x}} + \frac{1}{2\sqrt[3]{x^2}}$$
   (b)
   $$y' = \frac{1}{2\sqrt{x}} - \frac{1}{2\sqrt[3]{x}}$$
   (c)
   $$y' = \frac{1}{2\sqrt{x}} + \frac{1}{2\sqrt{x^3}}$$

(d)

$$y' = \frac{1}{2\sqrt{x}} - \frac{1}{2\sqrt{x^3}}$$

3. Is $f(x) = 2x^3 - 3x^2 - 120x + 15$ increasing, decreasing, or neither at $x = 0$?

   (a) Increasing
   (b) Decreasing
   (c) Neither
   (d) It cannot be determined without the graph

4. For the line $y = -\frac{2}{3}x + 4$

   (a) as x increases by 2, y decreases by 3
   (b) as x increases by 2, y increases by 3
   (c) as x increases by 3, y decreases by 2
   (d) as x increases by 3, y increases by 2

5. If $f(x) = x^2 - 3x$, then find $f'(x)$.

   (a)

$$\lim_{h \to 0} \frac{x^2 - 3x + h - (x^2 - 3x)}{h}$$

   (b)

$$\lim_{h \to 0} \frac{x^2 + 2xh + h^2 - 3x + h - (x^2 - 3x)}{h}$$

   (c)

$$\lim_{h \to 0} \frac{x^2 + 2xh + h^2 - 3x - h - (x^2 - 3x)}{h}$$

   (d)

$$\lim_{h \to 0} \frac{x^2 + 2xh + h^2 - 3x - 3h - (x^2 - 3x)}{h}$$

6. Does the graph of $f(x) = 2x^3 - 3x^2 - 120x + 15$ have a relative extremum at $x = -4$?

   (a) There is a relative minimum at $x = -4$.
   (b) There is a relative maximum at $x = -4$.

(c) The graph of $f(x)$ does not have a relative extremum at $x = -4$.

(d) It cannot be determined without the graph.

7. Find y' if $y = 10e^{3x+7}$.

(a) $y' = 10e^3$

(b) $y' = 30e^{3x+7}$

(c) $y' = 30e^3$

(d) $y' = \frac{3e^{3x+7}}{10}$

8. An object travels $d(t) = t^2 + 5t$ feet after t seconds. What is its instantaneous velocity at 3 seconds?

(a) 24 feet per second

(b) 11 feet per second

(c) 8 feet per second

(d) 17 feet per second

9. Is the graph of $f(x) = x^5 - 3x^2 + 4x - 10$ concave up, concave down, or neither at $x = 2$?

(a) Concave up

(b) Concave down

(c) Neither

(d) Concavity cannot be determined without the graph

10. For the line $y = x$

(a) as x increases by 1, y increases by 1

(b) as x increases by 1, y decreases by 1

(c) as x decreases by 1, y increases by 1

(d) the slope is 0

11. Evaluate

$$\int_3^5 \frac{2x}{x^2 - 4}\,dx$$

(a) $\ln 21 - \ln 5 \approx 1.4351$

(b) $\ln 5 - \ln 21 \approx -1.4351$

(c) $-\frac{16}{105}$

(d) The integral does not exist

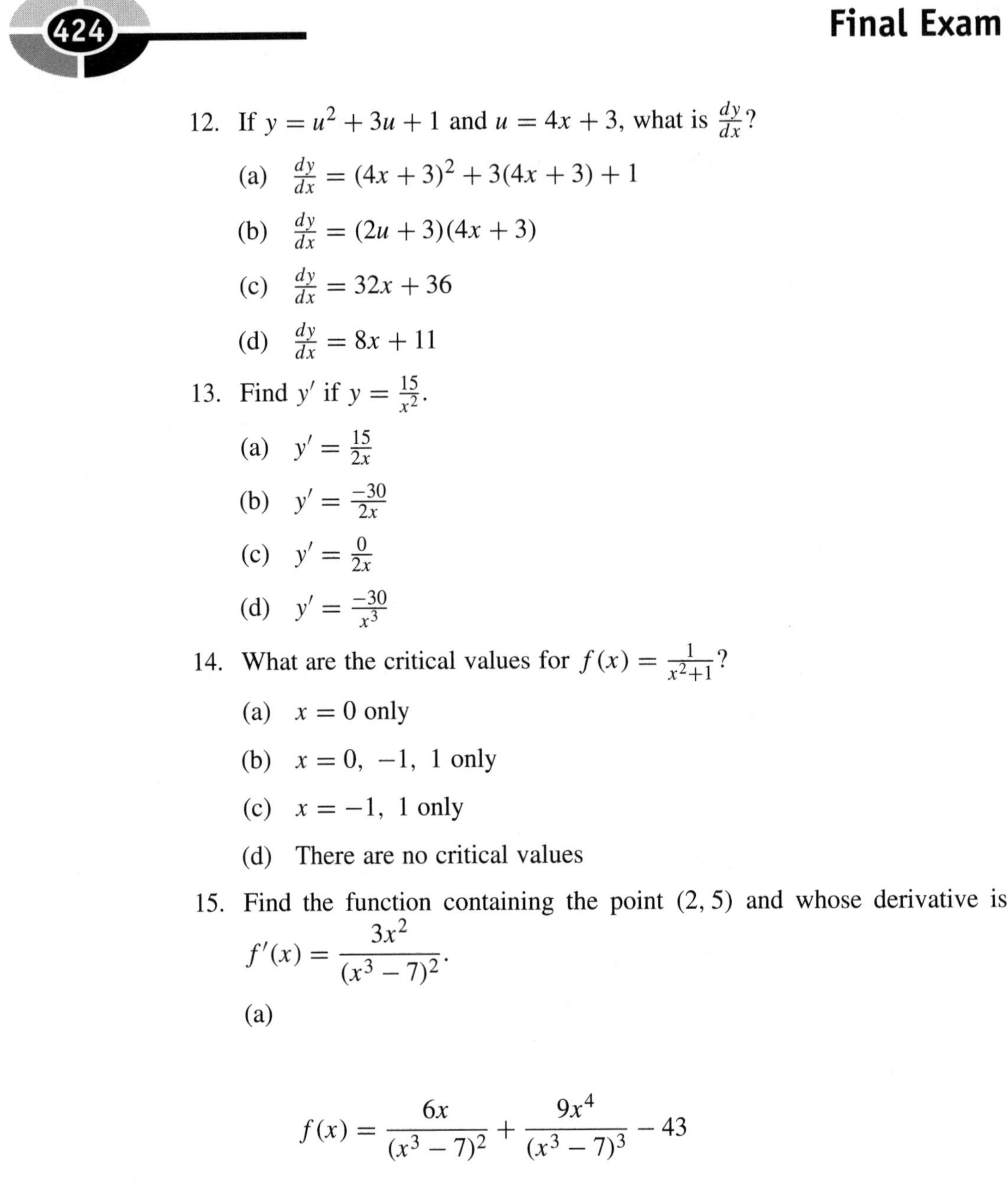

12. If $y = u^2 + 3u + 1$ and $u = 4x + 3$, what is $\frac{dy}{dx}$?

    (a) $\frac{dy}{dx} = (4x + 3)^2 + 3(4x + 3) + 1$

    (b) $\frac{dy}{dx} = (2u + 3)(4x + 3)$

    (c) $\frac{dy}{dx} = 32x + 36$

    (d) $\frac{dy}{dx} = 8x + 11$

13. Find y' if $y = \frac{15}{x^2}$.

    (a) $y' = \frac{15}{2x}$

    (b) $y' = \frac{-30}{2x}$

    (c) $y' = \frac{0}{2x}$

    (d) $y' = \frac{-30}{x^3}$

14. What are the critical values for $f(x) = \frac{1}{x^2+1}$?

    (a) $x = 0$ only

    (b) $x = 0, -1, 1$ only

    (c) $x = -1, 1$ only

    (d) There are no critical values

15. Find the function containing the point $(2, 5)$ and whose derivative is $f'(x) = \dfrac{3x^2}{(x^3 - 7)^2}$.

    (a)

    $$f(x) = \frac{6x}{(x^3 - 7)^2} + \frac{9x^4}{(x^3 - 7)^3} - 43$$

    (b)

    $$f(x) = -\frac{1}{x^3 - 7} + 6$$

(c)

$$f(x) = -\frac{3}{(x^3 - 7)^3} + 8$$

(d) $\ln|x^3 - 7| + 5$

16. Evaluate $\int \sqrt[4]{x}\, dx$.

(a) $\frac{4}{5}x^{5/4} + C$
(b) $\frac{5}{4}x^{5/4} + C$
(c) $\frac{5}{4}x^{4/5} + C$
(d) $\frac{4}{5}x^{4/5} + C$

17. The manager of an apartment complex is considering reducing the monthly rent. There are 100 apartments in the complex, and 80 of them occupied. The rent is now \$800 per month. The manager believes that for each \$16 decrease in the rent, two more apartments can be rented. What rent will maximize the revenue?

(a) \$720 per month
(b) \$700 per month
(c) \$680 per month
(d) \$660 per month

18. Evaluate approximately $\int_{-1}^{2} e^{2x+3}\, dx$.

(a) 1093.91
(b) 546.96
(c) 273.48
(d) 47.21

19. What is $\frac{dy}{dx}$ for $x^3y^2 - xy - 8x = 10$?

(a)

$$\frac{dy}{dx} = \frac{8 - 3x^2y^2 - 2x^3y + y}{-x}$$

(b)

$$\frac{dy}{dx} = \frac{8 - 3x^2y^2 - 2x^3y + y}{x}$$

(c)

$$\frac{dy}{dx} = \frac{8 - 3x^2y^2 + y}{2x^3y + x}$$

(d)

$$\frac{dy}{dx} = \frac{8 - 3x^2y^2 + y}{2x^3y - x}$$

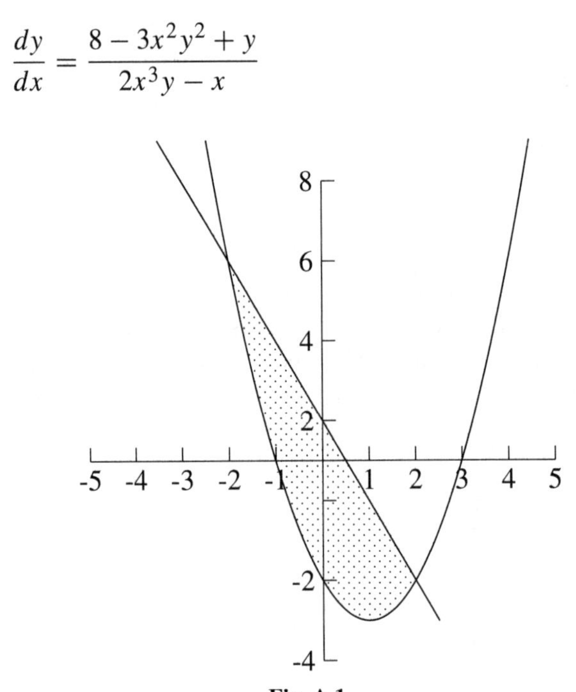

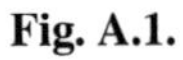

Fig. A.1.

20. Find the shaded area in Figure A.1. The line is $y = -2x + 1$, and the curve is $y = x^2 - 2x - 3$.

    (a) $\frac{20}{3}$
    (b) $-\frac{44}{3}$
    (c) $\frac{32}{3}$
    (d) 0

21. The population of a certain city can be approximated by $P(t) = 100{,}000e^{0.025t}$, t years after 1995. How fast is the population growing in the year 2005?

    (a) The population is growing at the rate of 12,840 people per year.
    (b) The population is growing at the rate of 2500 people per year.

(c) The population is growing at the rate of 3210 people per year.

(d) The population is growing at the rate of 2840 people per year.

22. Suppose $f'(5) = 0$ and $f''(5) = -6$. What does this mean?

(a) There is a relative maximum at $x = 5$.

(b) There is a relative minimum at $x = 5$.

(c) The relative maximum is -6.

(d) The relative minimum is -6.

23. Evaluate

$$\int \frac{1}{x^4}\,dx$$

(a) $\frac{1}{3}x^3 + C$

(b) $\frac{1}{5}x^5 + C$

(c) $-\frac{1}{3}x^{-3} + C$

(d) $-\frac{1}{5}x^{-5} + C$

24. Find the tangent line to $f(x) = x^3 - 5x^2 - x - 3$ at $x = -1$.

(a) $y = 12x + 10$

(b) $y = 12x + 4$

(c) $y = -2x - 4$

(d) $y = -2x - 5$

25. For the function $f(x) = -x^2 + 8x - 15$, which of the following is true?

(a) The maximum functional value is 1.

(b) The minimum functional value is 1.

(c) The maximum functional value is 4.

(d) The minimum functional value is 4.

26. Find y' if $y = \ln(6x^3 + 5x^2 + x)$.

(a)

$$y' = \frac{6x^3 + 5x^2 + x}{18x^2 + 10x + 1}$$

(b)

$$y' = \frac{(18x^2 + 10x + 1)}{\ln(6x^3 + 5x^2 + x)}$$

(c) $y' = \ln(18x^2 + 10x + 1)$

(d)

$$y' = \frac{18x^2 + 10x + 1}{6x^3 + 5x^2 + x}$$

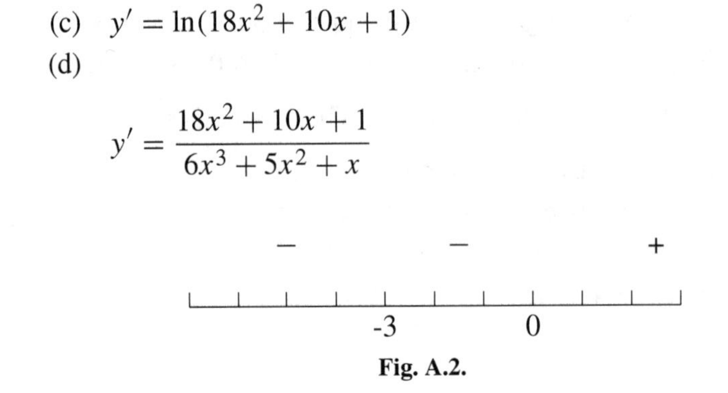

Fig. A.2.

27. The sign graph for a function $f(x)$ is given in Figure A.2. What can we conclude from this graph?

    (a) There is a relative minimum at $x = -3$ and a relative maximum at $x = 0$.
    (b) There is a relative maximum at $x = -3$ and a relative minimum at $x = 0$.
    (c) There is a relative minimum at $x = 0$ only.
    (d) There is a relative maximum at $x = 0$ only.

28. Evaluate $\int (6x^2 - 5x + 4)(4x^3 - 5x^2 + 8x + 3)^7\, dx$.

    (a) $\frac{1}{12}(4x^3 - 5x^2 + 8x + 3)^6 + C$
    (b) $\frac{1}{3}(4x^3 - 5x^2 + 8x + 3)^6 + C$
    (c) $\frac{1}{16}(4x^3 - 5x^2 + 8x + 3)^8 + C$
    (d) $\frac{1}{8}(4x^3 - 5x^2 + 8x + 3)^8 + C$

29. Find the average rate of change for the function $f(x) = x^2 - 6$ between $x = 2$ and $x = 5$.

    (a) -7
    (b) 0
    (c) $\frac{7}{1}$
    (d) $\frac{1}{7}$

30. The value of an investment for the first ten years can be approximated by $V(t) = 3.25t^4 - 83.9t^3 + 658.6t^2 - 1210t + 4989$, t years after purchase. What is happening to the investment's value at 5 years?

    (a) The value is increasing at the rate of \$708 per year.
    (b) The value is decreasing at the rate of \$708 per year.

(c) The value is increasing at the rate of \$6948 per year.
(d) The value is decreasing at the rate of \$6948 per year.

31. Evaluate

$$\int 4xe^{3x^2+5}\,dx$$

(a)

$$\frac{2}{3}e^{3x^2+5} + C$$

(b)

$$\frac{4}{3}e^{3x^2+5} + C$$

(c)

$$24e^{3x^2+5} + C$$

(d)

$$12e^{3x^2+5} + C$$

32. A square box is to be constructed so that it has a volume of six cubic feet. Material for the top costs \$0.40 per square foot; material for the bottom costs \$0.60 per square foot; and material for the sides costs \$0.45 per square foot. What is the height of the box that costs the least in material?

(a) About 1.75 feet
(b) About 1.81 feet
(c) About 1.95 feet
(d) About 3.43 feet

33. Find y' if $y = [(6x + 1)(x - 3)]^7$.

(a) $y' = 7[(6x + 1)(x - 3)]^6$
(b) $y' = (84x - 119)[(6x + 1)(x - 3)]^6$
(c) $y' = [6(x - 3) + (6x + 1)(1)]^7$
(d) $y' = 7[6(x - 3) + (6x + 1)(1)]^6$

34. Find y' if $y = \log_5(10x^2 - 7x)$.

(a)

$$y' = \frac{100x - 35}{10x^2 - 7x}$$

(b)

$$y' = \frac{\ln 5(20x - 7)}{10x^2 - 7x}$$

(c)

$$y' = \frac{20x - 7}{10x^2 - 7x}$$

(d)

$$y' = \frac{20x - 7}{\ln 5(10x^2 - 7x)}$$

35. Find the accumulated value of \$9000 that continuously flows into an account each year for 10 years. The account earns 6% annual interest, compounded continuously.

(a) \$95,400
(b) \$7399
(c) \$123,318
(d) \$161,176

36. Find the elasticity of demand function for a product whose demand function is $D(p) = 50 - 2p$.

(a)

$$E(p) = \frac{1}{25 - p}$$

(b) $E(p) = 25 - p$

(c)

$$E(p) = \frac{p}{25 - p}$$

(d)

$$E(p) = \frac{25 - p}{p}$$

37. Find

$$\lim_{x \to 6} \frac{x^2 - 5x - 6}{x^2 - 36}$$

(a) $\frac{7}{12}$
(b) $\frac{0}{0}$
(c) 0
(d) The limit does not exist

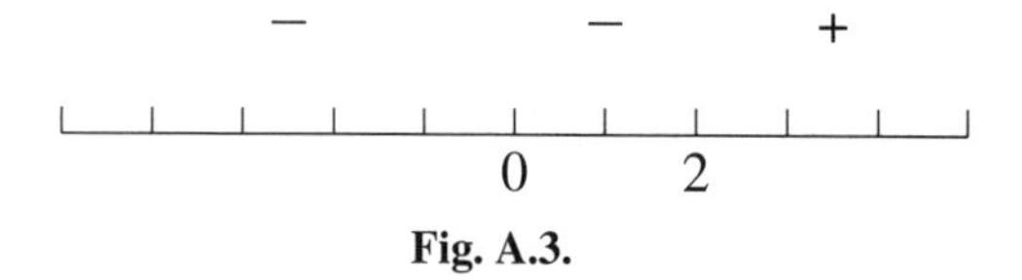

Fig. A.3.

38. The sign graph in Figure A.3 is the sign graph for the derivative of which function?

(a) $f(x) = 3x^3 - 8x^2 + 4$
(b) $f(x) = 3x^4 - 8x^3 + 4$
(c) $f(x) = 2x^3 - 6x^2 + 4$
(d) $f(x) = 2x^2 - 6x + 4$

39. $y = (5x^3 + 2x^2 + 3)^6$.

(a) $y' = (5x^3 + 2x^2 + 3)^5(15x^2 + 4x)$
(b) $y' = 6(15x^2 + 4x)^5$
(c) $y' = 6(5x^3 + 2x^2 + 3)^5$
(d) $y' = (90x^2 + 24x)(5x^3 + 2x^2 + 3)^5$

40. Find

$$\lim_{h \to 0} \frac{\frac{2}{x+h-1} - \frac{2}{x-1}}{h}$$

(a)

$$\lim_{h \to 0} \frac{-2}{(x + h - 1)(x - 1)}$$

(b)

$$\lim_{h\to 0}\frac{2h}{(x+h-1)(x-1)}$$

(c)

$$\lim_{h\to 0}\frac{-2h}{(x+h-1)(x-1)}$$

(d)

$$\lim_{h\to 0}\frac{2h}{(x-1)^2}$$

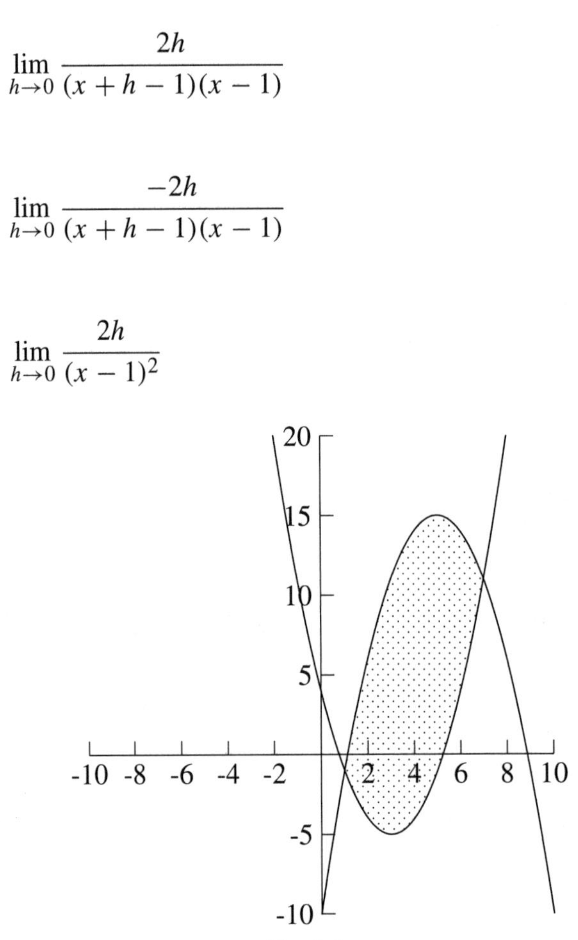

Fig. A.4.

41. Find the shaded area in Figure A.4. The curves are $y = -x^2 + 10x - 10$ and $y = x^2 - 6x + 4$.
    (a) 72
    (b) 60
    (c) 54
    (d) 78

42. Find the consumers' surplus for a product whose demand function is $D(x) = \frac{1000}{x+10}$ when $q = 90$ units are demanded.
    (a) \$2303
    (b) \$3203

(c) \$1403
(d) \$90

43. A grocery store sells 6000 ten-pound bags of pet food. Each bag costs \$1.20 to store for one year. Each order costs \$25. How many times per year should the store order the pet food?

(a) 10
(b) 12
(c) 14
(d) 16

44. Evaluate

$$\int \frac{4x+5}{6x^2+15x+2}\,dx$$

(a) $\frac{1}{3}\ln|6x^2+15x+2|+C$
(b) $3\ln|6x^2+15x+2|+C$
(c) $-\frac{3}{2}\ln|6x^2+15x+2|+C$
(d) The integral does not exist

45. What are the points of inflection for $f(x) = x^3 + 3x^2 - 24x + 4$?

(a) $(-4, 84)$ and $(2, -24)$ only
(b) $(-1, 30)$ only
(c) $(-4, 84)$, $(-1, 30)$, and $(2, -24)$
(d) There are no points of inflection

46. Find y' if $y = 10^{4x-x^2}$.

(a) $y' = \ln 10(4-2x)\cdot 10^{4x-x^2}$
(b) $y' = (4-2x)\ln 10^{4x-x^2}$
(c)

$$y' = \ln\left(\frac{4-2x}{4x-x^2}\right)$$

(d)

$$y' = \frac{4-2x}{\ln 10(4x-x^2)}$$

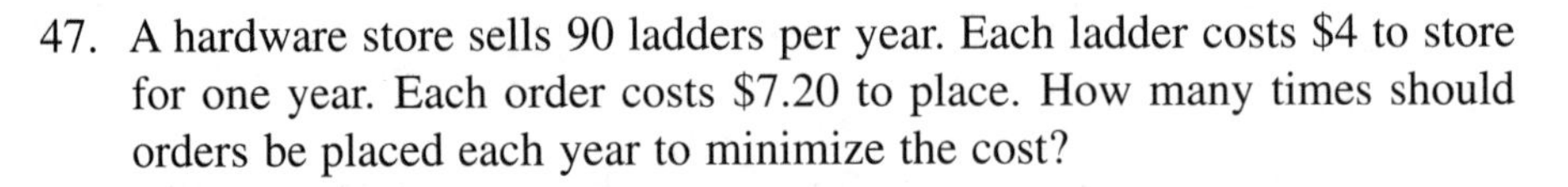

47. A hardware store sells 90 ladders per year. Each ladder costs \$4 to store for one year. Each order costs \$7.20 to place. How many times should orders be placed each year to minimize the cost?

    (a) 2 times per year
    (b) 3 times per year
    (c) 4 times per year
    (d) 5 times per year

48. Sales of a certain service depends on the sales budget. The number of orders in a month can be approximated by $s(a) = -0.001a^2 + 16a - 24{,}000$, where \$$a$ is the monthly sales budget. Currently, \$5000 is budgeted for sales each month. The company owner is planning on increasing the sales budget by \$500 per month. How will this affect the number of orders?

    (a) The sales level will increase at the rate of 3000 orders per month.
    (b) The sales level will increase at the rate of 2500 orders per month.
    (c) The sales level will increase at the rate of 2000 orders per month.
    (d) The sales level will increase at the rate of 1500 orders per month.

49. What is the absolute maximum of $f(x) = 2x^3 - 9x^2 - 24x + 5$ on the interval $[-2, 3]$?

    (a) The absolute maximum is 1 and the absolute minimum is -94.
    (b) The absolute maximum is 1 and the absolute minimum is -107.
    (c) The absolute maximum is 18 and the absolute minimum is -107.
    (d) The absolute maximum is 18 and the absolute minimum is -94.

50. The value of a piece of equipment can be approximated by $y = 20{,}000(0.90^x)$, x years after its purchase. How fast is its value decreasing four years after purchase?

    (a) Its value is decreasing at the rate of \$2000 per year.
    (b) Its value is decreasing at the rate of \$1380 per year.
    (c) Its value is decreasing at the rate of \$1460 per year.
    (d) Its value is decreasing at the rate of \$5830 per year.

51. Find $f'(x)$ if $f(x) = (4x^2 + 3x + 5)(x^2 + 2)$.

    (a) $f'(x) = (8x + 3)(x^2 + 2) + (4x^2 + 3x + 5)(2x)$
    (b) $f'(x) = (8x + 3)(2x)$
    (c) $f'(x) = (8x + 3)(x^2 + 2) - (4x^2 + 3x + 5)(2x)$
    (d) $f'(x) = (8x + 3)(x^2 + 2) - (4x^2 + 3x + 5)(2x + 2)$

52. Simplify $\ln(x-1) - \ln(2x+3)$.

   (a) $\ln[(x-1)(2x+3)]$
   (b) $\ln[(x-1) - (2x+3)]$
   (c) $\ln \dfrac{x-1}{2x+3}$
   (d) $\dfrac{\ln(x-1)}{\ln(2x+3)}$

53. Find $\frac{dy}{dx}$ if $y = \sqrt{x^2-4}$.

   (a) $\dfrac{dy}{dx} = \dfrac{1}{\sqrt{x^2-4}}$
   (b) $\dfrac{dy}{dx} = \dfrac{1}{2\sqrt{x^2-4}}$
   (c) $\dfrac{dy}{dx} = \dfrac{x}{\sqrt{x^2-4}}$
   (d) $\dfrac{dy}{dx} = \dfrac{x}{2\sqrt{x^2-4}}$

54. The revenue for a product t weeks after release during its first year can be approximated by $R(t) = -5t^2 + 333t + 50$. Find the average weekly revenue during the first year of the product's release.

   (a) \$210,469
   (b) \$17,539
   (c) \$4201
   (d) Losing \$187 per week

55. Find the price that has unit elasticity for a product whose demand function is $D(p) = 400e^{-0.02p}$.

   (a) \$20
   (b) \$30
   (c) \$40
   (d) \$50

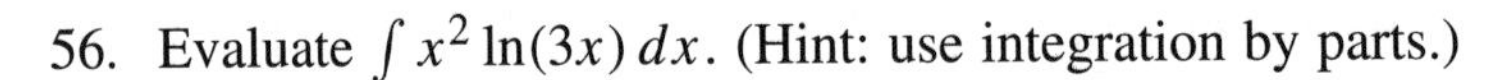

56. Evaluate $\int x^2 \ln(3x)\,dx$. (Hint: use integration by parts.)

    (a) $\frac{1}{3}x^3 \ln(3x) - \frac{1}{3}x^3 + C$
    (b) $\frac{1}{3}x^3 \ln(3x) - \frac{1}{9}x^3 + C$
    (c) $\frac{1}{3}x^3 \ln(3x) - \frac{1}{6}x^4 + C$
    (d) $\frac{1}{3}x^3 \ln(3x) - \frac{1}{2}x^4 + C$

57. The revenue for selling x units of a product is $R(x) = -0.01x^2 + 5x$. Find the marginal revenue for 20 units.

    (a) \$4.60
    (b) \$96
    (c) −\$3
    (d) \$1

58. A car traveling south on a highway averaged 64 mph. A small train passed underneath the highway at the same time the car was there. The train was traveling west, averaging 48 mph. An hour later, how fast was the distance between the car and train increasing?

    (a) About 100 mph
    (b) About 156 mph
    (c) About 62.5 mph
    (d) About 80 mph

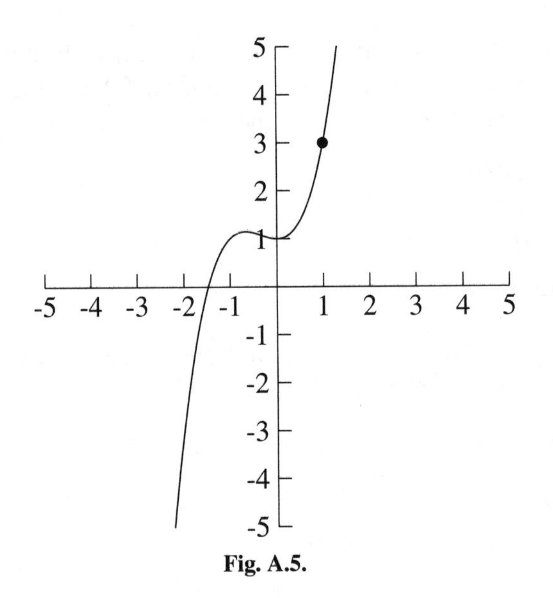

Fig. A.5.

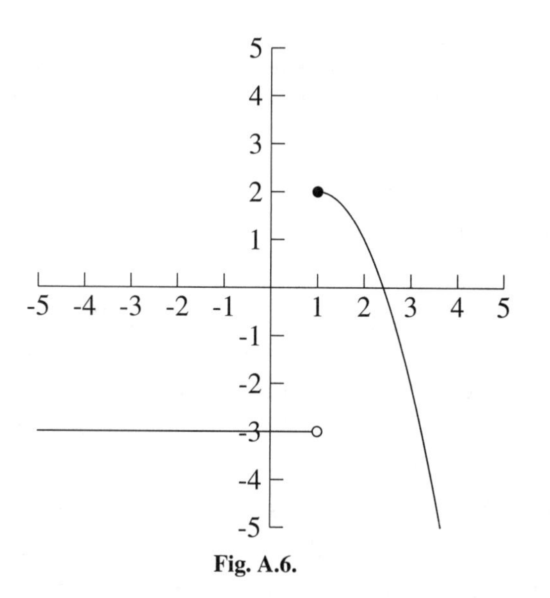

Fig. A.6.

59. The graph in Figure A.5 is the graph of $f(x)$. Find $\lim_{x \to 1} f(x)$.
    (a) 1
    (b) 2
    (c) 3
    (d) The limit does not exist

60. The graph in Figure A.6 is the graph of $g(x)$. Find $\lim_{x \to 1^-} g(x)$.
    (a) -3
    (b) 2
    (c) 1
    (d) The limit does not exist

61. Why is $g(x)$ (shown in Figure A.6) not continuous at $x = 1$?
    (a) $g(1)$ does not exist.
    (b) $\lim_{x \to 1} g(x)$ does not exist.
    (c) $\lim_{x \to 1} g(x)$ does exist but $\lim_{x \to 1} g(x) \neq g(1)$.
    (d) $\lim_{x \to 1^-} g(x)$ does not exist.

62. An open-topped box is to be constructed from a thin piece of metal that measures 15" × 18". After a square piece is cut from each corner, the

sides will be folded up to form the box. How much should be cut from each corner in order to maximize the volume of the box?

(a) About 2.08 inches
(b) About 2.72 inches
(c) About 3.16 inches
(d) About 8.28 inches

63. The number of newspaper subscribers in a small city can be approximated by $S(p) = 0.6p$, where p is the population. The population between the years 1980 and 2005 can be approximated by $p(t) = 2.15t^3 - 65.6t^2 + 897t + 39730$, t years after 1980. What is happening to the number of subscribers in the year 1990?

(a) The number of subscribers is increasing at the rate of 138 per year.
(b) The number of subscribers is increasing at the rate of 231 per year.
(c) The number of subscribers is increasing at the rate of 456 per year.
(d) The number of subscribers is increasing at the rate of 984 per year.

64. Find $f'(x)$ if

$$f(x) = \frac{16x + 3}{x^2 + 1}$$

(a)

$$f'(x) = \frac{16x(x^2 + 1) - (16x + 3)(2x)}{(x^2 + 1)^2}$$

(b)

$$f'(x) = \frac{16(x^2 + 1) + (16x + 3)(2x)}{(x^2 + 1)^2}$$

(c)

$$f'(x) = \frac{16(x^2 + 1) - (16x + 3)(2x)}{(x^2 + 1)^2}$$

(d)

$$f'(x) = \frac{16(x^2 + 1) + (16x + 3)(2x)}{x^2 + 1}$$

65. A cylindrical tank is being filled with a liquid solvent at the rate of 3 cubic feet per minute. The radius of the tank is 2 feet. How fast is the level of solvent rising?

    (a) About 28.27 feet per minute
    (b) About 9.42 feet per minute
    (c) About 0.48 feet per minute
    (d) About 0.24 feet per minute

66. Find $\frac{dy}{dx}$ for $(x + y)^2 = y^3$.

    (a)
    $$\frac{dy}{dx} = \frac{2x + 2y}{3y^2 - 2x - 2y}$$
    (b)
    $$\frac{dy}{dx} = \frac{2x + 2y}{3y^2}$$
    (c)
    $$\frac{dy}{dx} = \frac{3y^2}{2x + 2y}$$
    (d) $\frac{dy}{dx}$ does not exist

SOLUTIONS

1. d	2. d	3. b	4. c	5. d	6. b	7. b	8. b	9. a	10. a
11. a	12. c	13. d	14. a	15. b	16. a	17. a	18. b	19. d	20. c
21. c	22. a	23. c	24. b	25. a	26. d	27. c	28. c	29. c	30. a
31. a	32. c	33. b	34. d	35. c	36. c	37. a	38. b	39. d	40. a
41. a	42. c	43. b	44. a	45. b	46. a	47. d	48. a	49. d	50. b
51. a	52. c	53. c	54. c	55. d	56. b	57. a	58. d	59. c	60. a
61. b	62. b.	63. a	64. c	65. d	66. a				

INDEX

ABOUT THE AUTHOR

Rhonda Huettenmueller has taught mathematics at the college level for more than 15 years. Her ability to make higher math understandable and even enjoyable has earned her tremendous popularity and success with students. She incorporates many of her most effective teaching techniques in her books, including the best-selling *Algebra Demystified, College Algebra Demystified,* and *Precalculus Demystified.* She received her Ph.D. in mathematics from the University of North Texas.